# Industrial Chemistry: Principles and Applications

# Industrial Chemistry: Principles and Applications

Edited by
Cory Simmons

WILLFORD PRESS
www.willfordpress.com

Published by Willford Press,
118-35 Queens Blvd., Suite 400,
Forest Hills, NY 11375, USA

ISBN: 978-1-68285-593-5

**Cataloging-in-Publication Data**

Industrial chemistry : principles and applications / edited by Cory Simmons.
      p. cm.
Includes bibliographical references and index.
ISBN 978-1-68285-593-5
1. Chemistry, Technical. 2. Chemical engineering. I. Simmons, Cory.
TP145 .I53 2019
660--dc23

For information on all Willford Press publications
visit our website at www.willfordpress.com

WILLFORD PRESS

# Contents

# Preface

Over the recent decade, advancements and applications have progressed exponentially. This has led to the increased interest in this field and projects are being conducted to enhance knowledge. The main objective of this book is to present some of the critical challenges and provide insights into possible solutions. This book will answer the varied questions that arise in the field and also provide an increased scope for furthering studies.

Industrial chemistry is the study of applications of chemical processes for the development of consumer products from raw materials. Oil, metals, natural gas and minerals are some of the commonly used raw materials in such chemical processes. Industrial chemistry has applications across a range of other scientific fields and industries such as pharmaceuticals, food, cosmetics, polymer industry, among others. This book strives to present researches and studies that have transformed this discipline and aided its advancement. A number of key concepts and techniques central to the field of industrial chemistry are glanced at and their applications, as well as ramifications, are looked at in detail. From theories to research to practical applications, case studies related to all contemporary topics of relevance to this field have been included in this book. Students, researchers, experts and all associated with the discipline of industrial chemistry will benefit alike from this book.

I hope that this book, with its visionary approach, will be a valuable addition and will promote interest among readers. Each of the authors has provided their extraordinary competence in their specific fields by providing different perspectives as they come from diverse nations and regions. I thank them for their contributions.

**Editor**

# Acetylation of glycerol over bimetallic Ag–Cu doped rice husk silica based biomass catalyst for bio-fuel additives application

R. Jothi Ramalingam[1] · T. Radhika[2] · Farook Adam[2] · Tarekegn Heliso Dolla[3]

**Abstract** Acetylation of glycerol with acetic acid was carried out over bimetallic silver and copper deposited rice husk silica-alumina like ecofriendly green catalyst. Bio-additive like mono, di and tri acetyl glycerol synthesis from raw glycerol (one of the main product of biodiesel), which are gaining attention as additives for improving petroleum fuel properties towards biofuel additive development applications. Advantage of using bimetallic catalyst for glycerol acetylation due to possible synergistic effect between the metals and it enhances the catalytic conversion and selectivity compared to single metal catalyst. The prepared catalysts were characterised by XRD, FT-IR and TEM. Silver and copper incorporated RHS (rice husk silica)-alumina catalysts are shown higher activity and selectivity towards diacetin (di acetyl glycerol) and triacetins (tri acetyl glycerol) formation by catalytic acetylation of glycerol. Higher conversion (98 %) and good selectivity (51 %) is achieved.

**Keywords** Glycerol · Acetylation · Silver · Rice husk silica · Monoacetin · Triacetin · Biodiesel

## Introduction

Glycerol is a valuable by product during biodiesel production (10 % in weight). The utilization of glycerol for the synthesis of value added chemical is a topic of great industrial interest because glycerol can be formed in large amounts during the biodiesel production and represents as a waste byproduct. Its effective utilization will be a key factor that can promote biodiesel commercialization and further development. One of the most attractive outlets of glycerol is to produce glycols especially propanediols by selective hydrogenolysis of glycerol [1, 2]. This process provides a clean and economically competitive route for the production of these commodity chemicals from renewable glycerol instead of nonrenewable petroleum products. Consequently, many efforts have been attempted to facilitate this important reaction such as supported noble metals like Ru, Rh, and Pt are well-known active catalysts in the hydrogenolysis of glycerol [3]. Unfortunately, these catalysts are often promoting excessive C–C cleavage, resulting in a low selectivity to propanediols. As a less expensive alternative, copper-based catalysts have been reported to have a superior performance in this reaction due to their poor activity for C–C bond cleavage and high efficiency for C–O bond hydro-dehydrogenation [4, 5].

Karinen and Krause reported the etherification of glycerol with isobutene over acid exchange resin catalyst [6, 7]. Etherification of glycerol with tert-butanol over acid exchange resin or zeolite catalysts are also reported [8, 9]. Reddy et al. [10], and coworkers reported sulphated ceria-zirconia and ceria alumina type solid acid catalyst for glycerol acetylation to replace mineral acid based catalyst. Bagheri and Muhd [20], reviewed the various value added product formation from glycerol by bimetallic catalysts. Hence, utilization of glycerol by acetylating agent via

✉ R. Jothi Ramalingam
rjothiram@gmail.com

1 Surfactants research chair, Chemistry department, College of Science, King Saud University, Riyadh, Kingdom of Saudi Arabia

2 School of Chemical Sciences, Universiti Sains Malaysia, Penang 11800, Malaysia

3 College of Natural and Computational Sciences, Wolaita Sodo University, Wolaita Sodo, Ethiopia

catalytic method is another alternative methodology. The present study deals with bimetallic particle doped silica-alumina type solid acid catalyst preparation and tested for glycerol acetylation. The glycerol acetylaiton products such as mono, di and triacetyl esters have great industrial applications. The triacetylated derivative is known as tri-acetin and has application going from cosmetics to fuel additive [11, 12]. The mono and diacetylated esters (monoacetin and diacetin) are used in cryogenics and as raw material for production of biodegradable polyesters [13]. The aim of the present work deal with acetylation of glycerol over silver and copper incorporated rice husk silica-alumina solid acid catalyst to produce di and tri-acetylated glycerols with good selectivity under mild reaction conditions. The influence of various reaction parameters has also been studied.

## Experimental

### Catalyst preparation

Rice husk was collected from a rice mill in Penang, Malaysia. The rice husk silica was prepared by our previously reported procedure [14]. After washing and rinsing the rice husk (RH) several times with distilled water it was dried at room temperature. About 30.0 g of clean RH was stirred in 750 mL of 1 M $HNO_3$ at room temperature for 24 h to remove all metallic impurities and this acid treated RH was washed with distilled water until the pH of the rinse was constant (around 4.8–5.0) and dried in an oven at 373 K for 24 h, further kept in a muffle furnace at 873 K for 6 h for complete combustion. The white rice husk silica ash (RHA) thus obtained was used as source of silica for further studies.

About 10 g of the RHA was stirred continuously in 100 mL of 3 M NaOH to obtain the sodium silicate solution. To the obtained sodium silicate solution, Al $(NO_3)_3$·$9H_2O$ (Si:Al = 1:1) and cetyl trimethyl ammonium bromide (CTAB) were added and precipitated at pH = 12.3. The solid obtained was washed, dried at 383 K and calcined at 873 K. The xerogel obtained was ground to powder and labelled as RHS-Al. Similar procedure was repeated with the addition of $Cu(NO_3)_3$·$3H_2O$ to prepare Cu (10 wt%) doped RHS-Al catalyst (10Cu/RHS-Al). To this system appropriate amount of $AgNO_3$ was added and precipitated to obtain different amount of silver-copper deposited RHS-Al materials (1 %Ag–10 %Cu/RHS-Al and 5 % Ag–10 %Cu//RHS-Al).

### Characterization

The powder X-ray diffraction pattern of the catalysts was collected on Siemens Diffractometer D5000, Kristalloflex

operated at 40 kV and 10 mA with nickel filtered CuKα radiation, $c = 1.54$ Å. The $N_2$-sorption data and BET values were collected on NOVA 2200 type surface area and pore size analyzer. Pyridine adsorption analysis carried out by gaseous adsorption method using conventional hammet method. The FT-IR analysis was carried out on Perkin-Elmer System 2000 using KBr pellet method. The morphology and elemental analysis characterized by energy dispersive x-ray spectrometry (Edax Falcon System) and transmission electron microscopy (Philips CM12).

### Acetylation reaction

The acetylation of glycerol was carried out in a 50 mL double neck round bottom flask connected with a reflux condenser kept in an oil bath, temperature of which was controlled by a thermocouple. In a typical reaction, the catalyst powder (80 mg, dried at 383 K), was suspended in a mixture of glycerol and acetic acid (1:10) and kept in an oil-bath at the required reaction temperature (383 K). The system was subjected for stirring (700 rpm). About 5 μm of solution was withdrawn at regular intervals period (30 min) and analysed by GC using toluene as an internal standard. The products were analysed on Clarus 500 (PerkinElmer) gas chromatograph with a capillary column, Elite Wax (30 m length and 0.32 mm inner diameter) equipped with an FID detector using toluene as the internal standard. The set temperature programme was 323 K/ 3 min-temperature ramp of 283 K/min to 503 K with $N_2$ as carrier gas. Products were confirmed on GC–MS in Elite Wax column equipped with a mass selective detector.

## Results and discussion

### Characterisation of Ag–Cu–silica–alumina catalysts

The wide-angle powder X-ray diffraction pattern of as prepared catalysts with parent rice husk silica is shown in Fig. 1. The pattern depicts that RHS-Al is amorphous in nature with a broad peak in the $2\theta$ region of 20–30° at the calcination temperature of 773 K [16]. During copper incorporation in rice husk silica results in the formation of CuO nanoparticle and crystallized in monoclinic structure, which is confirmed by XRD pattern. The reflection observed at 35.5° and 38.7° corresponds to the formation of CuO nanoparticle phase formation [PDF 05-0661]. Similar observations are obtained for CuO-niobia/silica-alumina catalyst [17]. Small amount of silver particle incorporated on Cu/RHS-Al is not impact the crystal phase changes in the XRD pattern suggesting that the silver particle dispersion on the support and EDX analysis confirm the existence

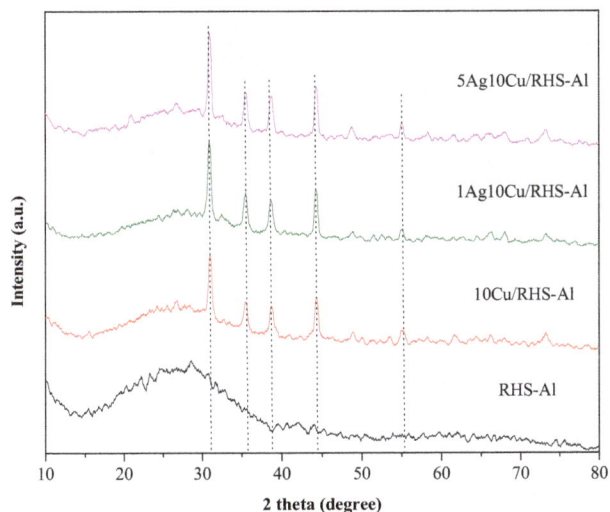

**Fig. 1** The powder X-ray diffraction pattern of prepared Ag–Cu/RHS-Al catalysts

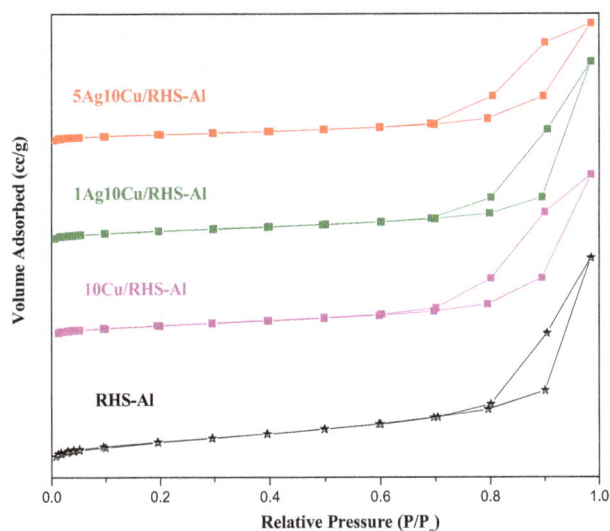

**Fig. 2** $N_2$ sorption–desorption analysis

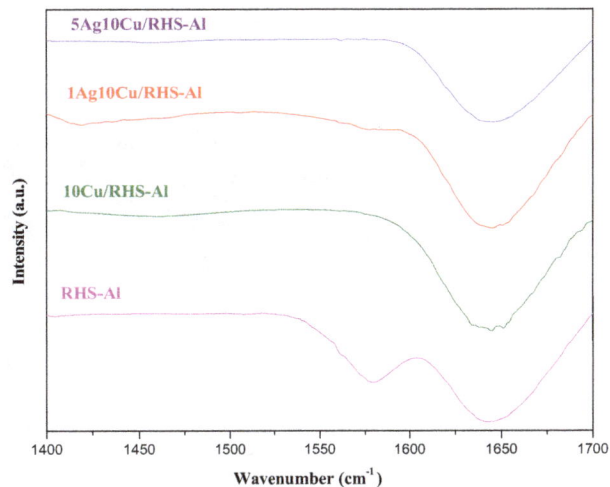

**Fig. 3** FT-IR spectra of pyridine-adsorbed Ag–Cu/RHS-Al catalysts

of Ag and Cu by elemental analysis. Nitrogen adsorption and desorption study is rice husk silica and silver, copper doped rice husk silica are shown in Fig. 2. The bimetallic particle incorporated rice husk silica shown the similar characteristic curve related with micro porous structure. Figure 3 shows the FT-IR spectra of representative samples recorded at 400–4000 $cm^{-1}$ rang and in all spectra, a broad absorption band appeared around $\sim 3456$ and 1635 $cm^{-1}$ corresponding to the vibrations of –OH and silanol group [14, 15]. The strong bands at $\sim 1011$ $cm^{-1}$ can be assigned to typical Si–O–Si vibration. Bands at $\sim 708$ and 452 $cm^{-1}$ are due to the Si–O vibration within Si–O–Si [14]. FT-IR spectra of samples after pyridine-adsorption

were recorded to study the nature of acidity exist in the catalyst (Fig. 3). For all catalysts, no prominent peaks were observed as seen in the spectra. In the entire sample a broad absorption band was observed at $\sim 1650$ $cm^{-1}$ corresponding to bronsted acid nature of the support [15]. In the case of RHS-A1 exhibits a band centred at $\sim 1575$ $cm^{-1}$ due to surface hydroxyl acidity. Therefore, bronsted acidity in nature exists predominantly in the as prepared Ag–Cu/RHS-Al solid acid catalyst. Morphologies of as synthesized bimetallic Cu–Ag-rice husk silica-alumina catalyst are shown in Fig. 4. The magnification of Fig. 4a is around 5 μm. The SEM images shown the aggregated particle morphology for as prepared bimetallic solid acid catalyst and copper oxide nanoparticle forms the cubic shape particle on RHS-alumina support, which is shown in arrow mark in Fig. 4a. The white colour cubic particle exists on silica-alumina due to formation of crystallized copper oxide particle after calcinations process (Fig. 4b). The amount of copper and silver is further determined by elemental analysis using EDX connected with SEM equipment. Figure 5 shows the elemental composition of as prepared bimetallic rice husk silica catalyst; edx spectra confirm the copper and silver particle existence in RHS support. TEM images are also confirms the aggregated particle morphology formation for as prepared solid catalysts and the size of the particle exists in the range of 30–40 nm and the scale value in the TEM image is 50 nm. The dense lengthy particle in the TEM image of Fig. 3b is due to existence copper oxide nanoparticle aggregation with RHS-alumina support. The length of the dense particle is more than 50 nm and below 80 nm. The pore size $N_2$ sorption analysis shows the uniform porosity and BET surface area of 1Ag–5Cu/RHS-Al to RHS-Al are shown in Fig. 2 and Table 1.

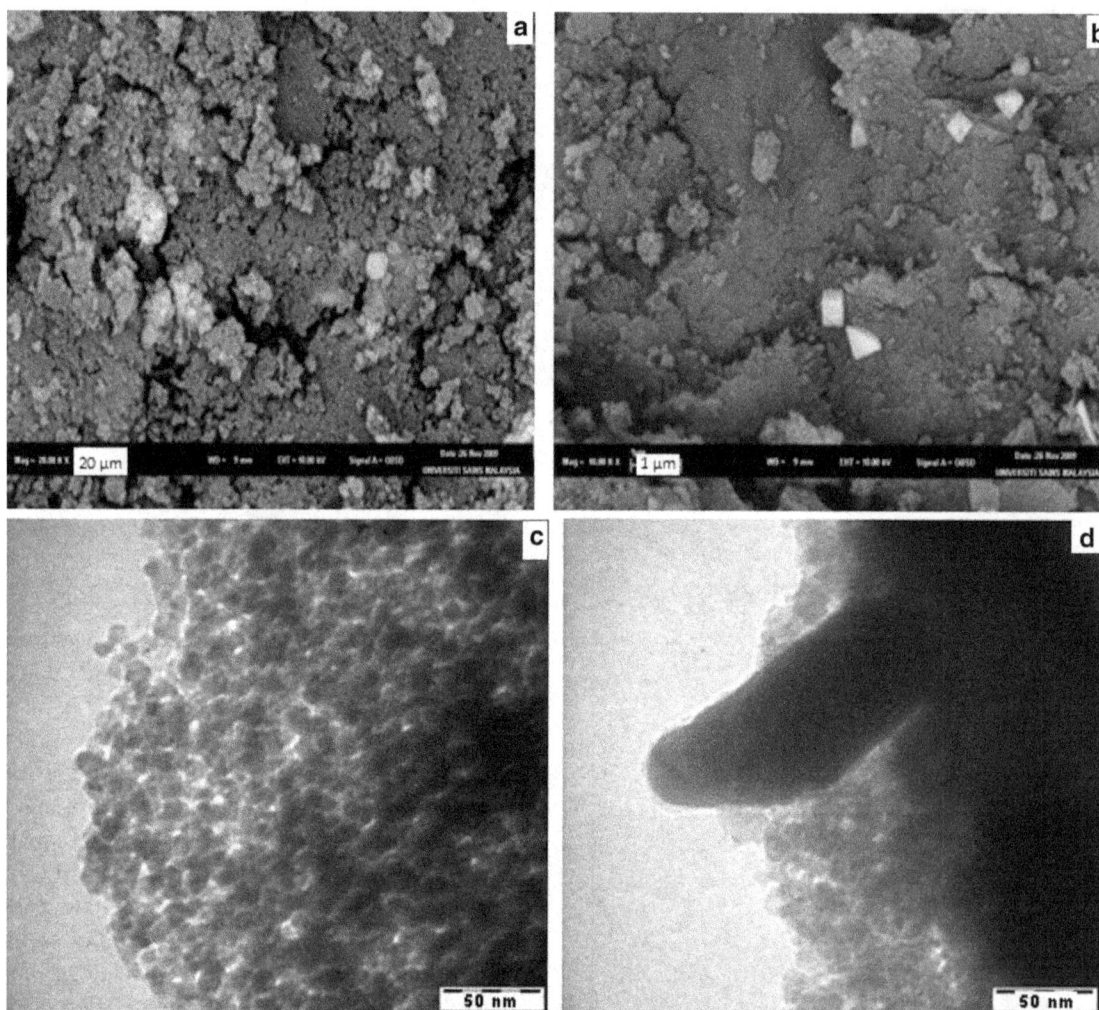

**Fig. 4** SEM images of **a** 1Ag10Cu/RHS-Al **b** 5Ag10Cu/RHS-Al and TEM images of **c** 1Ag10Cu/RHS-Al **d** 5Ag10Cu/RHS-Al

## Catalytic activity of Cu–Ag–silica–alumina catalysts

Acetylation of glycerol over different amount of copper and silver Incorporated rice husk silica catalyst is shown in Table 2. Silver and copper incorporated in the ratio of 1:10 and the reaction carried out at various temperatures. Conversion of glycerol acetylation over 1Ag–10Cu/RHS-Al catalyst is increased upon increase with reaction temperature. At higher temperature like 383 and 393 K shows the maximum conversion towards glycerol derivative formation. The selectivity towards monoacetin was decreased and diacetin and triacetin formation was increased. Tiracetin formation is increased at optimized temperature range between 353–373 K and above 373 K, triacetin formation was slashed. The selectivity of diacetin formation is very low at starting temperature and increased upon optimized reaction temperature. The di and tri-acetin formation was improved in the presence of bimetallic Ag–Cu

loaded silica-alumina catalyst at optimized reaction condition.

The effect of molar ratio between glycerol and acetic acid has also been studied in the presence of 1Ag10Cu-silica-alumina catalyst. The results of glycerol acetylation with different mole ratio of acetic acid and glycerol were shown in Table 3. In all analysis toluene is used as internal standard for determination of glycerol conversion in GC analysis. A Table 4 show the higher amount of silver incorporation on Cu/RHS-Al solid catalyst is increases the glycerol conversion and selectivity towards di and triacetin formation. Silver incorporation is acting as a promoter for copper-RHS-Al catalyst and increasing the metal particle deposition results in the higher amount of glycerol conversion occurred. From Table 3, Glycerol conversion increased gradually with increased amount of acetic acid, results in the formation of higher yield of diacetin and triacetin. At higher mole ratio between glycerol/acetic acid

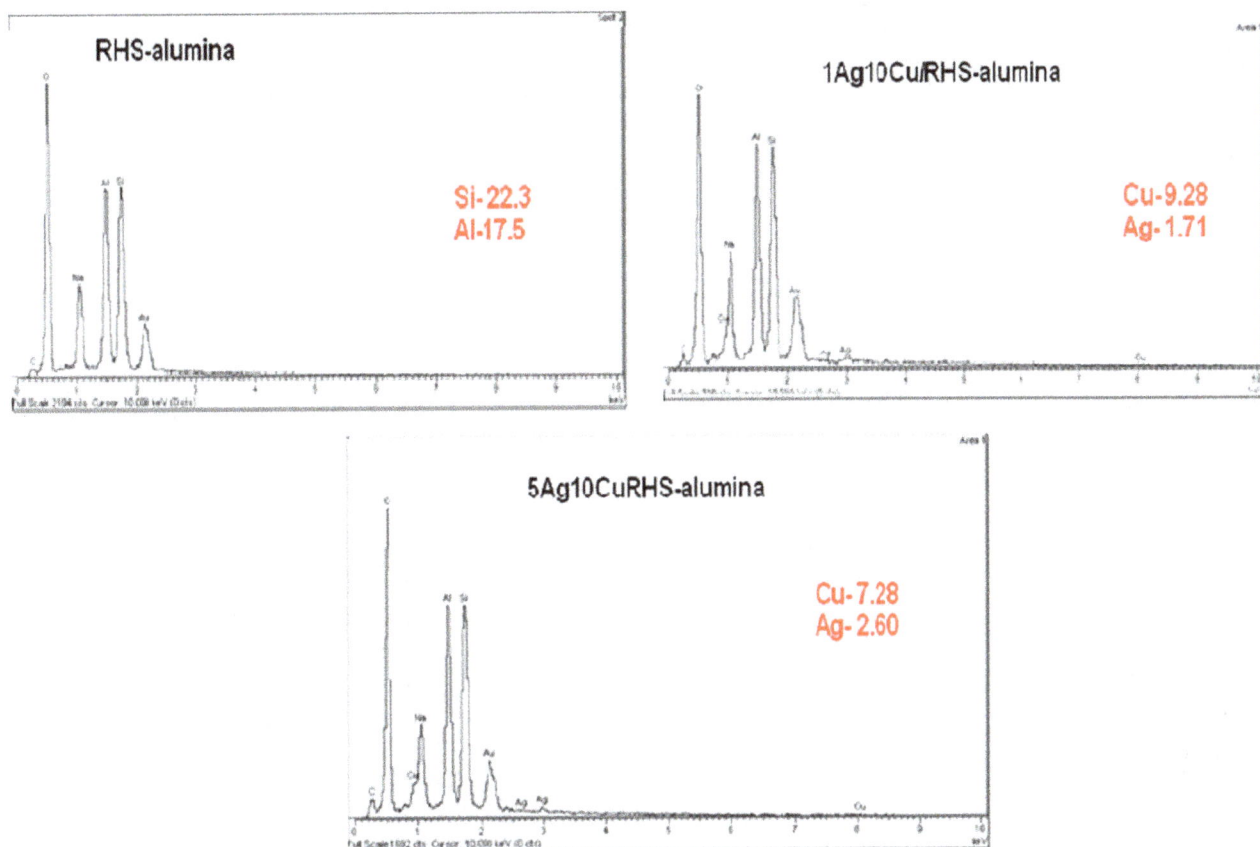

**Fig. 5** EDX analysis of selected catalysts

**Table 1** BET surface area of the catalysts

| Sample | BET surface area ($m^2g^{-1}$) | Crystallite size (nm) |
|---|---|---|
| RHS-Al | 79.4 | – |
| 10Cu/RHS-Al | 92.6 | – |
| 1Ag10Cu/RHS-Al | 108.1 | – |
| 5Ag10Cu/RHS-Al | 93.1 | 57.4 |

**Table 2** Influence of temperature for catalytic acetylation of glycerol

| Temperature (K) | Conversion (%) | Selectivity (%) | | |
|---|---|---|---|---|
| | | MAG | DAG | TAG |
| 333 | 69.4 | 100 | – | – |
| 353 | 78.9 | 71.1 | 5.1 | 23.8 |
| 373 | 80.9 | 72.8 | 2.5 | 24.6 |
| 383 | 83.4 | 68.3 | 18.5 | 13.2 |
| 393 | 80.5 | 65.6 | 16.8 | 17.6 |

Reaction conditions: catalyst = 1Ag 10Cu/RHS-Al; catalyst weight = 80 mg; molar ratio of glycerol to acetic acid = 1:10; reaction time = 5 h

is hinder the formation of monoacetin and higher selectivity obtained for triacetin formation in the presence of Ag–Cu–silica–alumina bimetallic catalyst. Table 3 shows

maximum conversion (80.4 and 85.7 %) with the higher selectivity towards triacetin formation (30.1 and 43.2 %) at increase amount of mole ratio such as 1:20 and 1:25. In eco-point of view and cost effective in large scale synthesis is optimized towards 1:10 more ratio usage of glycerol: acetic acid is the best for obtain MAG. Table 4 show the higher glycerol conversion with good selectivity at optimized glycerol and acetic acid mole ratio (1:10). Hence, bimetallic nanoparticle deposition on silica support and mole ratio variation between glycerol and acetic acid are playing important role in triacetin formation from glycerol. Addition of appropriate amount of silver and copper deposition on silica-alumina catalyst shows the considerable improvement in catalyst texture property towards high catalytic activity. The higher amount of bimetallic species incorporated catalysts (1Ag10Cu/RHS-Al and 5Ag10Cu/

**Table 3** Effect of molar ratio of glycerol to acetic acid for acetylation of glycerol

| Molar ratio (Glycerol: acetic acid)/mmol | Conversion (%) | Selectivity (%) | | |
|---|---|---|---|---|
| | | MAG | DAG | TAG |
| 1:5 | 76.3 | 84.8 | 12.3 | 2.9 |
| 1:10 | 83.4 | 68.3 | 18.5 | 13.2 |
| 1:15 | 81.2 | 55.3 | 28.0 | 17.7 |
| 1:20 | 80.4 | 43.9 | 26.0 | 30.1 |
| 1:25 | 85.7 | 32.5 | 24.3 | 43.2 |

Reaction conditions: catalyst = 1Ag 10Cu/RHS-Al; catalyst weight = 80 mg; temperature = 383 K; reaction time = 5 h

**Table 4** Acetylation of glycerol with acetic acid on influence of catalyst amount

| Catalyst | Conversion (%) | Selectivity (%) | | |
|---|---|---|---|---|
| | | MAG | DAG | TAG |
| RHS | – | – | – | – |
| RHS-Al | 76.8 | 89.4 | 10.6 | – |
| 10Cu/RHS-Al | 83.8 | 68.3 | 21.1 | 10.6 |
| 1Ag10Cu/RHS-Al | 97.5 | 3.7 | 58.4 | 37.9 |
| 5Ag10Cu/RHS-Al | 100 | 2.4 | 46.3 | 51.3 |

Reaction conditions: temperature = 383 K; 1Ag10Cu/RHS-Al; catalyst weight = 100 mg; molar ratio of glycerol to acetic acid = 1:10; reaction time = 5 h

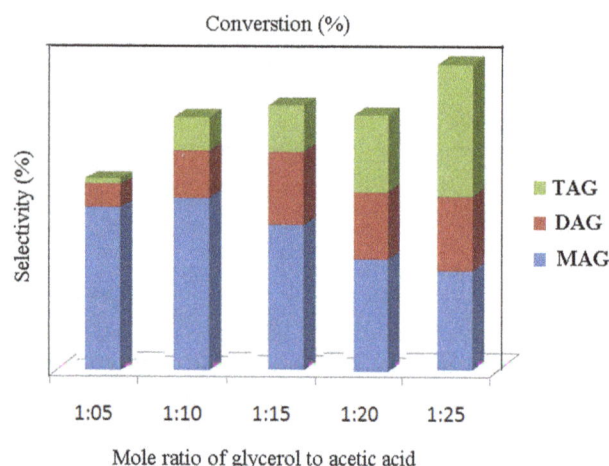

**Fig. 6** Acetylation of glycerol over different mole ratio of substrate and reactant

RHS-Al) shown the 100 % conversion for catalytic glycerol conversion and good product selectivity obtained towards diacetin and triacetin formation (Table 4). Figure 6 shows the pictorial representation of glycerol conversation and clear image of selectivity towards triacetin formation at higher mole ratio of substrate to reactant. One can envisage two possible mechanisms [18] for acetylation in strong acidic medium: the first one is the normal AAC2 mechanism, involving protonation of the carbonyl oxygen atom and nucleophilic attack in the carbonyl to form a tetrahedral intermediate, presenting a quaternary carbon atom (Scheme 1); the second mechanism is the AAC1, where protonation takes place in the oxygen atom attached to the carbonyl group, followed by formation of an acylium ion (Scheme 2). The first pathway is normally less energetic, because of the higher stability of the intermediate formed upon protonation in the carbonyl oxygen atom. On the other hand, formation of the tetrahedral intermediate is space demanding, and the second mechanism, involving the acylium ion, prevails in situations of steric constraints [19].

Table 5 shows the comparative catalytic activity of solid catalysts prepared by low cost methodology and conventional route prepared catalyst. The standard solid acid catalyst such as K-10, Niobium phosphate and Amberlyst 15 are shown complete conversion for glycerol acetylation with less selectivity towards triacetin formation. In our case, the prepared bimetallic catalysts like Ag–Cu–silica–alumina are showing higher selectivity towards diacetin (DAG) and triacetin (TAG) formation. The DAG and TAG are good additives for fuels such as biodiesel and gasoline due to their role in improve the viscosity. Another advantage of the present work is low cost methodology was adopted to prepare the bimetallic catalysts. The application of triacetin is very much vital for cosmetics formation and also used as additives for petroleum products.

## Conclusions

Rice husk silica-alumina and bimetallic Ag–Cu/RHS-Al solid acid catalysts were prepared by low cost methodology via sol–gel method and characterized by various physico-chemical techniques. Mono, Di, Triacetins are obtained as major products in the presence of Ag–Cu/RHS-Al catalyst and negligible conversion rate was obtained on raw RHS catalyst. The rod like aggregated particle morphology obtained for Ag–Cu/RHS-Al catalyst. Ag–Cu modified rice husk silica-alumina catalyst shows higher selectivity towards DAG and TAG formation compared to other conventional catalysts. Conversion and selectivity for

**Scheme 1** Acetylation mechanism with formation of a tetrahedral intermediate

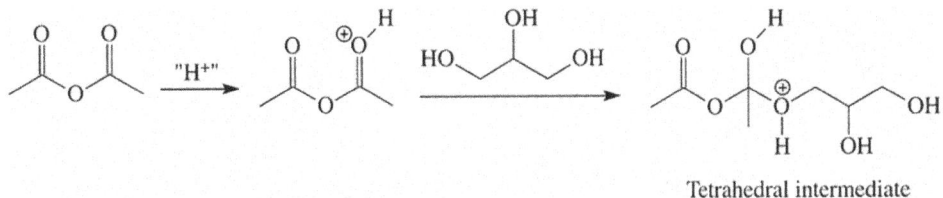

Tetrahedral intermediate

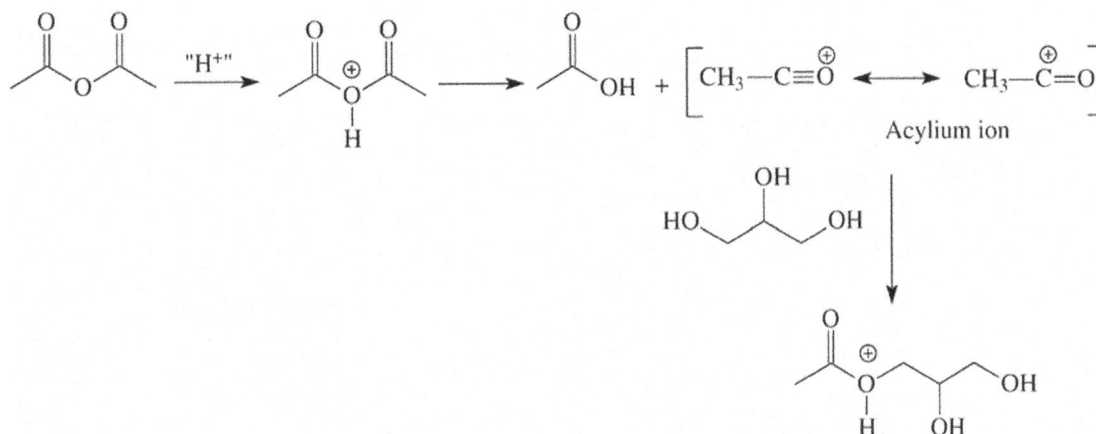

Acylium ion

**Scheme 2** Acetylation mechanism with formation of an acylium ion

**Table 5** Comparison of catalytic activity of various catalyst for glycerol acetylation

| Catalyst | Conversion (%) | Selectivity (%) | | |
|---|---|---|---|---|
| | | MAG | DAG | TAG |
| RHS-Al[#] | 76.8 | 89.4 | 10.6 | – |
| 10Cu/RHS-Al[#] | 83.8 | 68.3 | 21.1 | 10.6 |
| 1Ag10Cu/RHS-Al[#] | 97.5 | 3.7 | 58.4 | 37.9 |
| 5Ag10Cu/RHS-Al[#] | 100 | 2.4 | 46.3 | 51.3 |
| H-Beta* | 94 | 48 | 39 | 4 |
| K-10* | 94 | 36 | 52 | 6 |
| Niobium phosphate* | 94 | 38 | 49 | 7 |
| Amberlyst-15* | 97 | 18 | 55 | 24 |

* Ref. [17] Molar ratio glycerol to acetic acid = 1:4, reaction temperature is 393 K and reaction time 2 h

[#] Mole ratio of glycerol to acetic acid = 1:10, reaction temperature is 393 K and reaction time 5 h

glycerol acetylation reaction are improved greatly due to nature of active centers exists on bimetallic-rice husk silica catalyst.

**Acknowledgments** The research project was financially supported by King Saud University, Deanship of Scientific Research, Research Chairs.

**Authors Contributions** RJ performed all experiments, processed data, initial data analysis and first draft of manuscript; HAA guidance, TR secondary data analysis, subsequent drafts of manuscript FA, guidance, THD project design. All authors have read and approved the final manuscript.

## References

1. Dasari MA, Kiatsimkul P-P, Sutterlin WR, Suppes GJ (2005) Appl Catal A Gen. 281:225
2. Maris EP, Ketchie WC, Murayama M, Davis RJ (2007) J Catal 251:281
3. Feng J, Fu HY, Wang JB, Li RX, Chen H, Li XJ (2008) Catal Commun 9:1458
4. Runeberg J, Baiker A, Kijenski J (1985) Appl Catal. 17:309
5. Montassier C, Giraud D, Barbier J (1988) In: Guisnet M, Barrault J, Bouchoule C, Dupres D, Montassier C, Pérot G (eds.) Stud. Surf. Sci. Catal. Amsterdam: Elsevier, pp. 165
6. Mat R, Amin NAS, Ramli Z, Abu Bakar WA (2006) J Nat Gas Chem 15: 259
7. Karinen RS, Krause AOI (2003) Appl Catal A Gen. 306:128
8. Wessendorf R, Erdoel Kohle (1995) Erdgas Petrochemie 48: 138
9. Klepacova K, Mravec D, Bajus M (2005) Appl Catal A 294:141
10. Reddy PS, Sudarsanamm P, Raju G, Reddy BM (2012) J Indus Eng Chem. 18:648
11. Nomura S, Hyoshi T (1995) JP patent 203429
12. Lipkowski AW, Kijenski J, Walisiewicz-Niedbalska N (2005) Pol Chemik 58:238

13. Taguchi Y, Oishi A, Ikeda Y, Fujita K, Masuda T (2000) JP patent 298099
14. Adam F, Radhika T (2010) Chem Eng J 160:249
15. Muruvvet Y, Mehmet A, Tonbul Y, Kadir Y (1999) Turk J Chem 23:319
16. Fillipe AC Garcia, Valdeilson S. Braga, Júnia CM Silva, José A. Dias, Sílvia CL Dias, Jorge LB Davo (2007) Catal Lett 119:101
17. Silva LN, Gonçalves VLC, Mota CJA (2010) Catal Commun 11:1036
18. Lowry TH, Richardson KS (1981) Mechanism and theory in organic chemistry, 2nd edn. Harper and Row, New York, pp 652–658
19. Bender ML, Ladenheim H, Chen MC (1961) J Am Chem Soc 83:123–127
20. Bagheri S, Muhd Julkapl N, Yehye WA (2015) Renew Sustain Energy Rev 41:113–127

# 2

# Acetohydroxamic acid adsorbed at copper surface: electrochemical, Raman and theoretical observations

Juan Du[1] · Ye Ying[1] · Xiao-yu Guo[1] · Chuan-chuan Li[1] · Yiping Wu[1] · Ying Wen[1] · Hai-Feng Yang[1]

**Abstract** Corrosion inhibition effect of AHA film formed on the copper surface by self-assembled monolayers technique was estimated in 3 wt% NaCl solution by electrochemical impedance spectroscopy and polarization methods. Polarization data indicated that AHA was an anodic inhibitor. The maximum inhibition efficiency reached 93.5% in the case of assembly 3 h in 10 mM AHA solution. The adsorption of AHA on the copper surface fits Langmuir adsorption isotherm. Surface-enhanced Raman scattering together with quantum chemical studies demonstrated that N–O and C=O groups were attached to the copper surface, predicting the feasible adsorption centers and confirming the relationship between the molecular structures of AHA and its inhibition property.

**Keywords** Acetohydroxamic acid · EIS · Polarization · SERS · Anodic inhibitor

## Introduction

Highly electrical and thermal conductivities, as well as good mechanical workability of copper and its alloys enable them to have a diverse range of applications in pipelines for domestic and industrial water systems, shipbuilding, seawater desalination and heat exchanger [1–5].

However, copper and its alloys suffer serious corrosion in chloride environments, causing huge economy losses [6–8]. Thus, a considerable amount of efforts have been made to improve the corrosion resistance properties of copper using effective organic inhibitors with aromatic rings, and electronegative functional groups involving the heteroatoms of sulfur, nitrogen and oxygen, which may adsorb at the copper surface to form inhibitive coatings [9–15].

Unfortunately, among them, some are toxic and expensive. In recent years, various types of nontoxic organic compounds have been investigated to meet the recommendation using eco-friendly inhibitors as substitutes for restricted toxic inhibitors [16–21]. Some hydroxamic acids and their derivatives with biological activities, such as anti-inflammatory and anti-asthmatic, are used as pesticides and plant growth promoters [22]. In addition, hydroxamic acids could chelate with metal ions to form complexes [23, 24]. Thus, hydroxamic acid derivatives have already been reported as effective corrosion inhibitors for carbon steel and copper corrosion [25–27]. AHA as a potential corrosion inhibitor could form a protection film on the copper surface due to multiple adsorption centers of a nitrogen atom and two oxygen atoms in its structure (see Scheme 1). So far, the adsorption behavior of AHA on copper surface as well as its corrosion inhibition efficiency has not been observed in detail.

In this work, AHA was adsorbed on copper surface by self-assembled monolayers (SAMs) technique and the efficiency against corrosion in 3 wt% NaCl solution was estimated by electrochemical impedance spectroscopy (EIS) and polarization methods. SERS technique as a powerful tool to provide molecular fingerprint information was used to elucidate formation mechanism of AHA coating on the copper surface as well as Langmuir

✉ Ye Ying
yingye@shnu.edu.cn

✉ Hai-Feng Yang
haifengyang@yahoo.com

[1] The Education Ministry Key Lab of Resource Chemistry, Department of Chemistry, Shanghai Normal University, Shanghai 200234, People's Republic of China

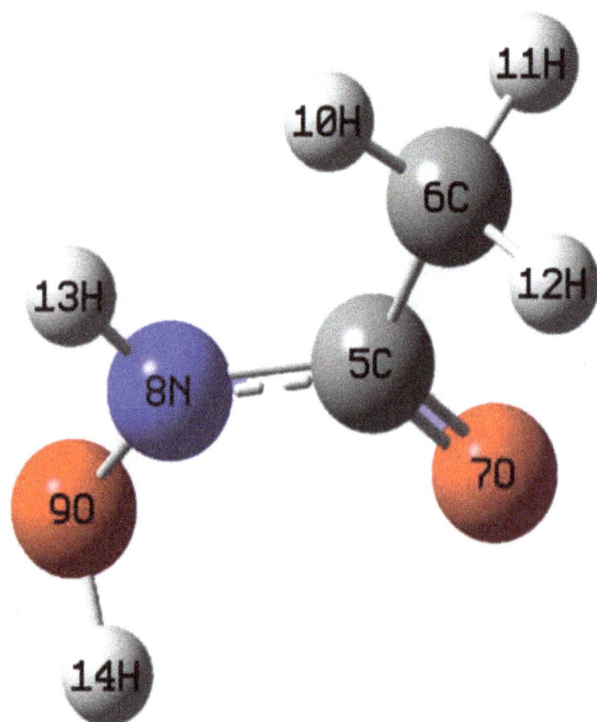

**Scheme 1** Optimized structure of acetohydroxamic acid

adsorption isotherm measurement. Furthermore, quantum chemical studies were used to predict the feasible adsorption centers and confirm the relationship between the molecular structures of AHA and its inhibition property.

## Experimental

### Materials and chemicals

Acetohydroxamic acid (AHA) was purchased from Sigma-Aldrich. Analytical grade NaCl was dissolved in ultrapure water (18 MΩ cm) to prepare 3 wt% NaCl corrosion media. Sulfuric acid and ethanol were analytical reagents, purchased from Sinopharm Chemical Reagents Company.

### Apparatus

Raman spectroscopic measurement was conducted using LabRam II confocal Laser Raman system (Dilor, France). A $1024 \times 800$ pixels charge-coupled device detector cooled by semiconductor was used, and the excitation source was a He–Ne laser at 632.8 nm with power of ca. 5 mW. The slit and pinhole were controlled at 100 and 1000 μm, respectively. The laser was focused onto the copper surface through a long-working-length of $50 \times$ objective. Each Raman spectrum was taken with 8 s

integration time and $3 \times$ repeats. The line of silicon positioned at 519 cm$^{-1}$ was used for spectral calibration.

The electrochemical measurements were carried out using CHI 750C electrochemistry workstation (CH Instruments, Inc.).

### Pretreatment for electrode

The copper electrode was constructed from polycrystalline copper (99.999%) rod inside of a Teflon sheath, and the exposure area of surface was 0.0314 cm$^2$. Before the Raman spectroscopic and electrochemical measurements, the copper electrode was sequentially abraded with 500- and 1000-grit papers, followed by 0.3 μm alumina powders to get a shiny mirror-like electrode surface, and then cleaned with Milli-Q water and pure ethanol in an ultrasonic bath to remove any remaining alumina particles and possible rust. For SERS detection, to obtain the necessary roughness of the copper surface, copper surface was first treated in 2 M H$_2$SO$_4$ solution using an oxidation–reduction cycle (ORC) process in the potential range from −0.55 to 0.45 V (vs. SCE) with scan rate at 0.02 V/s and 10 sweep segments and final potential was applied at −0.55 V (vs. SCE) for 60 s [28]. In a conventional three-electrode cell, the AHA-modified copper electrode (or bare copper specimen) and a platinum electrode were used as working electrode and the counter electrode, respectively. All potentials referred to in this paper are reported relative to the saturated calomel electrode (SCE), which was used as reference electrode.

### Coating the copper surface with AHA

The cleaned copper electrodes were immersed immediately into the deoxygenated AHA solutions with various concentrations. The assembly time effect on the formation of AHA film at the Cu surface was also considered. Before spectroscopic and electrochemical experiments, the AHA solution was removed and the electrode surface with AHA film was rinsed using Milli-Q water, and then dried by flowing nitrogen gas.

### Electrochemical measurements

The impedance spectra were performed in a three-electrode cell starting from open circuit potential (OCP) with the AC voltage amplitude of 5 mV (vs. SCE) in the frequency range from 0.01 Hz to 100 kHz. The impedance results were simulated with a compatible electronic equivalent circuit fitting. The electrochemical polarization curves were obtained from −0.1 V (vs. SCE) to −0.3 V (vs. SCE) with a scan rate of 1 mV s$^{-1}$.

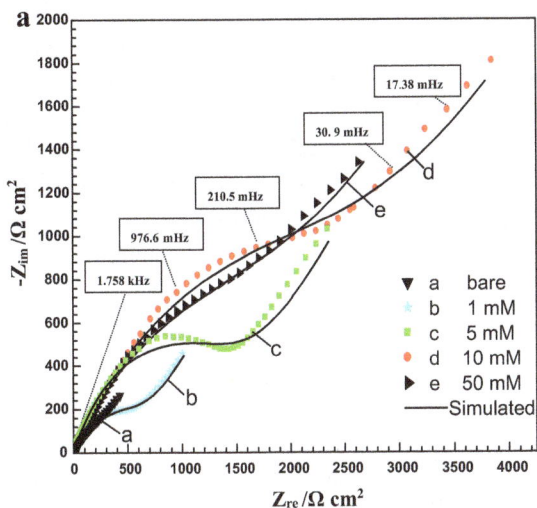

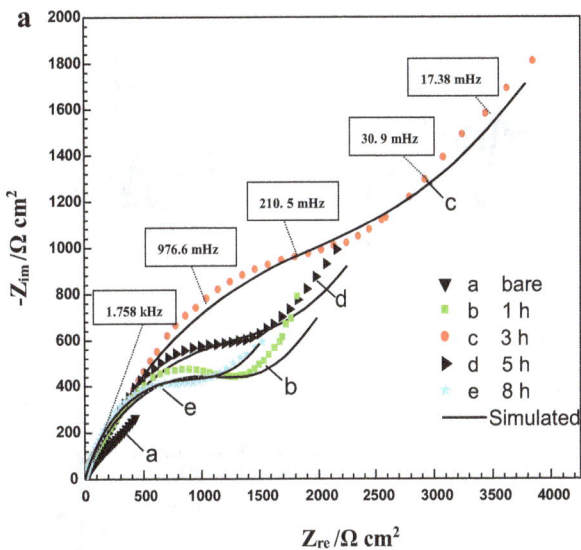

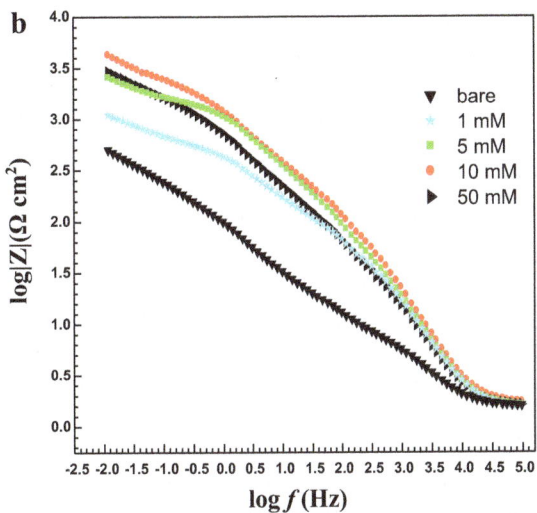

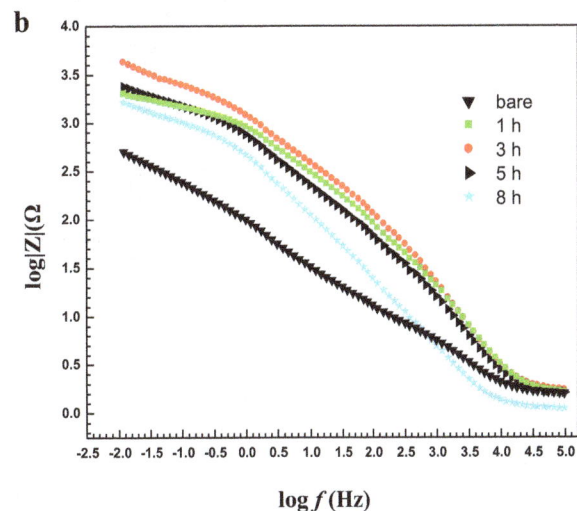

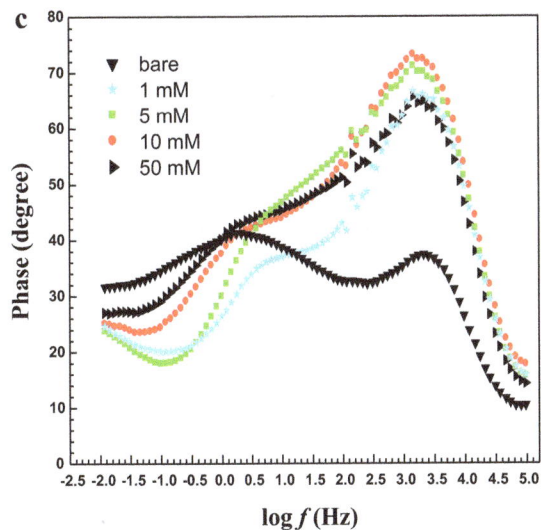

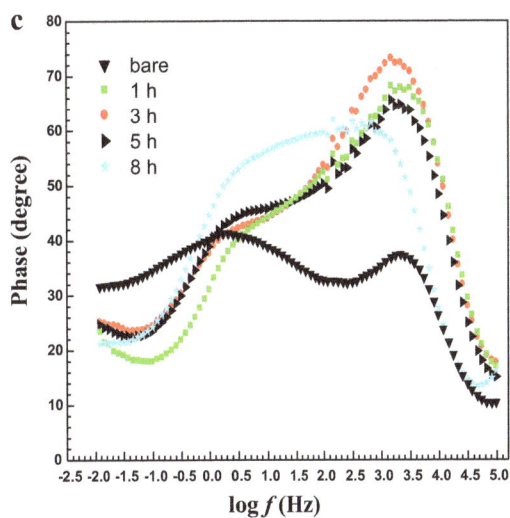

**Fig. 1** Nyquist (**a**), bode (**b**) and phase angle (**c**) plots of copper electrodes with acetohydroxamic acid films formed in different concentrations of acetohydroxamic acid solutions, acquired in 3 wt% NaCl solution

**Fig. 2** Nyquist (**a**), Bode (**b**) and phase angle (**c**) plots of copper electrodes in the absence and presence of film formed in 10 mM acetohydroxamic acid solution for different time, recorded in 3 wt% NaCl solution

**a**   $R(Q(RW))(QR)$

**b**   $R(C(R(Q(RW))))$

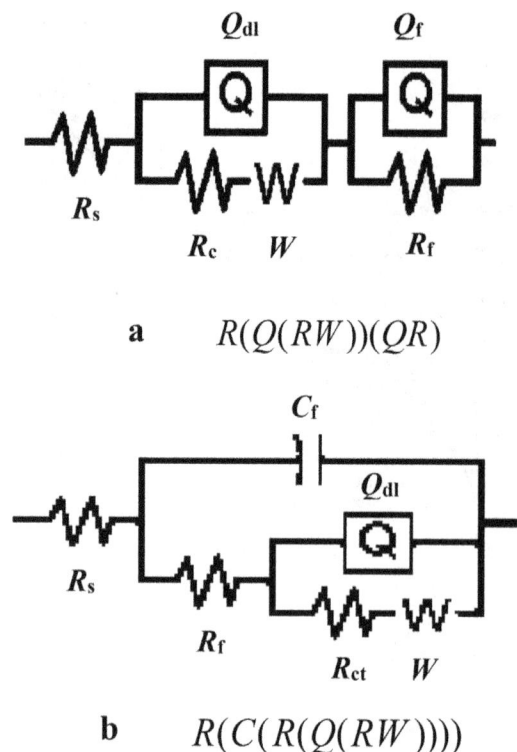

**Fig. 3** The electrical equivalent circuit models for impedance data of copper electrode without film (**a**) with AHA film (**b**)

**Theory computation details**

The geometry optimization of AHA molecule and its Raman spectral calculation were performed at the density functional theory (DFT) based on B3LYP/6-31G(d) level using Gaussian 03 package and the vibrations were

observed using GaussView 5.0. The SERS bands of AHA molecule were tentatively assigned according to the computation result obtained using B3LYP/Lanl2DZ method on the basis of AHA-Cu$_4$ mode. Additionally, the calculations of the frequencies have been scaled by a factor of 0.9940 for B3LYP/6-31G(d) level and a uniform scaling factor of 0.9762 for B3LYP/LanL2DZ method.

**Results and discussion**

**EIS spectra**

Electrochemical impedance spectroscopy for studying the inhibitor against corrosion does not result in any destruction of the film state on the electrode surface [29].

Nyquist plots along with bode and phase angle plots for the copper electrodes with AHA films formed in various AHA concentration solutions and under different assembly time corroded in 3 wt% NaCl solutions are shown in Figs. 1 and 2. The shape of the Nyquist plot for bare copper electrode is different from those of AHA-modified electrodes. The latter ones are composed of a high-frequency imperfect semicircle and a straight line seen as Warburg impedance in the low frequency [30].

Clearly, the Warburg impedance for copper is connected with the diffusion of many oxide species due to lacking protection film [31]. In Fig. 1a, compared with that of bare copper electrode, semicircle diameters of the electrodes with AHA increase visibly and it reaches a maximum when the film was formed in 10 mM AHA solution. It means that formed in an optimized assembly concentration of AHA

**Table 1** Electrochemical parameters calculated from EIS measurements for copper electrodes without the AHA films, in 3 wt% NaCl solution

| $C_{AHA}$ (mM) | $E_{OCP}$ (V vs. SCE) | $R_s$ ($\Omega$ cm$^2$) | $Q_f$ | | $R_f$ ($\Omega$ cm$^2$) | $Q_{dl}$ | | $R_{ct}$ ($\Omega$ cm$^2$) | $R_p$ ($\Omega$ cm$^2$) | $W$ | $\eta$ (%) |
|---|---|---|---|---|---|---|---|---|---|---|---|
| | | | $Y_0 \times 10^{-3}$ ($\Omega^{-1}$ cm$^{-2}$ s$^{n_1}$) | $n_1$ | | $Y_0 \times 10^{-4}$ ($\Omega^{-1}$ cm$^{-2}$ s$^{n_2}$) | $n_2$ | | | $Y_0$ ($\Omega^{-1}$ cm$^{-2}$ n$^{0.5}$) | |
| Bare | −0.193 | 1.45 | 5.37 | 0.565 | 227 | 1.08 | 0.858 | 4.84 | 232 | 0.0109 | – |

**Table 2** Electrochemical parameters calculated from EIS measurements for copper electrodes with the AHA films formed in different concentrations of AHA solutions, in 3 wt% NaCl solution

| $C_{AHA}$ (mM) | $E_{OCP}$ (V vs. SCE) | $R_s$ ($\Omega$ cm$^2$) | $C_f$ ($\mu$F cm$^{-2}$) | $R_f$ ($\Omega$ cm$^2$) | $Q_{dl}$ | | $R_{ct}$ ($\Omega$ cm$^2$) | $R_p$ ($\Omega$ cm$^2$) | $W$ | $\eta$ (%) |
|---|---|---|---|---|---|---|---|---|---|---|
| | | | | | $Y_0 \times 10^{-4}$ ($\Omega^{-1}$ cm$^{-2}$ S$^n$) | $n$ | | | $Y_0 \times 10^{-3}$ ($\Omega^{-1}$cm$^{-2}$ S$^{0.5}$) | |
| 1 | −0.200 | 1.59 | 7.61 | 24.5 | 5.93 | 0.515 | 764 | 789 | 5.92 | 70.7 |
| 5 | −0.225 | 1.62 | 7.24 | 39.0 | 1.77 | 0.611 | 1638 | 1677 | 2.83 | 86.2 |
| 10 | −0.236 | 1.72 | 6.31 | 54.7 | 2.46 | 0.540 | 3489 | 3544 | 1.68 | 93.5 |
| 20 | −0.224 | 1.64 | 8.31 | 32.4 | 3.81 | 0.536 | 1996 | 2028 | 2.21 | 88.6 |
| 40 | −0.225 | 1.62 | 6.48 | 29.0 | 3.05 | 0.485 | 1457 | 1486 | 4.00 | 84.4 |
| 50 | −0.230 | 1.62 | 8.07 | 16.2 | 4.36 | 0.536 | 2789 | 2805 | 2.16 | 91.7 |

**Table 3** Electrochemical impedance parameters for copper electrodes in 3 wt% NaCl solution, in the absence and presence of film formed in 10 mM AHA solution for different time

| Time (h) | $E_{OCP}$ (V vs. SCE) | $R_s$ ($\Omega$ cm$^2$) | $C_f$ ($\mu$F cm$^{-2}$) | $R_f$ ($\Omega$ cm$^2$) | $Q_{dl}$ $Y_0 \times 10^{-4}$ ($\Omega^{-1}$ cm$^{-2}$ S$^n$) | $n$ | $R_{ct}$ ($\Omega$ cm$^2$) | $R_p$ ($\Omega$ cm$^2$) | $W$ $Y_0 \times 10^{-3}$ ($\Omega^{-1}$ cm$^{-2}$ S$^{0.5}$) | $\eta$ (%) |
|---|---|---|---|---|---|---|---|---|---|---|
| 1 | −0.212 | 1.62 | 5.98 | 26.6 | 2.44 | 0.576 | 1586 | 1613 | 4.18 | 85.6 |
| 3 | −0.236 | 1.72 | 6.31 | 54.7 | 2.46 | 0.540 | 3489 | 3544 | 1.68 | 93.5 |
| 5 | −0.222 | 1.61 | 7.68 | 20.9 | 3.66 | 0.569 | 2008 | 2029 | 3.30 | 88.6 |
| 8 | −0.251 | 1.05 | 13.7 | 0.111 | 5.02 | 0.655 | 1270 | 1270 | 5.39 | 81.8 |

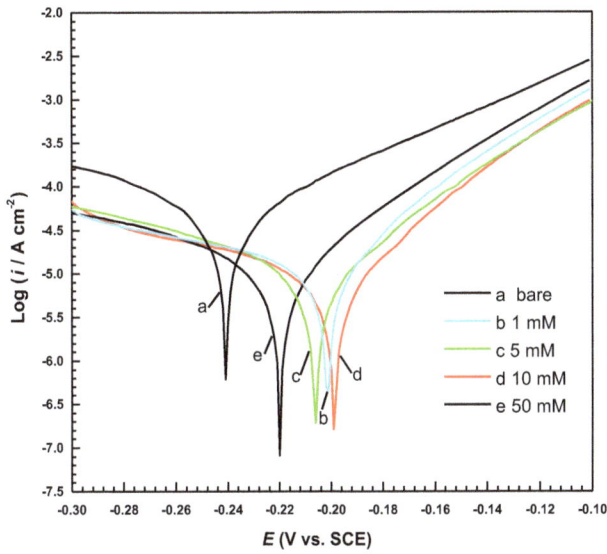

**Fig. 4** Anodic and cathodic polarization curves of the copper without and with AHA film formed at different concentrations for 3 h: *a* bare, *b* 1 mM, *c* 5 mM, *d* 10 mM, *e* 50 mM, in 3 wt% NaCl solution

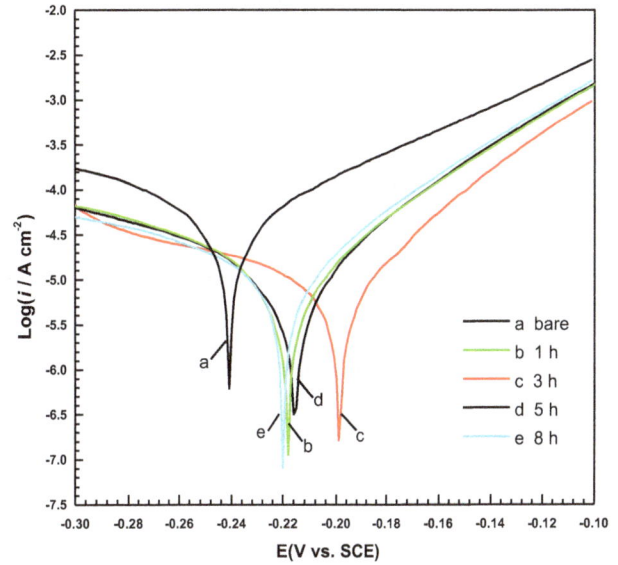

**Fig. 5** Anodic and cathodic polarization curves for copper electrodes without and with AHA films formed in 10 mM AHA solution for different time: *a* bare, *b* 1 h, *c* 3 h, *d* 5 h, *e* 8 h in 3 wt% NaCl solution

solution, molecular adsorption layer on the copper surface is compact for enhancing corrosion protection [32]. Similarly, the $Z_{mod}$ value of the bode plot (Fig. 1b) and phase angle value from phase angle plot (Fig. 1c) for the film-modified electrodes formed in 10 mM AHA solution are highest, indicating a superior protection performance. In Fig. 2a, semicircle diameter of the electrode with AHA formed in 10 mM AHA solution for 3 h assembly time comes to the greatest value; the same results have been observed in the bode plots (Fig. 2b) and phase angle plots (Fig. 2c). If the assembly time is less than 3 h, the AHA layer adsorbed on the electrode might not be dense enough while more than 3 h, accumulation of AHA molecules occurs on the surface, which also affects the compact structure of film and hinders inhibition of copper corrosion.

The equivalent circuit models for analyzing impedance characteristics of electrodes with and without AHA are displayed in Fig. 3. Such equivalent circuit models were selected, considering the possible adsorption fashion and the film structure of AHA at the copper surface as well as

evaluating the least error for each parameter routinely less than 10% and the Chi-square values ($\chi^2$) less than $1 \times 10^{-3}$.

In Fig. 3a, the equivalent circuit model of R(Q(RW))(QR) is fitting the Nyquist plot of the bare copper while R(C(R(Q(RW)))) equivalent circuit in Fig. 3b is for the AHA-modified copper.

$R_s$ is the solution resistance, and $R_{ct}$, the charge transfer resistance is attributed to the corrosion reaction at the electrode/solution interface [33]. $R_f$ represents the resistance of the film formed on the copper surface, and $W$ indicates the Warburg impedance. $Q_{dl}$ and $Q_f$ are defined as constant phase elements (CPE), representing film capacitance and a modified double-layer capacitance [34], described by the following equation [35]:

$$Z_{CPE} = Y_0^{-1}(j\omega)^{-n} \tag{1}$$

where $Y_0$ is the modulus, $j$ is the imaginary root, $\omega$ is the angular frequency and $n$ is the phase ($-1 \leq n \leq +1$).

**Table 4** Corrosion parameters obtained from potentiodynamic polarization curves for copper surface without and with AHA film formed in different concentrations of AHA solutions for 3 h assembly time, in 3 wt% NaCl solutions

| $C_{AHA}$ (mM) | $E_{corr}$ (V vs. SCE) | $i_{corr}$ (uA cm$^{-2}$) | $\beta_c$ (V dec$^{-1}$) | $\beta_a$ (V dec$^{-1}$) | $\eta$ (%) |
|---|---|---|---|---|---|
| Bare | −0.241 | 78.89 | 4.323 | 12.84 | – |
| 1 | −0.201 | 17.25 | 6.748 | 18.21 | 78.13 |
| 5 | −0.206 | 13.29 | 6.712 | 17.61 | 83.15 |
| 10 | −0.199 | 10.13 | 9.460 | 19.61 | 87.16 |
| 50 | −0.197 | 11.12 | 8.770 | 18.91 | 85.90 |

**Table 5** Polarization parameters for the copper without and with AHA film formed in 10 mM AHA solution for different time, in 3 wt% NaCl solution

| Time (h) | $E_{corr}$ (V vs. SCE) | $i_{corr}$ (uA cm$^{-2}$) | $\beta_c$ (v dec$^{-1}$) | $\beta_a$ (v dec$^{-1}$) | $\eta$ (%) |
|---|---|---|---|---|---|
| Bare | −0.241 | 78.89 | 4.323 | 12.84 | – |
| 1 | −0.218 | 13.2 | 4.207 | 17.81 | 83.27 |
| 3 | −0.199 | 10.13 | 9.460 | 19.61 | 87.16 |
| 5 | −0.216 | 12.2 | 5.454 | 17.61 | 84.54 |
| 8 | −0.220 | 17.09 | 3.055 | 17.71 | 78.34 |

**Table 6** Comparison of the inhibition efficiencies of different copper inhibitors

| Inhibitor | Concentration (mM) | Medium | $\eta$ (%) | References |
|---|---|---|---|---|
| AHA | 10 | 3 wt% NaCl | 93.5 | This work |
| DMTD | 10 | 0.5 M HCl | 84.3 | Qin et al. [40] |
| PU | 10 | 1 M NaCl | 76.0 | Scendo [41] |
| MPTT | 20 | 0.5 M NaCl | 94.4 | Chen et al. [42] |
| BBTD | 1 | 3 wt% NaCl | 87.6 | Zhang et al. [5] |
| AAP | 10 | 3 wt% NaCl | 90.6 | Song et al. [30] |

Relying on the different values of n, *CPE* may be inductance ($n = -1$, $Q = L$), resistance ($n = 0$, $Q = R$), Warburg impedance ($n = 0.5$, $Q = W$) or capacitance ($n = 1$, $Q = C$).

Electrochemical parameters calculated from EIS measurements for copper electrodes without and with AHA films are listed in Table 1 (bare copper), Table 2 (with films formed in different AHA concentrations solutions) and Table 3 (with films under different assembly time).

All the n values are over 0.5, indicating the relatively low corrosion of the electrode [36]. The inhibition efficiency ($\eta$%) is described in the following equation [37]:

$$\eta(\%) = \frac{R_p - R_p^0}{R_P} \times 100 \quad (2)$$

$R_p^0$ refers the polarization resistance of the bare copper, and $R_p$ is the polarization resistance of the AHA-modified electrode ($R_p$ is the sum of $R_{ct}$ and $R_f$) [38].

Observation from the three Tables reveals that the $R_p$ values for the AHA-modified copper electrodes increase visibly and the W values tend to decrease obviously compared to the bare copper electrode. Under the above-mentioned optimized condition, $R_p$ reaches the maximum value, W reaches the minimum value and the maximum inhibition efficiency reaches 93.5%. The RSD (relative standard deviation) of the inhibition efficiencies calculated from electrochemical impedance parameters for three copper electrodes with film is 0.13%. In all, impedance data suggest that the AHA monolayer has a remarkable protection behavior for copper.

**Polarization measurements**

Polarization curves of the electrodes with different AHA concentrations and different assembly time recorded in 3 wt% NaCl solutions are given in Figs. 4 and 5,

**Fig. 6** Micrographs of copper surfaces with AHA film formed in 10 mM AHA solution for 3 h assembly time (**a**), and with AHA film formed in same condition as a and exposed to 3 wt% NaCl for 5 h (corrosion time) at 298 K (**b**). Bare copper before (**c**) and after (**d**) exposed to 3 wt% NaCl for 5 h (corrosion time) at 298 K

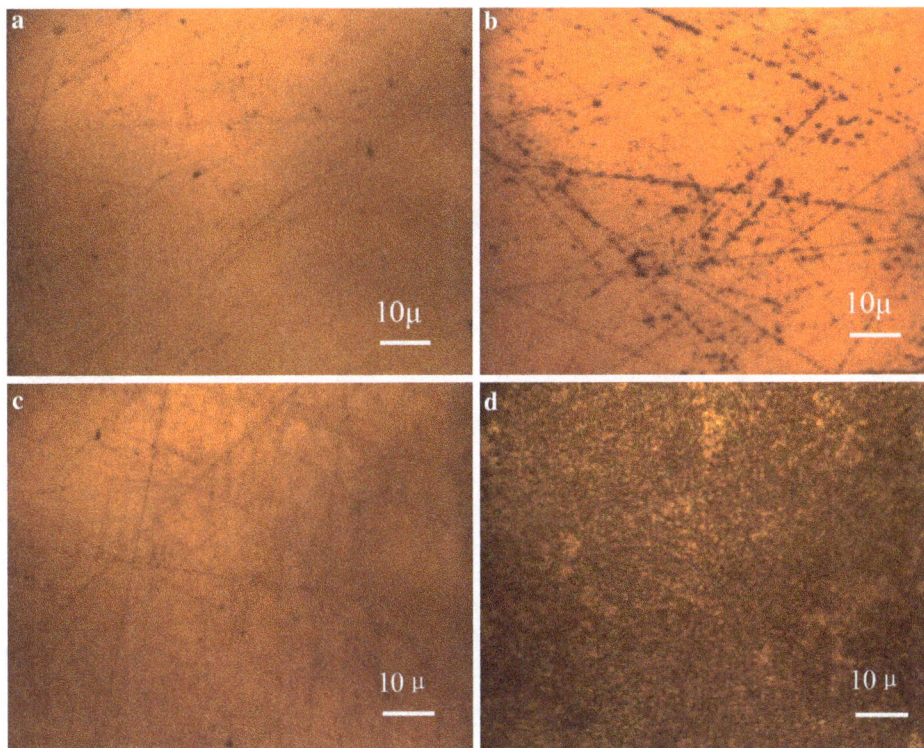

respectively. The reaction of the cathodic oxygen reduction is depicted as follows:

$$O_2 + 4e^- + 2H_2O \rightarrow 4OH^- \tag{3}$$

Also, in the presence of $Cl^-$, the anodic dissolution process of copper consists of the following steps (the ionization of copper with $Cl^-$ and the diffusion of $CuCl_2^-$ to the bulk solution) [39]:

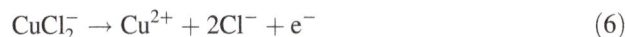

$$Cu + Cl^- \rightarrow CuCl + e^- \tag{4}$$

$$CuCl + Cl^- \rightarrow CuCl_2^- \tag{5}$$

$$CuCl_2^- \rightarrow Cu^{2+} + 2Cl^- + e^- \tag{6}$$

The related electrochemical parameters obtained from the polarization curves, such as cathodic and anodic Tafel slopes ($\beta_c$ and $\beta_a$), corrosion potential ($E_{corr}$), corrosion current density ($i_{corr}$) and the inhibition efficiency ($\eta\%$), are given in Tables 4 and 5. $E_{corr}$ and $i_{corr}$ were estimated by the method of Tafel extrapolation.

It can be seen from Figs. 4 and 5 that the Tafel slopes greatly change after the addition of AHA. Besides, current densities shift to lower values obviously at the same potential, compared with the bare electrode. Additionally, corrosion potential shifts to the positive direction significantly, indicating that AHA has more pronounced influence in the anodic dissolution process of copper with respect to the cathodic oxygen reduction. It might point out that in Fig. 4 the plot c (5 mM) looks tricky in comparison with others. We repeated the each experiment more than 5× and the results were quite similar. A possible reason was that

the best uniform and compact film of AHA at the copper surface just formed under optimal concentration and assembly time.

The comparison of this work with other inhibitors, such as 4-amino-antipyrine (AAP) [30], as well as 2,5-dimercapto-1,3,4-thiadiazole (DMTD) [40], bis-(1-benzotriazolymethylene)-(2,5-thiadiazoly)-disulfide (BBTD) [5], purine (PU) [41], 5-Mercapto-3-phenyl-1,3,4-thiadiazole-2-thione potassium (MPTT) [42], is listed in Table 6. By contrast, it can be seen that the AHA is an effective corrosion inhibitor for copper in 3 wt% NaCl.

## Microscopic analysis

Figure 6 shows the surface morphologies of AHA-modified copper and bare copper electrode corroded in 3 wt% NaCl solutions for 5 h. In Fig. 6d, after 5 h immersion in high-salt media, the porous structure and rough surface of the bare copper indicate serious corrosion occurrence while the corrosion behavior of the AHA-modified copper surface is less severe under the same condition (see Fig. 6b). The above observation confirms that AHA is an effective inhibitor for copper corrosion, acquired from electrochemical measurement.

## Adsorption isotherm

For further shedding insight on adsorption mechanism of inhibitor interaction with the copper surface, many

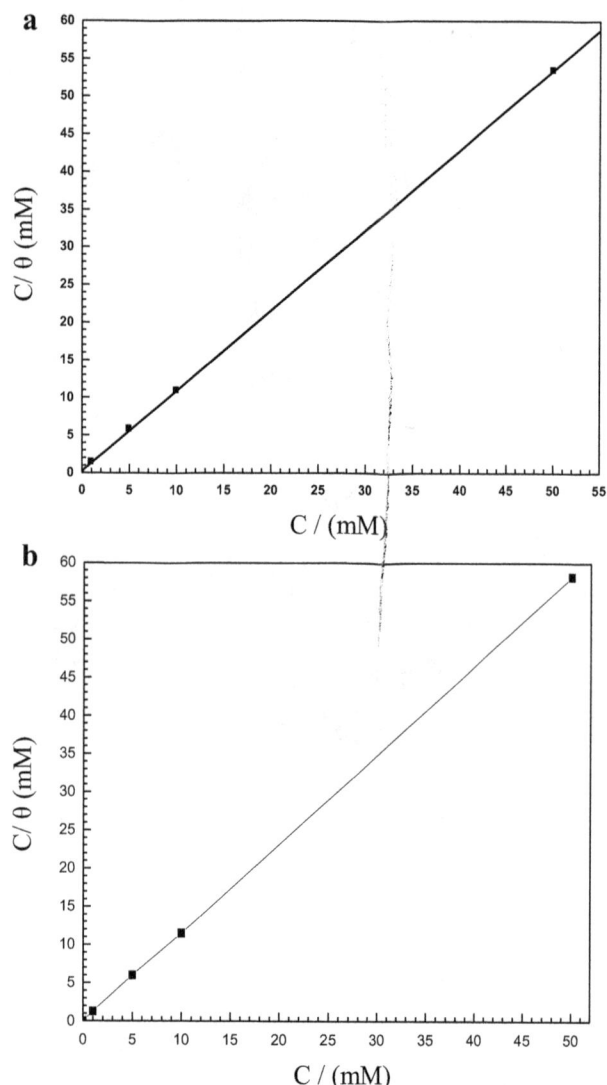

**Fig. 7** Langmuir adsorption isotherm plot for AHA film formed under the optimized condition at the copper surface in 3 wt% NaCl solution at 298 K according to **a** EIS results and **b** potentiodynamic polarization results

adsorption modes such as the Langmuir, Temkin and Frumkin isotherms could be investigated [43]. Therefore, the surface coverage ($\theta$) for different concentrations of AHA modification film was calculated from EIS parameters in Table 1 ($\eta(\%) = 100 \times \theta$).

It is found that the adsorption of AHA on the copper surface can be described by Langmuir adsorption isotherm equation [44, 45]:

$$\frac{c}{\theta} = \frac{1}{K_{ads}} + c \tag{7}$$

where $K_{ads}$ is the equilibrium constant, $\theta$ is the degree of surface coverage on the metal surface and $c$ is the AHA concentration. According to EIS experimental results, a plot of $c/\theta$ against $c$ (Fig. 7a) showed a straight line

($y = 1.06463x + 0.2659$), with the linear correlation coefficient $R^2 = 0.9999$, and its nearly unit slope confirms that the adsorption of AHA on the copper surface fits Langmuir adsorption isotherm. Additionally, adsorption isotherm drawn from potentiodynamic polarization results in Table 4 is also given in Fig. 7b. A plot of $c/\theta$ against $c$ shows a straight line ($y = 1.16282x + 0.0452$), with the linear correlation coefficient $R^2 = 0.9999$.

In addition, the standard free energy of adsorption, $\Delta G^0_{ads}$, is obtained from the following equation [46]:

$$\Delta G^0_{ads} = -RT \ln(55.5 K_{ads}) \tag{8}$$

where $R$ is the general gas constant (8.314 J mol$^{-1}$ K$^{-1}$), the temperature in Kelvin (298.15 K) is the absolute temperature, and the value of 55.5 mol L$^{-1}$ is the concentration of water in pure solution. The calculated $K_{ads}$ values are $3.76 \times 10^3$ M$^{-1}$ and $2.21 \times 10^4$ M$^{-1}$, based on EIS and potentiodynamic polarization results, respectively. The relevant $\Delta G^0_{ads}$ is $-30.4$ kJ mol$^{-1}$(EIS results) or $-34.8$ kJ mol$^{-1}$(potentiodynamic polarization results). The large negative value of $\Delta G^0_{ads}$ indicates that AHA was strongly adsorbed on the copper surface [47, 48].

## Raman studies

Figure 8a, b illustrates the normal Raman spectrum of AHA powder and SERS spectrum of the AHA-modified copper electrode formed in 10 mM AHA solution for 3 h assembly time, respectively. Table 7 shows density functional theory (DFT) calculation results for Raman and SERS analysis. The corresponding assignments for Raman spectral analysis were performed on the basis of B3LYP/6-31G(d) calculation. Also, to assign SERS spectrum of AHA, DFT calculation for geometry optimization and vibration modes based on B3LYP/LanL2DZ was performed with model AHA-Cu$_4$ [49]. The optimized geometry model AHA-Cu$_4$ can be seen in Fig. 9.

It should be mentioned that in inset of Fig. 8 with spectral range from 200 to 750 cm$^{-1}$, the Raman bands around 528 and 618 cm$^{-1}$ are from the oxide species of copper, indicating that the oxide layers were inevitably generated in ORC pretreatment process for SERS activity of copper [50].

Combined with the calculation results in Table 7, we can better understand the vibrational modes in Fig. 8a, b. In Fig. 8a, the Raman peaks at 967 and 989 cm$^{-1}$ represent N–O–H bending in plane and C–H bending in plane. The stretching vibration modes of N–O, C–C, C=O and C–H appear at 1088, 1366, 1619 and 2998 cm$^{-1}$. The asymmetric stretching vibration mode of C–H appears at 2941 cm$^{-1}$. The Raman peaks at 1390 cm$^{-1}$ represent OH bending and C-N bending out of plane.

**Fig. 8** *a* Normal Raman spectrum of AHA in powder, *b* SERS spectrum of AHA film at the copper electrode formed in 10 mM AHA solution for 3 h assembly time, *inset* showing the SERS spectrum (200–750 cm$^{-1}$) of the oxide layers

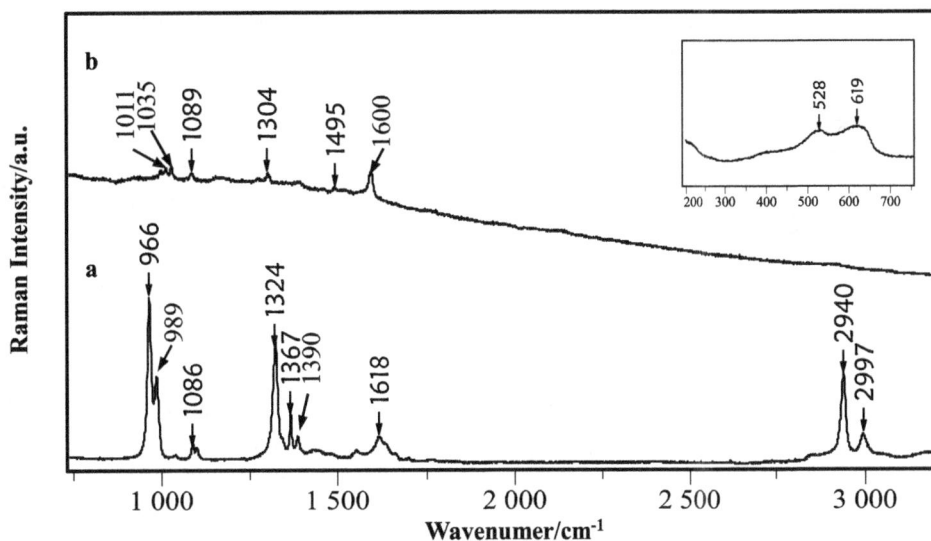

**Table 7** Assignment for Raman vibrational modes of AHA and SERS vibrational modes of AHA-Cu$_4$ based on DFT calculation

| Raman (cm$^{-1}$) | Calculated Raman (cm$^{-1}$) (B3LYP/6-31G(d)) | Approximate assignment | SERS (cm$^{-1}$) | Calculated SERS (cm$^{-1}$) (AHA-Cu$_4$) | Approximate assignment |
|---|---|---|---|---|---|
| 967$^s$ | 957 | N–O–H$^{ip.bend}$ | | 940 | C–N–H$^{str.}$ |
| 989$^s$ | 1011 | C–H$^{ip.bend}$ | 1011$^m$ | 1011 | C–H$^{ip.bend}$ |
| | 1022 | C–H$^{op.bend}$ | 1035$^m$ | 1020 | C–H$^{op.bend}$ |
| 1088$^w$ | 1096 | N–O$^{str.}$ | 1088$^s$ | 1085 | N–O$^{str.}$ |
| | | | | 1122 | O–H$^{str.}$ |
| | 1286 | N–H$^{bend}$ | 1305$^m$ | 1309 | N–H$^{bend}$ |
| 1327$^s$ | 1355 | CH$_3^{deformation}$ | | | |
| 1366$^m$ | 1371 | C–C$^{str.}$ | | | |
| 1390$^m$ | 1410 | OH$^{bend}$ | | 1410 | OH$^{bend}$ |
| | | C–N$^{op.bend}$ | | | C–N$^{op.bend}$ |
| | 1507 | C–N$^{str.}$ | 1495$^m$ | 1503 | C–N$^{str.}$·N–H$^{bend}$ |
| 1619$^m$ | 1667 | C=O$^{str.}$ | 1600$^s$ | 1627 | C=O$^{str.}$ |
| 2941$^s$ | 2945 | C–H$^{as.str.}$ | | 2967 | O–H$^{str.}$ |
| 2998$^m$ | 3019 | C–H$^{str.}$ | | 3066 | CH$_3^{deformation}$ |

Wavenumber is given in cm$^{-1}$

*w* weak, *m* medium, *s* strong, *as* asymmetric, *str.* stretching, *ip* in plane, *op* out plane, *bend* bending

In Fig. 8b, the peaks at 1011 and 1035 cm$^{-1}$ represent C–H bending in plane and C–H bending out of plane. The bands with medium intensities at 1305 and 1495 cm$^{-1}$ could be assigned to N–H bending and C-N stretching, while the strong bands at 1088 and 1600 cm$^{-1}$ are from N–O stretching and C=O stretching. According to the surface selection rule for SERS [51, 52] and the SERS mechanism [53, 54], the enhanced bands in the SERS spectrum might correspond to either the vibrational moieties attached to the surface or the vibration direction perpendicular to the metal surface; in contrast, the intensities of vibrational modes with parallel polarized components to the surface will be decreased. Thus, the N–O and C=O groups might be perpendicular to the surface, due to their high SERS intensities. The proposed adsorption model of AHA at the copper surface is shown in Fig. 10.

## Quantum chemistry calculations

As mentioned in SERS spectrum analysis, AHA molecule may adsorb at the copper surface by transferring electrons from the N–O and C=O groups to the unfilled hybrid orbital of copper; quantum chemical calculations are used to predict the feasible adsorption centers of a free single molecule on the bare metal surface to confirm the relationship between the AHA molecular structures and its inhibition property. In addition, note that all the calculated parameters obtained in the gas phase and the solvent effects were ignored.

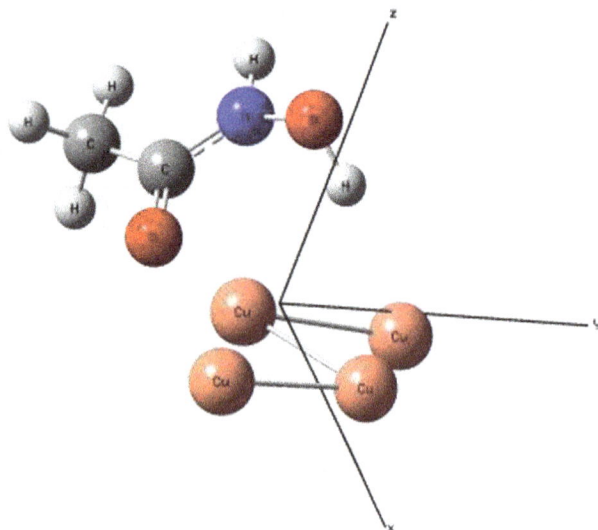

**Fig. 9** The geometry model of AHA-Cu₄ optimized by B3LYP/LANI2DZ method

**Table 8** Quantum chemical parameter comparison of some molecules in literature with the AHA molecule calculated with DFT method

| | $E_{LUMO}$ (kJ mol$^{-1}$) | $E_{HOMO}$ (kJ mol$^{-1}$) | $\Delta E$ (kJ mol$^{-1}$) | $\mu$ (Debye) |
|---|---|---|---|---|
| AHA | −175.9 | −427.6 | 251.8 | 5.413 |
| MPTT | −0.072 | −0.186 | 0.114 | 9.834 |
| AAP | −0.505 | −5.296 | 4.791 | 4.257 |

The geometry of the AHA is fully optimized based on the method of B3LYP/LANI2DZ. The illustration of the highest occupied molecular orbital (HOMO) and the lowest unoccupied molecular orbital (LUMO) is given in Fig. 11. The quantum chemical parameters, such as $E_{HOMO}$, $E_{LUMO}$, dipole moment, $\mu$, the energy gap, $\Delta E$ ($\Delta E = E_{LUMO} - E_{HOMO}$), are listed in Table 8.

Clearly in Table 8, the high value of $E_{HOMO}$ ($E_{HOMO} = -427.64$ kJ mol$^{-1}$) indicates the strong ability of AHA molecule donating electrons to form covalent bond with unoccupied d-orbitals of metal [55], while the low value of $E_{LUMO}$ ($E_{LUMO} = -175.86$ kJ mol$^{-1}$) indicates that the AHA molecule has a tendency to accept electrons from d-orbitals of metal to form back-donating bond. The values of the high dipole moment ($\mu = 5.4125$ D) and the low energy gap ($\Delta E = 251.78$ kJ mol$^{-1}$) facilitate adsorption to the copper surface and, therefore, enhance the inhibition efficiency [56], which is in good agreement with the experimental results.

It can be found in Fig. 11 that the HOMO is located within the region around the N–O and C=O groups, which could be regarded as the feasible sites for interaction with the copper surface. Additionally, the nitrogen atom and two oxygen atoms have large electron densities and their Mulliken atomic charges are −0.302, −0.357 and −0.363, respectively. Thus, AHA can adsorb on the copper surface by donating the electrons from the two O atoms to the d-orbitals of copper.

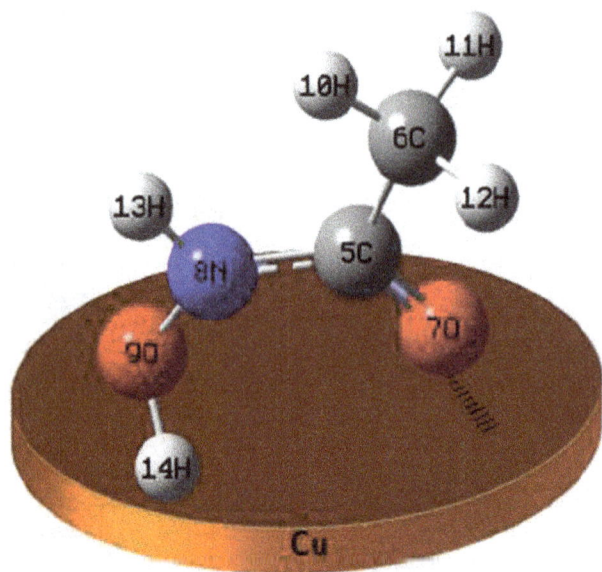

**Fig. 10** The proposed adsorption model of AHA on the copper surface

**Fig. 11** Molecular orbital plots for acetohydroxamic acid

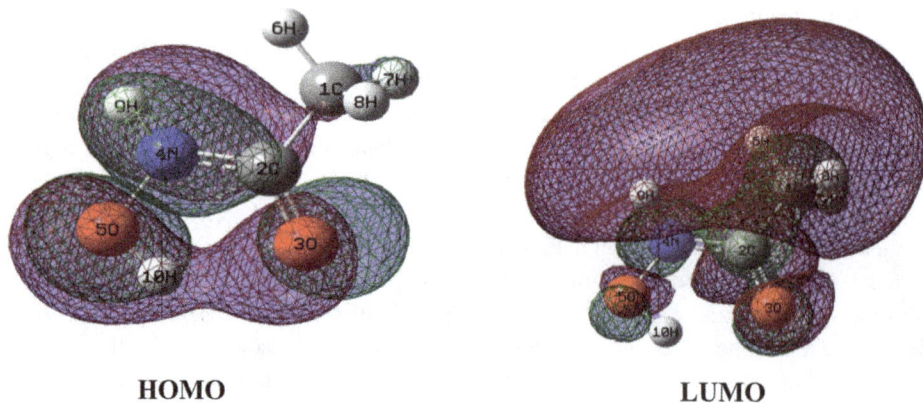

**HOMO**　　　　　**LUMO**

# Conclusions

Consequently, the above investigations confirm that the application of acetohydroxamic acid as an effective corrosion inhibitor for copper in 3 wt% NaCl solution is promising. Some conclusions could be drawn:

1. A relatively compact AHA film could be formed on the copper surface in 10 mM AHA solution for 3 h assembly time and its inhibition efficiency in 3 wt% NaCl solution reached 93.5%, supported by surface morphologies images recorded when the copper with optimized AHA film corroded in 3 wt% NaCl solution for 5 h.
2. Adsorption of AHA on the copper surface fits Langmuir adsorption isotherm.
3. AHA molecules adopted the N–O and C=O groups to attach onto the surface for constructing the protection film.

**Acknowledgements** This work is supported by the National Natural Science Foundation of China (No. 21475088) and International Joint Laboratory on Resource Chemistry (IJLRC), Shanghai Key Laboratory of Rare Earth Functional Materials, and Shanghai Municipal Education Committee Key Laboratory of Molecular Imaging Probes and Sensors.

# References

1. Issaadi S, Douadi T, Chafa S (2014) Adsorption and inhibitive properties of a new heterocyclic furan Schiff base on corrosion of copper in HCl 1 M: experimental and theoretical investigation. Appl Surf Sci 316:582–589
2. Lei YH, Ohtsuka T, Sheng N (2015) Corrosion protection of copper by polypyrrole film studied by electrochemical impedance spectroscopy and the electrochemical quartz microbalance. Appl Surf Sci 357:1122–1132
3. Song P, Guo XY, Pan YC, Shen S, Sun YQ, Wen Y, Yang HF (2013) Insight in cysteamine adsorption behaviors on the copper surface by electrochemistry and Raman spectroscopy. Electrochim Acta 89:503–509
4. Tian H, Li W, Cao K, Hou B (2013) Potent inhibition of copper corrosion in neutral chloride media by novel non-toxic thiadiazole derivatives. Corros Sci 73:281–291
5. Zhang DQ, Gao LX, Zhou GD (2004) Inhibition of copper corrosion by bis-(1-benzotriazolymethylene)-(2,5-thiadiazoly)-disulfide in chloride media. Appl Surf Sci 225:287–293
6. Yu YZ, Zhang DQ, Zeng HJ, Xie B, Gao LX, Lin T (2015) Synergistic effects of sodium lauroyl sarcosinate and glutamic acid in inhibition assembly against copper corrosion in acidic solution. Appl Surf Sci 355:1229–1237
7. Tian HW, Cheng YF, Li WH, Hou BR (2015) Triazolyl-acylhydrazone derivatives as novel inhibitors for copper corrosion in chloride solutions. Corros Sci 100:341–352
8. Ghelichkhah Z, Sharifi-Asl S, Farhadi K, Banisaied S, Ahmadi S, Macdonald DD (2015) L-cysteine/polydopamine nanoparticle-coatings for copper corrosion protection. Corros Sci 91:129–139
9. Wang D, Xiang B, Liang YP, Song S, Liu C (2014) Corrosion control of copper in 3.5 wt% NaCl solution by domperidone: experimental and theoretical study. Corros Sci 85:77–86
10. Hegazy MA, Hasan AM, Emara MM, Bakr MF, Youssef AH (2012) Evaluating four synthesized Schiff bases as corrosion inhibitors on the carbon steel in 1 M hydrochloric acid. Corros Sci 65:67–76
11. Mahdavian M, Ashhari S (2010) Corrosion inhibition performance of 2-mercaptobenzimidazole and 2-mercaptobenzoxazole compounds for protection of mild steel in hydrochloric acid solution. Electrochim Acta 55:1720–1724
12. Avci G (2008) Corrosion inhibition of indole-3-acetic acid on mild steel in 0.5 M HCl. Colloids Surf A Physicochem Eng Asp 317:730–736
13. Ansari KR, Quraishi MA, Singh A (2014) Schiff's base of pyridyl substituted triazoles as new and effective corrosion inhibitors for mild steel in hydrochloric acid solution. Corros Sci 79:5–15
14. Behpour M, Mohammadi N (2012) Investigation of inhibition properties of aromatic thiol self-assembled monolayer for corrosion protection. Corros Sci 65:331–339
15. Lokesh KS, Keersmaecker MD, Elia A, Depla D, Dubruel P, Vandenabeele P, Vlierberghe SV, Adriaens A (2012) Adsorption of cobalt(II) 5,10,15,20-tetrakis(2-aminophenyl)-porphyrin onto copper substrates: characterization and impedance studies for corrosion inhibition. Corros Sci 62:73–82
16. Golestani Gh, Shahidi M, Ghazanfari D (2014) Electrochemical evaluation of antibacterial drugs as environment-friendly inhibitors for corrosion of carbon steel in HCl solution. Appl Surf Sci 308:347–362
17. Zhang DQ, Gao LX, Zhou GD (2005) Inhibition of copper corrosion in aerated hydrochloric acid solution by amino-acid compounds. J Appl Electrochem 35:1081–1085
18. Ramezanzadeh B, Vakili H, Amini R (2015) The effects of addition of poly(vinyl) alcohol (PVA) as a green corrosion inhibitor to the phosphate conversion coating on the anticorrosion and adhesion properties of the epoxy coating on the steel substrate. Appl Surf Sci 327:174–181
19. Yang HF, Feng J, Liu YL, Yang Y, Zhang ZR, Shen GL, Yu RQ (2004) Electrochemical and surface enhanced Raman scattering spectroelectrochemical study of phytic acid on the silver electrode. J Phys Chem B 108:17412–17417
20. Dost K, Tokul O (2006) Determination of phytic acid in wheat and wheat products by reverse phase high performance liquid chromatography. Anal Chim Acta 558:22–27
21. El-Sayed AR, Harm U, Mangold KM, Fürbeth W (2012) Protection of galvanized steel from corrosion in NaCl solution by coverage with phytic acid SAM modified with some cations and thiols. Corros Sci 55:339–350
22. Reddy AS, Kumar MS, Reddy GR (2000) A convenient method for the preparation of hydroxamic acids. Tetrahedron Lett 41:6285–6288
23. Farkas E, Enyedy ÉA, Csóka H (1999) A comparison between the chelating properties of some dihydroxamic acids, desferrioxamine B and acetohydroxamic acid. Polyhedron 18:2391–2398
24. Farkas E, Enyedy ÉA, Zekany L, Deak G (2001) Interaction between iron(II) and hydroxamic acids: oxidation of iron(II) to iron(III) by desferrioxamine B under anaerobic conditions. J Inorg Biochem 83:107–114
25. Alagta A, Felhösi I, Kálmán E (2007) Hydroxamic acid corrosion inhibitor for steel in aqueous solution. Mater Sci Forum 537–538:81–88
26. Ezznaydy G, Shaban A, Telegdi J, Ouaki B, Hajjaji SE (2015) Inhibition of copper corrosion in saline solution by mono-hydroxamic acid. J Mater Environ Sci 6(7):1819–1823
27. Telegdi J, Rigó T, Kálmán E (2005) Molecular layers of hydroxamic acids in copper corrosion inhibition. J Electroanal Chem 582:191–201
28. Brown GM, Hope GA (1995) In-situ spectroscopic evidence for the adsorption of $SO_4^{2-}$ ions at a copper electrode in sulfuric acid solution. J Electroanal Chem 382:179–182

29. Gao X, Zhao CC, Lu HF, Gao F, Ma H (2014) Influence of phytic acid on the corrosion behavior of iron under acidic and neutral conditions. Electrochim Acta 150:188–196

30. Hong S, Chen W, Luo HQ, Li NB (2012) Inhibition effect of 4-amino-antipyrine on the corrosion of copper in 3 wt% NaCl solution. Corros Sci 57:270–278

31. Pan YC, Wen Y, Xue LY, Guo XY, Yang HF (2012) Adsorption behavior of methimazole monolayers on a copper surface and its corrosion inhibition. J Phys Chem C 116:3532–3538

32. Hong S, Chen W, Zhang Y, Luo HQ, Li M, Li NB (2013) Investigation of the inhibition effect of trithiocyanuric acid on corrosion of copper in 3.0 wt% NaCl. Corros Sci 66:308–314

33. Reznik VS, Akamsin VD, Khodyrev YP, Galiakberov RM, Efremov YY, Tiwari L (2008) Mercaptopyrimidines as inhibitors of carbon dioxide corrosion of iron. Corros Sci 50:392–403

34. Skale S, Doleček V, Slemnik M (2007) Substitution of the constant phase element by Warburg impedance for protective coatings. Corros Sci 49:1045–1055

35. Soltani N, Behpour M, Oguzie EE, Mahluji M, Ghasemzadeh MA (2015) Pyrimidine-2-thione derivatives as corrosion inhibitors for mild steel in acidic environments. RSC Adv 5:11145–11162

36. Li CC, Guo XY, Shen S, Song P, Xu T, Wen Y, Yang HF (2014) Adsorption and corrosion inhibition of phytic acid calcium on the copper surface in 3 wt% NaCl solution. Corros Sci 83:147–154

37. Sherif EM, Park SM (2006) Effects of 2-amino-5-ethylthio-1,3,4-thiadiazole on copper corrosion as a corrosion inhibitor in aerated acidic pickling solutions. Electrochim Acta 51:6556–6562

38. Pan YC, Wen Y, Guo XY, Song P, Shen S, Du YP, Yang HF (2013) 2-Amino-5-(4-pyridinyl)-1,3,4-thiadiazole monolayers on copper surface: observation of the relationship between its corrosion inhibition and adsorption structure. Corros Sci 73:274–280

39. Li W, Hu LC, Zhang SG, Hou BR (2011) Effects of two fungicides on the corrosion resistance of copper in 3.5% NaCl solution under various conditions. Corros Sci 53:735–745

40. Qin TT, Li J, Luo HQ, Li M, Li NB (2011) Corrosion inhibition of copper by 2,5-dimercapto-1,3,4-thiadiazole monolayer in acidic solution. Corros Sci 53:1072–1078

41. Scendo M (2007) The effect of purine on the corrosion of copper in chloride solutions. Corros Sci 49:373–390

42. Chen W, Hong S, Li HB, Luo HQ, Li M, Li NB (2012) Protection of copper corrosion in 0.5 M NaCl solution by modification of 5-mercapto-3-phenyl-1,3,4-thiadiazole-2-thione potassium self-assembled monolayer. Corros Sci 61:53–62

43. Zhang F, Tang YM, Cao Z, Jing WH, Wu ZL, Chen YZ (2012) Performance and theoretical study on corrosion inhibition of 2-(4-pyridyl)-benzimidazole for mild steel in hydrochloric acid. Corros Sci 61:1–9

44. Zhou X, Yang HY, Wang FH (2011) [BMIM] BF4 ionic liquids as effective inhibitor for carbon steel in alkaline chloride solution. Electrochem Acta 56:4268–4275

45. Abd El Rehim SS, Sayyah SM, El-Deeb MM, Kamal SM, Azooz RE (2016) Adsorption and corrosion inhibitive properties of P(2-aminobenzothiazole) on mild steel in hydrochloric acid media. Int J Ind Chem 7:39–52

46. Benahmed M, Djeddi N, Akkal S, Laouar H (2016) Saccocalyx satureioides as corrosion inhibitor for carbon steel in acid solution. Int J Ind Chem 7:109–120

47. Madkour LH, Elshamy IH (2016) Experimental and computational studies on the inhibition performances of benzimidazole and its derivatives for the corrosion of copper in nitric acid. Int J Ind Chem 7:195–221

48. Morad MS (2008) Corrosion inhibition of mild steel in sulfamic acid solution by S-containing amino acids. J Appl Electrochem 38:1509–1518

49. Tao S, Yu LJ, Pang R, Huang YF, Wu DY, Tian ZQ (2013) Binding interaction and Raman spectra of p–π conjugated molecules containing $CH_2/NH_2$ groups adsorbed on silver surfaces: a DFT study of wagging modes. J Phys Chem C 117:18891–18903

50. Shen S, Guo XY, Song P, Pan YC, Wang HQ, Wen Y, Yang HF (2013) Phytic acid adsorption on the copper surface: observation of electrochemistry and Raman spectroscopy. Appl Surf Sci 276:167–173

51. Moskovits M (1982) Surface selection rules. J Chem Phys 77:4408–4416

52. Moskovits M, Suh JS (1984) Surface selection rules for surface-enhanced Raman spectroscopy: calculations and application to the surface-enhanced Raman spectrum of phthalazine on silver. J Phys Chem 88:5526–5530

53. Félidj N, Aubard J, Lévi G, Krenn JR, Salerno M, Schider G, Lamprecht B, Leitner A, Aussenegg FR (2002) Controlling the optical response of regular arrays of gold particles for surface-enhanced Raman scattering. Phys Rev B 65:075419

54. McFarland AD, Young MA, Dieringer JA, VanDuyne RP (2005) Wavelength-scanned surface-enhanced Raman excitation spectroscopy. J Phys Chem B 109:11279–11285

55. Akalezi CO, Oguzie EE (2016) Evaluation of anticorrosion properties of *Chrysophyllum albidum* leaves extract for mild steel protection in acidic media. Int J Ind Chem 7:81–92

56. Yan Y, Li WH, Cai LK, Hou BR (2008) Electrochemical and quantum chemical study of purines as corrosion inhibitors for mild steel in 1 M HCl solution. Electrochem Acta 53:5953–5960

# 3

# Sorption behavior of some lanthanides on polyacrylamide stannic molybdophosphate as organic–inorganic composite

E. A. Abdel-Galil[1] · A. B. Ibrahim[1] · M. M. Abou-Mesalam[1]

**Abstract** Sorption behavior of some lanthanide ions such as lanthanum and samarium ions on polyacrylamide stannic molybdophosphate {PASnMoP} as organic–inorganic composite has been investigated. The distribution coefficients of $La^{3+}$ and $Sm^{3+}$ ions in different pH media on PASnMoP were determined with the selectivity order $La^{3+} > Sm^{3+}$. Capacity of PASnMoP for $La^{3+}$ and $Sm^{3+}$ ions was determined and found 24.66 and 15.38 mg g$^{-1}$ for $La^{3+}$ and $Sm^{3+}$ ions, respectively. The resin dosage dependence for the sorption behavior for $La^{3+}$ and $Sm^{3+}$ ions on polyacrylamide stannic molybdophosphate was conducted. The adsorption isotherms were described by means of Langmuir and Freundlich isotherms for $La^{3+}$ and $Sm^{3+}$ ions. The Langmuir model represented the adsorption process better than the Freundlich model. The kinetic data were tested using Logergren-first-order and Pseudo-second-order kinetic models. The data correlated well with the Pseudo-second-order kinetic model, indicating that the chemical adsorption was the rate limiting step. Thermodynamic parameters $\Delta G°$, $\Delta H°$ and $\Delta S°$ were also calculated and the data showed that the ion exchange of $La^{3+}$ and $Sm^{3+}$ ions on PASnMoP was spontaneous and endothermic in nature.

**Keywords** Distribution coefficient · Lanthanides · Sorption isotherm · Polyacrylamide stannic molybdophosphate · Composite · Thermodynamic

## Introduction

Rare earth elements (REE) have been increasingly used in the field of chemical engineering, nuclear energy, optical, magnetic, luminescence and laser materials, high-temperature superconductors, secondary batteries and catalysis [1–6]. Lanthanum, one of the most abundant of the lanthanides, is an important element of mischmetal and hydrogen-absorbing alloy [3], Neodymium is the raw material used in high-strength permanent magnets (Nd-B-Fe), making it less expensive than samarium-cobalt permanent magnets [7]. Yttrium is an important element and in great demand in astronavigation, luminescence, nuclear energy and metallurgical industries [8]. Organic polymers as ion exchangers are well known for their uniformity, chemical stability and control of their ion-exchange properties through synthetic methods. The inorganic ion-exchange materials besides other advantages are important in being more stable to high temperature and radiation field than the organic ones [9]. To obtain a combination of these advantages associated with polymeric and inorganic materials as ion exchangers, attempts have been made to develop polymeric–inorganic composite ion exchangers by incorporation of organic monomers in the inorganic matrix [10]; few such excellent ion-exchange materials have been developed and are successfully being used in chromatographic techniques [11–13].

In the present work, polyacrylamide stannic molybdophosphate was synthesized as reported earlier [14]. Distribution coefficient and separation factors ($\propto$) of $La^{3+}$ and $Sm^{3+}$ on polyacrylamide stannic molybdophosphate were determined. Capacity, sorption isotherms and thermodynamic parameters have been calculated for the sorption of $La^{3+}$ and $Sm^{3+}$ ions on polyacrylamide stannic molybdophosphate.

✉ E. A. Abdel-Galil
ezzat_20010@yahoo.com

[1] Atomic Energy Authority, Hot Labs. Center, P.O. 13759, Cairo, Egypt

## Materials and methods

All reagents and chemicals were of analytical grade purity and used without further purification. pH measurements were performed using pH meter of the bench, model 601A, USA. The concentration of $La^{3+}$ and $Sm^{3+}$ ions in solutions was measured by UV spectrophotometer UV-1700.

Polyacrylamide stannic molybdophosphate {PASnMoP} as organic–inorganic composite was prepared as described earlier by Abdel-Galil [14]. Polyacrylamide was prepared by mixing equal volume of acrylamide and potassium persulfate. A viscous solution was obtained by heating the mixture gently at 70 °C with continuous stirring. Inorganic precipitate of Sn(IV) molybdophosphate was prepared at 25 °C by mixing equal volumes of the solutions of stannic chloride (0.1 M), ammonium molybdate (0.1 M) and orthophosphoric acid (1 M). The yellow precipitate was obtained when the pH of the mixture was adjusted to 1.05 by adding aqueous ammonia ($NH_4OH$). The viscous solution of polyacrylamide was added to the yellow inorganic precipitate of Sn(IV) molybdophosphate and mixed thoroughly with continuous stirring. The resultant yellow colored slurry was refluxed for 3 h at a temperature of $70 \pm 5$ °C and the color of the slurry changed from yellow to greenish color. The resultant greenish colored slurry was kept for 24 h at room temperature for digestion. The supernatant liquid was decanted and gel was filtered using a centrifuge (about $10^4$ rpm) and dried at $50 \pm 1$ °C. The product was crashed to obtain small granules and converted to $H^+$-form by treating with 1 M $HNO_3$ for 24 h with occasional shaking intermittently replacing the supernatant liquid with fresh acid and the color of the product became yellow. The excess acid was removed after several washing with DMW, dried at 50 °C and sieved to obtain particles of particular size range (0.115–0.375 mm). The percentage of yield and physical appearance of beads were selected for further studies.

## Sorption studies

The distribution coefficients ($k_d$) of $La^{3+}$ and $Sm^{3+}$ ions on polyacrylamide Sn(IV) molybdophosphate were determined by batch equilibration technique, as a function of different pH values. 0.1 g of {PASnMoP} was shaken with 10 ml of 50 ppm of $La^{3+}$ and $Sm^{3+}$ ions solution at V/m ratio of 100 ml/g. The mixture was placed overnight (sufficient to attain the equilibrium) in a shaker thermostat adjusted at 25, 45 or $65 \pm 1$ °C. After equilibrium, the solutions were separated by centrifugation and the concentration of $La^{3+}$ and $Sm^{3+}$ ions in the solution was determined using UV spectrophotometer. The pH values of the solutions were measured before and after equilibrium

using pH meter. The distribution coefficient ($k_d$) and separation factor ($\propto$) were evaluated by;

$$k_d = \frac{A_o - A_f}{A_f} \times \frac{V}{m} (ml/g) \tag{1}$$

$$\propto = \frac{(k_d)A}{(k_d)B} \tag{2}$$

where $A_o$ is the concentration of the ions in solution before equilibrium, $A_f$ is the concentration of the ions in solution after equilibrium, $V$ is the volume of the solution (ml), $m$ is the weight of the exchanger (g), and $\propto$ is the separation factor between two neighboring ions A and B.

## Capacity measurements

The capacity of polyacrylamide stannic(IV) molybdophosphate for $La^{3+}$ and $Sm^{3+}$ ions was determined by the repeated batch technique, by equilibrating 0.05 g of PASnMoP ion exchanger with 5 ml of 100, 200, 400, 600, 800 and/or 1000 ppm of $La^{3+}$ and/or $Sm^{3+}$ ion solutions on a shaker thermostat adjusted at $25 \pm 1$ °C. After equilibrium, the solution was separated and repeated until no further sorption occurs. The capacity was calculated using the following equation:

$$Capacity = Uptake \times \frac{V}{m} \times C_o \ mg/g \tag{3}$$

where $C_o$ is the initial ion concentration in solution, $V$ is the solution volume (ml) and $m$ is the weight of the exchanger (g).

## Adsorption isotherm

The adsorption isotherms were done, as it is well known, by a gradual increase in the concentration of sorbate ion in solution and measuring the amount sorbed at each equilibrium concentration. The degree of sorption showed therefore is a function of the concentration of sorbate ions only. The adsorption isotherms were carried out with different initial concentrations varying from 100 to 1000 ppm at different reaction temperatures 25, 45 and/or $65 \pm 1$ °C and at constant V/m value of 100 ml/g and pH 3. After equilibrium, the respective mixture was filtered and then the concentration of $La^{3+}$ and $Sm^{3+}$ was measured.

## Kinetic measurements

The kinetic analysis of the adsorption process for $La^{3+}$ and $Sm^{3+}$ ions on polyacrylamide stannic(IV) molybdophosphate was carried out by mixing the exchanger with metal ions solution at 50 ppm with a V/m ratio of 100 ml/g in a shaker thermostat at $25 \pm 1$ °C. The solution was separated at different time intervals and analyzed to determine the metal ion concentration in solution using UV spectrophotometer for $La^{3+}$ and $Sm^{3+}$. The extent of sorption was determined from the equation.

$$\text{Sorption} \% = \{(A_i - A_f)/A_i\} \times 100 \tag{4}$$

where $A_i$ and $A_f$ are the initial and final concentrations of metal ions in solution.

## Results and discussion

Polyacrylamide stannic molybdophosphate as organic–inorganic cation exchange material was prepared as described earlier by Abdel-Gelil [14].

The influence of adsorbent dosage on the adsorption of $La^{3+}$ and $Sm^{3+}$ was studied and the data are shown in Fig. 1. The adsorption of the metals increased with increasing the dosage of the exchanger. This increasing may be related to the increase in the surface area of the exchanger by increasing the adsorbent dosage, so more available active sides on the adsorbent and thus making easier penetration of metal ions to the sorption sides [15]. The data also indicate that the adsorption was almost constant at higher dosage than 0.1 g and the optimum dosage used in all investigation was 0.1 g with batch factor 100 ml/g.

The cation exchange process between $H^+$ (solid phase) and $H^+$ in solution can be represented by the following reaction:

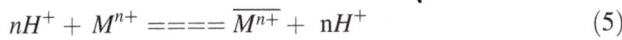

$$nH^+ + M^{n+} ==== \overline{M^{n+}} + nH^+ \tag{5}$$

where a bar over a character denotes the concentration of $M^{n+}$ on the solid and unbar denotes the concentration of $M^{n+}$ in the solution phase.

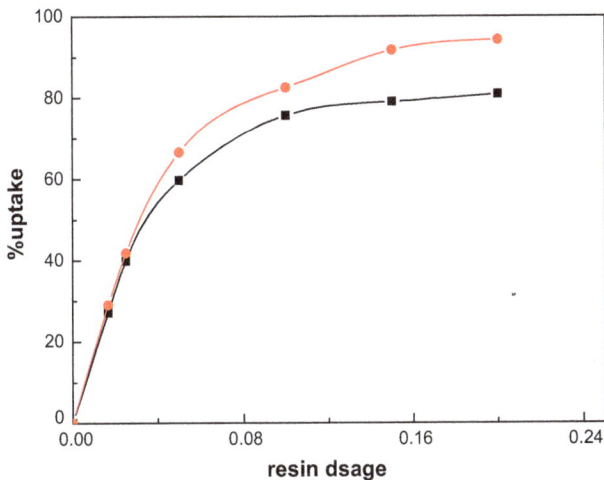

The distribution coefficient ($k_d$) values are defined by the following relation:

$$k_d = \frac{[\overline{M^{n+}}]}{[M^{n+}]} \tag{6}$$

The selectivity $K_H^M$ can be defined by the following equation:

$$K_H^M = \frac{[M^{n+}][\overline{H^+}]^n}{[\overline{H^+}][M^{n+}]} \tag{7}$$

From Eqs. (6 and 7), the distribution coefficient ($k_d$) can be written in the simple form as follows:

$$k_d = K_H^M \times \frac{[\overline{H^+}]^n}{[H^+]^n} \tag{8}$$

(or)

$$\log k_d = \log K_H^M \times [\overline{H^+}]^n - n \log[H^+] \tag{9}$$

From very dilute solutions, the activity coefficient in solution is very small and can be neglected, then

$$[\overline{M^{n+}}] << [\overline{H^+}] \quad \text{and} \quad [M^{n+}] << [H^+]$$

Then, the term $\log K_H^M \times [\overline{H^+}]$ is considered as constant and, thus, Eq. (9) can be reduced in the following form:

$$\text{Log } k_d = -n \log[H^+] = n \text{ pH} \tag{10}$$

When $\log k_d$ values are plotted against log pH, a straight line with slope n obtained, where n is refer to the valance of the sorbed ions. Figure 2 and Table 1 show the dependency of $k_d$ values of $La^{3+}$ and $Sm^{3+}$ ions onto PASnMoP on the pH of the ion medium. The linear relationships between log $k_d$ and pH values were observed for $La^{3+}$ and $Sm^{3+}$ ions with slopes 0.246 and 0.182, respectively. These slopes did not equal to the valence of the metal ions sorbed which

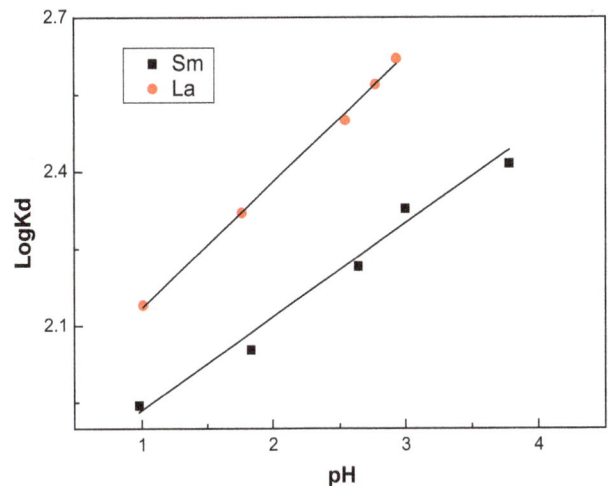

Fig. 1 Effect of ion exchanger dosage on sorption of $Sm^{3+}$ and $La^{3+}$ on {PASnMoP}

Fig. 2 Log $k_d$ of $La^{3+}$ and $Sm^{3+}$ ions versus pH on {PASnMoP}

**Table 1** $K_d$ and separation factor values for $La^{3+}$ and $Sm^{3+}$ ions on PASnMoP

| pH | $k_d$, ml/g (separation factor, $\propto$) | |
|---|---|---|
| | $La^{3+}$ | $Sm^{3+}$ |
| 0.98 | 134.89 | 87.096 |
| | | (1.548) |
| 1.69 | 199.53 | 114.82 |
| | | (1.737) |
| 2.49 | 316.23 | 158349 |
| | | (1.995) |
| 2.72 | 363.08 | 177.83 |
| | | (2.042) |
| 2.90 | 398.11 | 190.55 |
| | | (2.089) |

prove the non-ideality of the exchange reaction between $La^{3+}$ and $Sm^{3+}$ ions and PASnMoP. These findings cannot be explained only in terms of electrostatic interaction between the hydrated cations and the anionic sites in the exchanger. It may, therefore, be considered that the dependence of $k_d$ for cations cannot be understood by a purely columbic interaction with the anionic sites, but also may be due to the formation of a covalent bond similar to a weakly acidic resin; such interaction would be closely related to the ionic potential of the cations [14].

The $k_d$ values and separation factors of $La^{3+}$ and $Sm^{3+}$ ions in different pH on PASnMoP are summarized in Table 1. From the data in Table 1, we found that the selectivity order of the investigated cations on PASnMoP in the same conditions has the following sequence; $La^{3+} > Sm^{3+}$. This selectivity sequence is in accordance with the ionic radii; the ions with smaller ionic radii easily exchanged and move faster than the ions with greater ionic radii [14]. Also from Table 1, the separation factors between $La^{3+}$ and $Sm^{3+}$ metal ions on PASnMoP are relatively high and predict some selective separation of these ions which were available on PASnMoP.

Figures 3 and 4 show the $k_d$ values for $La^{3+}$ and $Sm^{3+}$ ions, respectively, in different media at different reaction temperatures. The effect of reaction temperature on the adsorption was carried out in the temperature range $25–65 \pm 1$ °C. From these figures, we found that the $K_d$ values are increased for $La^{3+}$ and $Sm^{3+}$ ions on PASnMoP with increasing the reaction temperature from 25 to $65 \pm 1$ °C. This behavior may be due to the endothermic nature of the system for the reaction of $La^{3+}$ and $Sm^{3+}$ ions on PASnMoP. Also, the increasing $k_d$ with increasing the reaction temperatures may be attributed to the increasing mobility of $La^{3+}$ and $Sm^{3+}$ ions with increasing the reaction temperature [16]. Similar results were obtained by Abou-Mesalam et al. for the distribution coefficients of

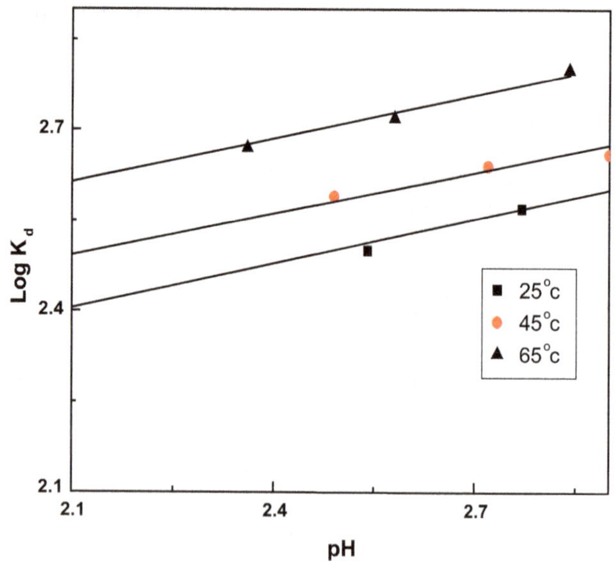

**Fig. 3** Log $k_d$ of $La^{3+}$ ion as a function of pH on PASnMoP at different reaction temperatures

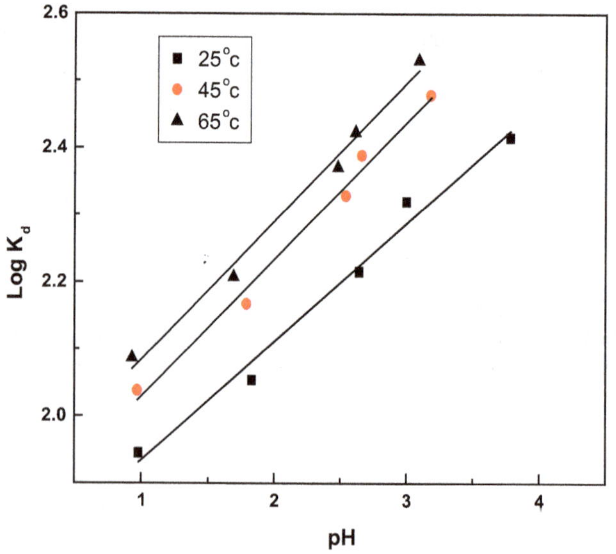

**Fig. 4** Log $k_d$ of $Sm^{3+}$ ion as a function of pH on PASnMoP at different

$^{22}$Na, $^{60}$Co and $^{152,154}$Eu ions on titanium antimonate [17–19].

Figure 5 shows the linear relation between in $k_d$ of $La^{3+}$ and $Sm^{3+}$ ions on PASnMoP and 1/T according to the Van't Hoff relation [17];

$$\ln k_d = \frac{\Delta S^\circ}{T} - \frac{\Delta H^\circ}{RT} \qquad (11)$$

where $\Delta S^\circ$ is the entropy change of adsorption, $\Delta H^\circ$ is the enthalpy change of adsorption, $R$ is the gas constant, and $T$ is the absolute temperature.

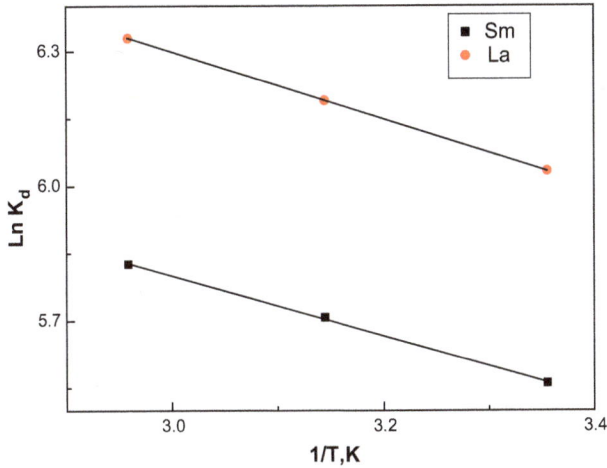

**Fig. 5** Van't Hoff plot of the adsorption of $La^{3+}$ and $Sm^{3+}$ ions on PASnMoP

**Table 2** Thermodynamic parameters for adsorption of $La^{3+}$ and $Sm^{3+}$ ions on PASnMoP

| Metal | Temp. | $\Delta G^\circ$ | $\Delta H^\circ$ | $\Delta S^\circ$ |
|-------|-------|--------|--------|--------|
| $La^{3+}$ | 298 | −14.93 | 14.32 | 98 |
|  | 318 | −16.36 |  | 96 |
|  | 338 | −17.78 |  | 94.9 |
| $Sm^{3+}$ | 298 | −13.77 | 12.79 | 89 |
|  | 318 | −15.08 |  | 87 |
|  | 338 | −16.37 |  | 86 |

It was found that the distribution coefficient of $La^{3+}$ and $Sm^{3+}$ ions increased with increasing temperature from 298 to 338°K (i.e., the distribution coefficient decreased with increasing $1/T$) as shown in Fig. 5. This increase in the extent of adsorption with the increase in temperature was attributed to acceleration of some originally slow adsorption steps and creation of some new active sites on the adsorbent surfaces [20, 21]. From the slopes and intercepts of these straight lines represented in Fig. 5, the enthalpy change of adsorption ($\Delta H^\circ$) and entropy change of adsorption ($\Delta S^\circ$) were evaluated and represented in Table 2.

As shown in Table 2, the positive values of ($\Delta H^\circ$) indicate the endothermic nature of the adsorption process [14]. The positive values of $\Delta S^\circ$ indicate that the increased randomness of solid solution interface during the adsorption of these cations on PASnMoP [16]. The data in Table 2 indicate that the values of $\Delta H^\circ$ for $La^{3+}$ and $Sm^{3+}$ ions on PASnMoP are greater than 12.6 $KJmol^{-1}$ which indicated the presence of other mechanism for the adsorption of $La^{3+}$ and $Sm^{3+}$ ions on PASnMoP beside ion-exchange mechanism [22, 23]. These results are supported also from the data of $K_d$ where the slope of the

**Table 3** Capacity of PASnMoP for $La^{3+}$ and $Sm^{3+}$ ions at $25 \pm 1$ °C

| Concentration, mg/L | Capacity, mg/g | |
|---------------------|----------|----------|
|  | $La^{3+}$ | $Sm^{3+}$ |
| 100 | 19.49 | 17.34 |
| 200 | 37.88 | 31.29 |
| 400 | 63.48 | 54.98 |
| 600 | 82.74 | 69.86 |
| 800 | 97.07 | 77.85 |
| 1000 | 98.85 | 77.69 |

linear relationship between $K_d$ and pH is not equal to the valence of $La^{3+}$ and $Sm^{3+}$ ions.

The free energy change of specific adsorption $\Delta G^\circ$ was calculated using the relation:

$$\Delta G^\circ = \Delta H^\circ - T\Delta S^\circ \quad (12)$$

and

$$\Delta G^\circ = RT \ln k_d \quad (13)$$

The negative values of free energy change $\Delta G^\circ$ represented in Table 2 indicate that the adsorption process is spontaneous and indicates the preferable adsorption of these cations on PASnMoP compared with $H^+$ ion [14, 16].

The capacity of PASnMoP for $La^{3+}$ and $Sm^{3+}$ ions was studied and the data are tabulated in Table 3. From Table 3, it is clear that the capacity of polyacrylamide Sn(IV) molybdophosphate samples for $La^{3+}$ and $Sm^{3+}$ has the following order: $La^{3+} > Sm^{3+}$. This sequence is in accordance with the hydrated radii of the exchanged ions. The ions with smaller hydrated radii enter the pores of the exchanger, resulting in higher adsorption [24, 25].

## Sorption isotherm

Sorption equilibrium is usually described by an isotherm equation whose parameters express the surface properties and affinity of the sorbent, at a fixed temperature and pH. An adsorption isotherm describes the relationship between the amount of adsorbate on the adsorbent and the concentration of dissolved adsorbate in the liquid at equilibrium [26]. Langmuir and Freundlich isotherms are Komman kinds of several isotherm equations that were tested to fit the obtained sorption data.

The Langmuir adsorption model assumes that molecules are adsorbed at fixed number of well-defined sites, each of which can only hold one molecule and no trans-migration of adsorbate in the plane of the surface. These sites are also assured to be energetically equivalent and distant to each other, so there are no interactions between the molecules adsorbed to adjacent sites. The linear form of the Langmuir isotherms is represented by the following equation [27]:

$$\frac{C_e}{q_e} = \frac{C_e}{q_m} + \frac{1}{K_L q_m} \quad (14)$$

where $C_e$ is the equilibrium concentration of the metal (mg/L) and $q_e$ is the amount of the metal adsorbed (mg) by per unit of the adsorbent (g). $q_m$ and $K_L$ are Langmuir constants relating adsorption capacity (mg/g) and the energy of adsorption (L/g), respectively, and evaluated from slope and intercept of the linear plots of $C_e/q_e$ versus $C_e$, respectively.

The linearized Langmuir adsorption of $La^{3+}$ and $Sm^{3+}$ is given in Figs. 6 and 7. The Langmuir adsorption constants evaluated from isotherms and their correlation coefficient are presented in Table 4; it is clear that the Langmuir isotherm model provides an excellent fit to the equilibrium adsorption data, giving correlation coefficient of 0.994 for $La^{3+}$ and 0.980 for $Sm^{3+}$, respectively.

Based on the further analysis of Langmuir equation, the dimensionless parameter of the equilibrium or adsorption intensity ($R_L$) can be expressed by

$$R_L = \frac{1}{1 + K_L C_o} \quad (15)$$

where $C_o$ (mg $L^{-1}$) is the initial amount of adsorbate.

The $R_L$ parameter is considered as more reliable indicator of the adsorption. There are four probabilities for the $R_L$ value: (1) for favorable adsorption, $0 < R_L < 1$, (2) for unfavorable adsorption, $R_L > 1$, (3) for linear adsorption, $R_L = 1$, and (4) for irreversible adsorption, $R_L = 0$.

The variation of $R_L$ with the initial metal concentration of solution is shown in Fig. 8. $R_L$ values were found to be between 0 and 1 for all concentrations of $La^{3+}$ and $Sm^{3+}$

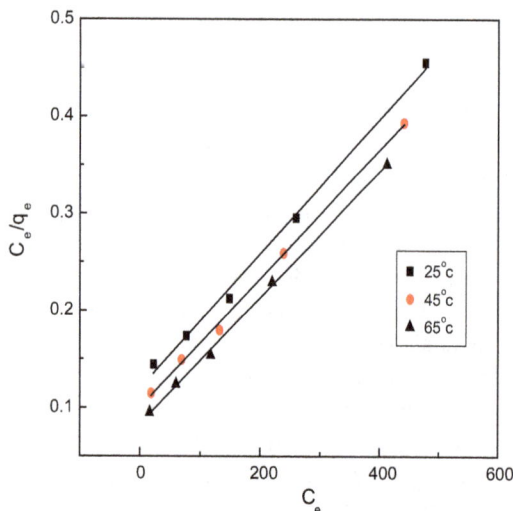

**Fig. 7** Langmuir isotherm plot for the adsorption of $Sm^{3+}$ onto {PASnMoP}

**Table 4** Parameters of Langmuir isotherm for ion exchange of $La^{3+}$ and $Sm^{3+}$ on PASnMoP

| Metal | T °K | $q_m$ | $R_L$ | $K_L$ | Ln b | $R^2$ |
|-------|------|-------|-------|-------|------|-------|
| $La^{3+}$ | 298 | 14.43 | 0.147 | 0.0057 | −5.15 | 0.9946 |
| | 318 | 15.1 | 0.131 | 0.0066 | −5.01 | 0.9977 |
| | 338 | 15.4 | 0.114 | 0.0077 | −4.85 | 0.9984 |
| $Sm^{3+}$ | 298 | 4.922 | $4.34 \times 10^{-6}$ | 230.34 | 5.43 | 0.9800 |
| | 318 | 5.40 | $4.11 \times 10^{-6}$ | 243.01 | 5.49 | 0.9910 |
| | 338 | 5.98 | $4.21 \times 10^{-6}$ | 237.39 | 5.46 | 0.9890 |

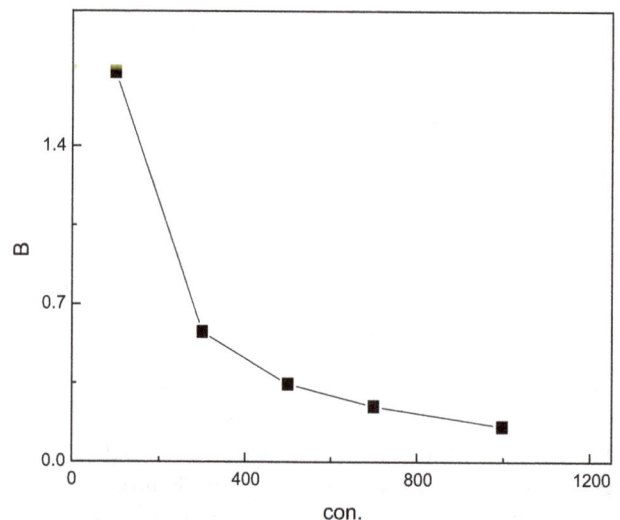

**Fig. 6** Langmuir isotherm plot for the adsorption of $La^{3+}$ onto {PASnMoP}

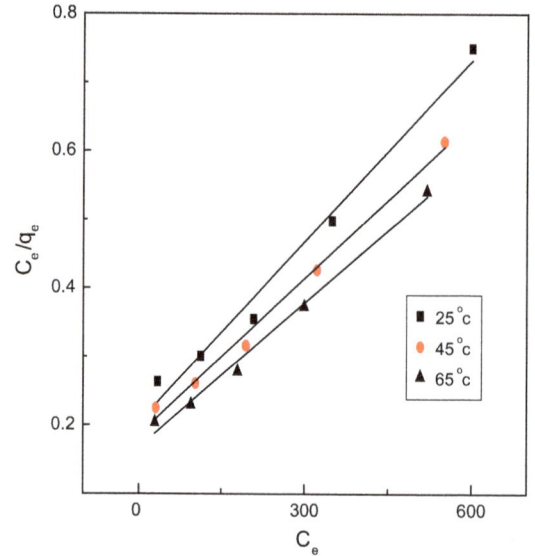

**Fig. 8** Variation of adsorption intensity ($R_L$) with initial $La^{3+}$ ion concentration ($C_o$ ppm)

and, therefore, ion exchange of both $La^{3+}$ and $Sm^{3+}$ is favorable. From Fig. 8, we can also see that the $R_L$ values decreased with the increasing initial concentration. This indicates that ion exchange is more favorable for the higher initial $La^{3+}$ and $Sm^{3+}$ concentration than for the lower one.

Freundlich isotherm is an empirical equation that encompasses the heterogeneity of sites and the exponential distribution of sites and their energies. The sorption data have been analyzed using the logarithmic form of the Freundlich isotherm as shown below:

$$Log\ q_e = \log K_f + \frac{1}{n} \log C_e \qquad (16)$$

where $K_f$ (mg/g) and $n$ are Feundlich constants incorporating all factors affecting the adsorption process such as adsorption capacity and intensity of adsorption. These constants are determined from the slope and intercept of linear plot of log $q_e$ versus log $C_e$, respectively.

The constants $K_f$ and $n$ of the Freundlich model are, respectively, obtained from the intercept and the slope of the linear plot of log $q_e$ versus log $C_e$ according to Figs. 9 and 10. The constants $K_f$ can be defined as an adsorption coefficient which represented the quantity of adsorbed metal ion for a unit equilibrium concentration (i.e., $C_e = 1$). Higher values of $K_f$ indicate higher affinity for $La^{3+}$ and $Sm^{3+}$. The slope $1/n$ is a measure of the adsorption intensity or surface heterogeneity [28, 29].

The Freundlich adsorption constants evaluated from isotherms and their correlation coefficient are presented in Table 5. For $1/n = 1$, the partition between the two phases is independent of the concentration; the situation $1/n < 1$ is the most common and corresponds to a normal L-type

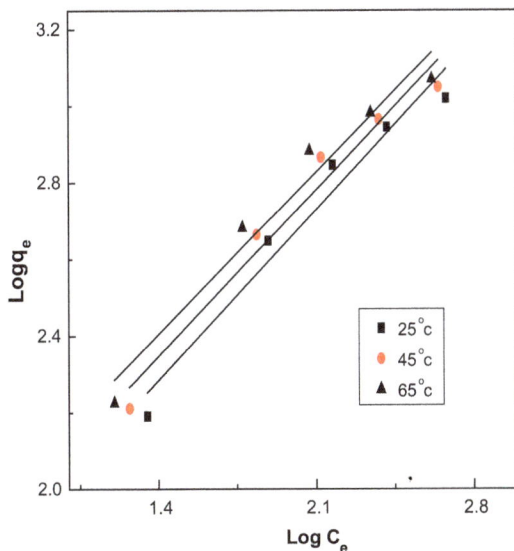

Fig. 10 Freundlich isotherm plot for the adsorption of $Sm^{3+}$ onto PASnMoP

Table 5 Parameters of Freundlich isotherm for ion exchange of $La^{3+}$ and $Sm^{3+}$ on {PASnMoP}

| Metal | Temp. | $1/n$ | $K_f$ | $R^2$ |
|---|---|---|---|---|
| $La^{3+}$ | 298 | 0.636 | 24.66 | 0.946 |
| | 318 | 0.622 | 29.85 | 0.954 |
| | 338 | 0.606 | 35.80 | 0.951 |
| $Sm^{3+}$ | 298 | 0.647 | 15.38 | 0.931 |
| | 318 | 0.662 | 15.84 | 0.951 |
| | 338 | 0.672 | 16.97 | 0.950 |

Langmuir isotherm, while $1/n > 1$ is indicative of a cooperative adsorption which involves strong interaction between the molecules of adsorbate. Values of $1/n < 1$ show favorable ion exchange of metals on ion-exchange resin, as shown in Table 5.

## Kinetic investigation

The two important physicochemical factors for parameter evaluation of the adsorption process as a unit operation are the kinetics and the equilibrium. Kinetics of adsorption describing the solute uptake rate, which in turn governs the residence time of adsorption reaction, is one of the important characteristics defining the efficiency of adsorption. Hence, in the present study, the kinetics of metal removal has been carried out at 298–338°K to understand the behavior of this exchanger.

In the study, Lagergren-first-order equation and Pseudo-second-order equation were used to test the experimental data. The lagergren-first-order equation is expressed as [30, 31]:

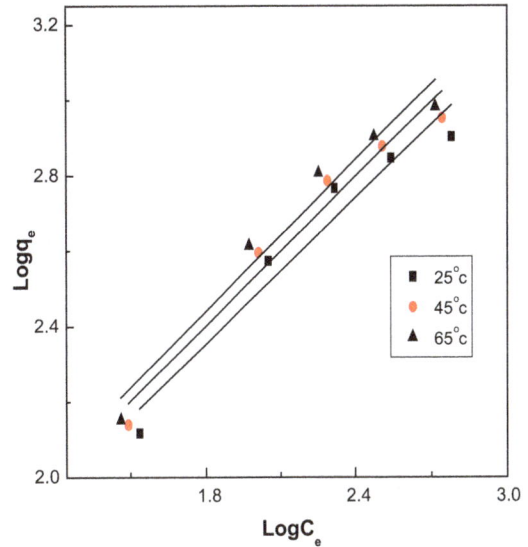

Fig. 9 Freundlich isotherm plot for the adsorption of $La^{3+}$ onto PASnMoP

**Table 6** Comparison between adsorption rate constants, estimated $q_e$ and coefficients of correlation associated with the Lagergren-first-order and the Pseudo-second-order kinetic models

| Metal | T °K | First-order kinetic model | | | Second-order kinetic model | | | | |
|---|---|---|---|---|---|---|---|---|---|
| | | $q_e$ (mg g$^{-1}$) | $k_1 \times 10^{-2}$ (min$^{-1}$) | $R^2$ | h (mg g$^{-1}$ min$^{-1}$) | $k_2 \times 10^{-3}$ (mg g$^{-1}$ min$^{-1}$) | $q_e$ (mg g$^{-1}$) | $T_{1/2}$ (min.) | $R^2$ |
| La$^{3+}$ | 298 | 35.54 | −9.1 | 0.8698 | 5.11 | 7.69 | 81.56 | 15.93 | 0.999 |
| | 318 | 31.05 | −13.9 | 0.9846 | 9.31 | 13.34 | 83.54 | 8.96 | 0.999 |
| | 338 | 22.19 | −13.4 | 0.9959 | 13.77 | 18.95 | 85.25 | 6.18 | 0.999 |
| Sm$^{3+}$ | 298 | 20.41 | −3.1 | 0.9574 | 4.22 | 7.8 | 73.36 | 17.38 | 0.999 |
| | 318 | 19.27 | −2.7 | 0.9460 | 4.94 | 8.58 | 75.93 | 15.35 | 0.999 |
| | 338 | 15.19 | −1.8 | 0.8236 | 5.83 | 6.9 | 77.55 | 13.36 | 0.999 |

$$\ln(q_e - q_t) = \ln q_e - k_1 t \qquad (17)$$

where $k_1$ (min$^{-1}$) is the rate constant of first-order adsorption, $q_e$ is the amount of metal adsorbed at equilibrium and $q_t$ is the amount adsorbed at time "$t$". Plotting $\ln(q_e - q_t)$ against "$t$" at (298–338) °K provided first-order adsorption rate constant ($k_1$) and $q_e$ values from the slope and intercept (Table 6).

The Pseudo-second-order equation [32, 33].

$$\frac{t}{q_e} = \frac{1}{k_2 q_e^2} + \frac{1}{q_e} \qquad (18)$$

The product $k_2 q_e^2$ is the initial adsorption rate "h" (mg g$^{-1}$ min$^{-1}$):

$$h = k_2 q_e^2 \qquad (19)$$

The half adsorption time is the time required to uptake half of the maximal amount of adsorbate at equilibrium. It

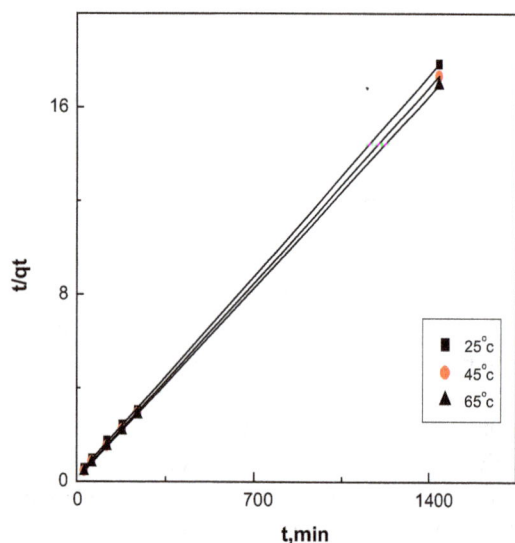

**Fig. 12** Test of Pseudo-second-order equation for adsorption of Sm$^{3+}$ from {PASnMoP} at different reaction temperatures

characterizes the adsorption rate as well. In case of Pseudo-second-order process, its value is given by the following relationship:

$$t_{1/2} = \frac{1}{k_2 q_e} \qquad (20)$$

where $k_2$ (mg g$^{-1}$ min$^{-1}$) is the Pseudo-second-order rate constant, $q_e$ is the amount adsorbed at equilibrium, $t_{1/2}$ is the half adsorption time and $q_t$ is the amount of metal adsorbed at time "$t$". Plotting $t/q_t$ against "$t$" at (298–338 °K) (Figs. 11, 12) provided second-order adsorption rate constant ($k_2$) and $q_e$ values from the slope and intercept (Table 6). The values of correlation coefficient indicate a better fit of Pseudo-second-order model with the experimental data compared to the Lagergren-first-order model at all studied temperatures. The same type results were also given in same works.

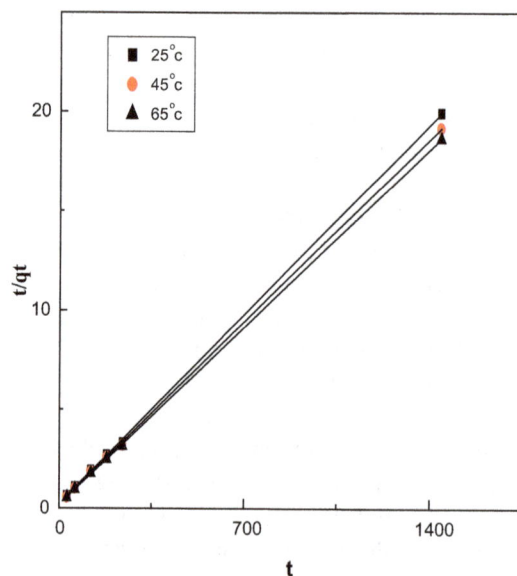

**Fig. 11** Test of Pseudo-second-order equation for adsorption of La$^{3+}$ from {PASnMoP} at different reaction temperatures

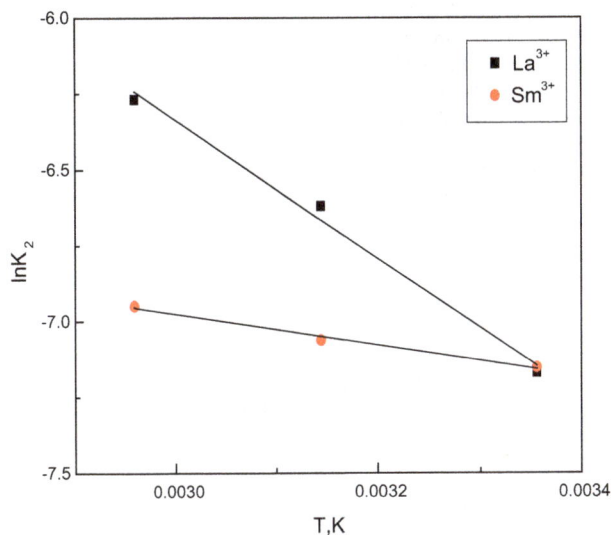

**Fig. 13** Plots of Ln K vs. 1/T for adsorption of $La^{3+}$ and $Sm^{3+}$ ions onto PASnMoP

**Table 7** Thermodynamic parameters of $La^{3+}$ and $Sm^{3+}$ adsorption onto {PASnMoP}

| Metal | $\Delta H^{o}$ kJ mol$^{-1}$ K$^{-1}$ | $\Delta S^{o}$ kJ mol$^{-1}$ K$^{-1}$ | $\Delta G^{o}$ kJ mol$^{-1}$ | | |
|---|---|---|---|---|---|
| | | | 298 °K | 318 °K | 338 °K |
| $La^{3+}$ | 18.94 | 4.13 | 17764.19 | 17475.8 | 17591.42 |
| $Sm^{3+}$ | 4.51 | 45.14 | 17714.63 | 18665.5 | 19502.31 |

## Thermodynamic parameters evaluation

Temperature dependence of the adsorption process is associated with several thermodynamic parameters. Thermodynamic considerations of an ion-exchange process are necessary to conclude whether the process is spontaneous or not. Thermodynamic parameters such as Gibbs free energy ($\Delta G^{\circ}$), enthalpy change ($\Delta H^{\circ}$) and entropy change ($\Delta S^{\circ}$) can be estimated using equilibrium constants changing with temperature. The Gibbs free energy change of the adsorption reaction is given by the following Eq. (23):

$$\overline{\Delta G^{\circ}} = - RT \ln K \qquad (21)$$

where $R$ is universal gas constant (8.314 mol$^{-1}$ k$^{-1}$), $T$ is the absolute temperature ($K$) and $K(q_e/C_e)$ is the distribution coefficient [15, 34].

Relation between $\Delta G^{\circ}$, $\Delta H^{\circ}$ (enthalpy change) and $\Delta S^{\circ}$ (entropy change) can be expressed by the following equation [35, 36]:

$$\Delta G^{\circ} = \Delta H^{\circ} - T\Delta S^{\circ} \qquad (22)$$

Equation (22) can be written as:

$$\ln K = \frac{\Delta S^{\circ}}{R} - \frac{\Delta H^{\circ}}{RT} \qquad (23)$$

where values of $\Delta H^{\circ}$ and $\Delta S^{\circ}$ can be determined from the slope and the intercept of the plot between ln $K$ versus 1/$T$ (Fig. 13).

The values $\Delta G^{\circ}$, $\Delta H^{\circ}$ and $\Delta S^{\circ}$ along with relation coefficient are given in Table 7. The magnitude of $\Delta G^{\circ}$ decreased with rising the temperature; from Table 7, the values of $\Delta H^{\circ}$ were positive, indicating that the ion-exchange reaction is endothermic. This is also supported by the increase in value of uptake capacity of the adsorbent with rising the temperature.

The positive values of $\Delta S$ show the increasing randomness at the solid/liquid interface during the adsorption of $La^{3+}$ and $Sm^{3+}$ on PASnMoP. Obviously, it is shown from the results reported in Table 7 that the temperature affects the adsorption process of metal ion adsorption onto the resin in which the higher temperature provided more energy to enhance the adsorption rate.

## Conclusion

The distribution coefficients of the prepared polyacrylamide Sn(IV) molybdophosphate (PASnMoP) have been investigated for $La^{3+}$ and $Sm^{3+}$ ions and the values of thermodynamic parameters were determined and the overall adsorption processes were found to be spontaneous and endothermic. The linear Langmuir and Freundlich isotherm models were used to represent the experimental data, and the experimental data could be relatively well intercept by the Langmuir isotherm. $R_L$ values between 0 and 1.0 further indicate a favorable adsorption of $La^{3+}$ and $Sm^{3+}$. By applying the kinetic models to the experimental data, it was found that the adsorption of $La^{3+}$ and $Sm^{3+}$ onto PASnMoP followed the Pseudo-second-order rate kinetics. The negative $\Delta G^{\circ}$ values showed that the ion exchange of $La^{3+}$ and $Sm^{3+}$ was spontaneous. The positive values of $\Delta S^{\circ}$ revealed the increased randomness at the solid solution interface.

## References

1. Nabi SA, Naushad MU (2007) Studies of cation exchange thermodynamics for alkaline earths and transition metal ions on a new crystalline cation exchanger: aluminium tungstate and distribution coefficient values of metal ions in surfactant media. Coll Surf A Phys Eng Asp 293:175–184
2. Nabi SA, Naushad MU, Khan AM (2006) Sorption studies of metal ions on napthol blue-black modified Amberlite IR-400 anion exchange resin. Separation and determination of metal ion contents of pharmaceutical preparation. Coll Surf A Phys Eng Asp 280:66–70

3. Nabi SA, Naushad MU, Ganai SA (2010) Preparation and characterization of a new inorganic cation-exchanger: zirconium (IV) iodosilicate: Analytical applications for metal content determination in pharmaceutical sample and synthetic mixture. Desalination Water Technol 16:29–38

4. Al-Othman ZA, Inamuddin Mu (2011) Naushad, Forward ($M^{2+}$–$H^+$) and reverse ($H^+$–$M^{2+}$) ion exchange kinetics of the heavy metals on polyaniline Ce(IV) molybdate: a simple practical approach for the determination of regeneration and separation capability of ion exchanger. Chem Eng J 171:456–463

5. Al-Othman ZA, Naushad MU, Inamuddin (2011) Organic–inorganic type composite cation exchanger poly-o-toluidine Zr(IV)-tungstate: Preparation, physicochemical characterization and its analytical application in separation of heavy metals. Chem Eng J 172:369–375

6. Al-Othman ZA, Naushad M, Ali R (2013) Kinetic, equilibrium isotherm and thermodynamic studies of Cr(VI) adsorption onto low-cost adsorbent developed from peanut shell activated with phosphoric acid. Environ Sci Poll Res 20(5):3351–3365

7. Zhou J, Duan W, Zhou X, Zhang C (2007) Application of annular centrifugal contractors in the extraction flow sheet for producing high purity yttrium. Hydrometallurgy 85:154–162

8. Qureshi M, Varshney KG (1991) Inorganic ion exchangers. In: Chemical analysis, CRC Press. Boca Ralon

9. Clearfield A (2000) Solvent extraction and ion exchanger. 18:655–678

10. Khan AA, Inamuddin, Alam MM (2005) Preparation, characterization and analytical applications of a new and novel electrically conducting fibrous type polymeric–inorganic composite material: polypyrrole Th(IV) phosphate used as a cation-exchanger and Pb(II) ion-selective membrane electrode. J Mater Res Bull 40:289–305

11. Khan AA, Alam MM (2003) Synthesis, characterization and analytical applications of a new and novel 'organic inorganic' composite material as a cation exchanger and Cd(II) ion-selective membrane electrode: polyaniline Sn(IV) tungstoarsenate. J React Funct Polym 55:277–290

12. Khan AA, Khan A, Inamuddin (2007) Preparation and characterization of a new organic–inorganic nano-composite poly-o toluidine Th(IV) phosphate: its analytical applications as cation-exchanger and in making ion-selective electrode. J Talanta 72:699–710

13. Khan AA, Paquiza L (2011) Characterization and ion-exchange behavior of thermally stable nano-composite polyaniline zirconium titanium phosphate: its analytical application in separation of toxic metals. J Desalination 265:242–254

14. El-Naggar IM, Mowafy EA, Abdel-Galil EA, El-Shahat MF (2010) Synthesis, characterization and ion-exchange properties of a novel 'organic–inorganic' hybrid cation-exchanger: polyacrylamide Sn(IV) molybdophosphate. Global J Phys Chem 1:91–106

15. Sari A, Tuzen M, Cıtak D, Soylak M (2007) Adsorption characteristics of Cu (II) and Pb(II) onto expanded perlite fromaqueous solution. J Hazard Mater 148:387–394

16. Abdel-Galil EA (2006) Chemical studies for sorption of some radionuclides on Silico(IV) titanate as cation exchanger" M.Sc. Thesis, Chemistry Dept., Fac. of Sci., Zagazig Univ

17. Abou-Mesalam MM, Shady SA (2004) Chemical in situ precipitation and immobilization technologies of radioactive liquid waste using titanium(iv) antimonate ion exchanger. Arab J Nucl Sci Appl 37:101–111

18. Clark A (1970) Theory of adsorption and catalysis. Academic Press, New York, p 54

19. Abou-Mesalam MM (2012) Evaluation of Crystalline size and Lattice Strain in Nano Particles of Transition Metals Hexacyano Ferrate. International Journal of advanced Chemical Technology 2(1)

20. Abou-Mesalam MM (2011) Hydrothermal synthesis and characterization of a novel zirconium oxide and its application as an ion exchanger. Adv Chem Eng Sci 1:20–25

21. Mishra SP, Singh UK, Tiwari D (1996) Inorganic particles in removal of toxic metal ions, IV. Efficient removal of zinc ions from aqueous solution by hydrous zirconium oxide. J Radiat Chem 210:207–211

22. Helfferich F (1962) Ion exchange. McGraw Hill, New York

23. Abou-Mesalam MM (2003) Sorption kinetics of copper, zinc, cadmium and nickel ions on synthetized silico-antimonate ion exchanger. J Coll Surf 215:205–211

24. Nabi SA, Usmani S, Rahman N (1996) Synthesis, characterization and analytical Applications Of an ion exchange material: zirconium (IV) iodophosphate. Ann Chim Fr 21:521–530

25. Shady SA (2009) Selctivity of cesium from fission radionuclides using resorcinol—formaldehyde and zirconyl—mplybdopyrophosphate as ion exchangers. J Hazard Mater 167:947–952

26. Paric J, Trago M, Medvidovic NV (2004) Removal of zinc, copper and lead by natural zeolite—a comparison of adsorption isotherms. Water Res 38:1839–1899

27. Langmuir I (1916) The constitution and fundamental properties of solids and liquids. Part I. solids. J Am Chem Soc 38:2221–2295

28. Gopal V, Elango KP (2007) Equilibrium, kinetic and hermodynamic studies of adsorption of fluoride onto plaster of paris. J Hazard Mater 141:98–105

29. Chabani M, Amrane A, Bensmaili A (2006) Kinetic modeling of the adsorption of nitrates by ion exchange resin. J Chem Eng 125:111–117

30. Agrawal A, Sahu KK (2006) Kinetic and isotherm studies of Cadmium adsorption on manganes nodule residue. J Hazard Mater 137:915–924

31. Uysal M, Ar I (2007) Removal of Cr4+ from industrial wastewaters by adsorption. Part1. Determination of optimum conditions. J Hazard Mater 149:282–291

32. Prasanna Kuma Y, King P, Prasad VSRK (2006) Equilibrium and Kinetic studies for the biosorption system of Copper$^{2+}$ ion from aqueous solution using Tectona grandis LF. leaves powder. J Hazard Mater 137:1211–1217

33. Chen CL, Li XL, Zhao DI., Tan XI., Wang XK (2007) Adsorption Kinetic, thermodynamics and adsorption studies of Th$^{4+}$ on oxidized multi –wall Carbon nanotubes. Coll Surf A 302:449–954

34. Sari A, Mendil D, Tuzen M, Soylak M (2008) Biosorption of Cd (II) and Cr(III) from aqueous solution by moss (Hylocomium-splendens) biomass: equilibrium, kinetic and thermodynamic studies. J Chem Eng 144:1–9

35. Donat R, Akdogan A, Erdem E, Cetisli H (2005) Themodynamics of Pb$^{2+}$ and Ni$^{2+}$ adsorption onto natural bentonite from aqueous solutions. J Coll Interface Sci 286:43–52

36. Khani MH, Keshtkar AR, Ghannadi M, Pahlavanzadeh H (2008) Equilibrium, kinetic and thermodynamic study of the biosorption of uranium onto Cystoseriaindica algae. J Hazard Mater 150:612–618

**4**

# Industrial-scale purging of ammonia by using nitrogen before environmental discharge

Ahmet Ozan Gezerman[1]

**Abstract** For efficient discharge and storage of refrigerated chemicals such as ammonia, special processes must be developed, and several related parameters must be checked and evaluated. In this study, pressure changes in ammonia storage systems that are purged by nitrogen gas, during filling by gaseous ammonia, were calculated and an environment-friendly technique for discharging ammonia gas was developed. In addition, exergy analysis for the system was performed, and the nitrogen discharge rate in the system was calculated. The total exergy loss was determined to be 43.18 %, and the nitrogen discharge rate was determined to be 38,995 $dm^3\ h^{-1}$ for the proposed system.

**Keywords** Purging · Ammonia · Nitrogen · Cryogenic · Multistage compressor

## List of symbols

| | |
|---|---|
| $e_{cp}$ | Exergy loss in the compressor (kJ $kg^{-1}$) |
| $e_{cd}$ | Exergy loss in the condenser (kJ $kg^{-1}$) |
| $e_{exp}$ | Exergy loss in the expansion valve (kJ $kg^{-1}$) |
| $e_{pump}$ | Exergy loss in the ammonia pump (kJ $kg^{-1}$) |
| $e_{vessel}$ | Exergy loss in the ammonia storage vessel (kJ $kg^{-1}$) |
| $e$ | Total exergy loss of the compression system (%) |
| $\eta$ | Exergy efficiency of the compression system (%) |
| $s$ | Specific entropy (kJ $kg^{-1}\ K^{-1}$) |
| $T$ | Temperature (K) |
| $T_a$ | Ambient temperature (K) |
| $w$ | Compression work per unit refrigerated mass (kJ $kmol^{-1}$) |
| $\psi_r$ | Exergy of reaction at 1 atm, 25 °C (kJ $kmol^{-1}$) |
| $g_{f_i}$ | Chemical exergy of a component (kJ $kmol^{-1}$) |
| $v_i$ | Volume fraction ($cm^3\ cm^{-3}$) |
| $n$ | Mole fraction (mole $mole^{-1}$) |
| $T_{cd}$ | Temperature in the condenser (K) |
| $T_{cp}$ | Temperature in the multistage compressor (K) |
| $T_{exp}$ | Temperature in the expansion valve (K) |
| $T_{pump}$ | Temperature in the ammonia pump (K) |

✉ Ahmet Ozan Gezerman
ahmet_ozan@yahoo.com;
ahmetozangezerman78@hotmail.com

[1] Department of Chemical Engineering, Faculty of Chemical and Metallurgical Engineering, Yildiz Technical University, Istanbul, Turkey

## Introduction

For storing refrigerated gasses like ammonia, maintenance of systems in which these gasses are used is very important. Since ammonia is a pungent, cryogenic, and corrosive refrigerated liquid, several problems are encountered in systems in which it is used. For example, ammonia has a low boiling point (−33 °C), so the process lines should be able to prevent temperature variations. Therefore, special methods are needed to purge ammonia from the process lines. Based on this demand, Shi et al. [1] designed a purging tower and investigated the tower's performance in ammonia storage systems.

In addition, a different method was investigated by Oberski et al. [2], where ammonia gas was used as a reductant to purge the system.

Another patented approach toward purging ammonia during the storage process was demonstrated by Ishizaka et al. [3]. Here, titanium tetrachloride was used as the inert material, but this process has not been implemented on an industrial scale. Another investigation on the purging of refrigerated chemicals such as ammonia from process lines was performed by Shaikh et al. [4]. In this work, air was used to purge the ammonia from the storage system. However, the air was introduced into the system by a device that increased the operation cost. Further, Sun et al. investigated the purging of refrigerated and cryogenic materials from process lines [5]. In these studies, ammonia was treated using urea, which was introduced by a dosing system.

Another investigation of an ammonia storage system and its periodic maintenance was performed by Mayer et al. [6], and this work is useful because it tackled purging on an industrial scale.

There are several experimental works on the application of ammonia storage systems. Le Lostec et al. [7] reported promising experimental-scale results for an ammonia–water absorption chiller system.

Wright [8] also developed a patent for the storage of refrigerated materials and purging of the systems; however, the operation cost is high. Other methods have also been designed based on computational fluid dynamics (CFD) for safe purging of the pipelines of refrigerated systems [9]. Maekawa [10] applied another method for purging ammonia storage systems, using fluorine as the cleaning gas. Another gas-purging application is a refrigeration cycle system, developed by Dincer and Kanoglu [11].

Ammonia, which is the main chemical considered in the present work, is widely used in several industries. In processes involving ammonia, appropriate methods of discharge are imperative for long-term maintenance of the process equipment and production/compressor lines. Hence, purging is indispensable, and for this purpose, the physical form of ammonia must be controlled. Ammonia is generally used in anhydrous form (99.9 %) in industrial processes. However, because of its pungent odor, ammonia stored at cryogenic temperatures should not be discharged into the environment in anhydrous form.

In the present work, I propose that during ammonia storage, the empty tank should be first filled with ammonia gas and then with liquid ammonia. During the filling operation, the main objective is to condition the tank to afford an ambient storage environment for liquid ammonia. During the operation, the liquid anhydrous ammonia storage tank is partially filled (i.e., only three-fourth of the tank is filled), and the unfilled portion is used to accommodate gasified ammonia. A part of this gasified ammonia is used for conditioning the other empty tank, while the rest is sent to the multistage compressor for liquefaction, in order to maintain the level of liquid in the liquid ammonia storage tank. During the operation, the gaseous ammonia obtained from liquid ammonia is compressed and then liquefied

by the multistage compressor. This approach helps in mitigating thermodynamic losses, as confirmed by a thermodynamic assessment of the proposed process based on exergy analysis.

## System description and calculations

For increasing ammonia gas concentrations in the tank, the compressor suction valve of tank B is slowly opened. The purpose of this action is to purge nitrogen gas into tank B, by taking the ammonia gas from tank A, so that the ammonia gas concentration in tank B is increased and nitrogen gas is discharged. Ammonia gas from tank A reaches tank B through the shortest possible route in the pipeline. With increasing ammonia concentration in tank B, the pressure on the nitrogen gas mixture increases, and hence, nitrogen is forced out of the tank (Fig. 1).

Tank A is filled with liquid ammonia, and tank B is filled with nitrogen gas. The purpose of this arrangement is to purge nitrogen gas in tank B, in preparation for filling with liquid ammonia. Because of the ambient temperature, the liquid ammonia in tank A (1 atm, $-33\,°C$) may vaporize and reach the compressor suction by pipeline 1. The liquefied ammonia is repeatedly directed to tank A from the compressor discharge via pipelines 2–5. The conditioning work involving prefilling tank B with liquid ammonia, by using pipeline 1, was accomplished by the vaporized ammonia in tank A.

The changes in pressure with time during the purging of ammonia gas using nitrogen gas are shown in Fig. 2. The green line denotes the pressure change depending on ammonia gas concentration during 8 h; the blue line indicates the pressure change of nitrogen gas in 8 h. The letters on the green line denote the changes in pressure made to purge ammonia gas. Accordingly, if the pressure difference between tanks A and B is known, the purge ratio from the discharge point in tank B can be calculated. If $(P_A - P_B) = \Delta P = 0.014$ atm, according to Fig. 4, the nitrogen gas concentration can be 13 % when the ammonia gas concentration in tank is 87 %. Then, the nitrogen purge rate is [12, 13]

$$P \times V = n \times R \times T,$$

$$P_A = 0.048 \text{ atm}, \ P_B = 0.034 \text{ atm},$$

$$(P_A - P_B) = \Delta P = 0.014 \text{ atm},$$

$V$ = Volume of the ammonia storage tank in the process : $23,000 \text{m}^3$,

$$V = 23,000 \text{ m}^3 \times \left(\frac{13}{100}\right) = 2990 \text{ m}^3 \times 1000 \text{ dm}^3$$
$$= 2,990,000 \text{ dm}^3,$$

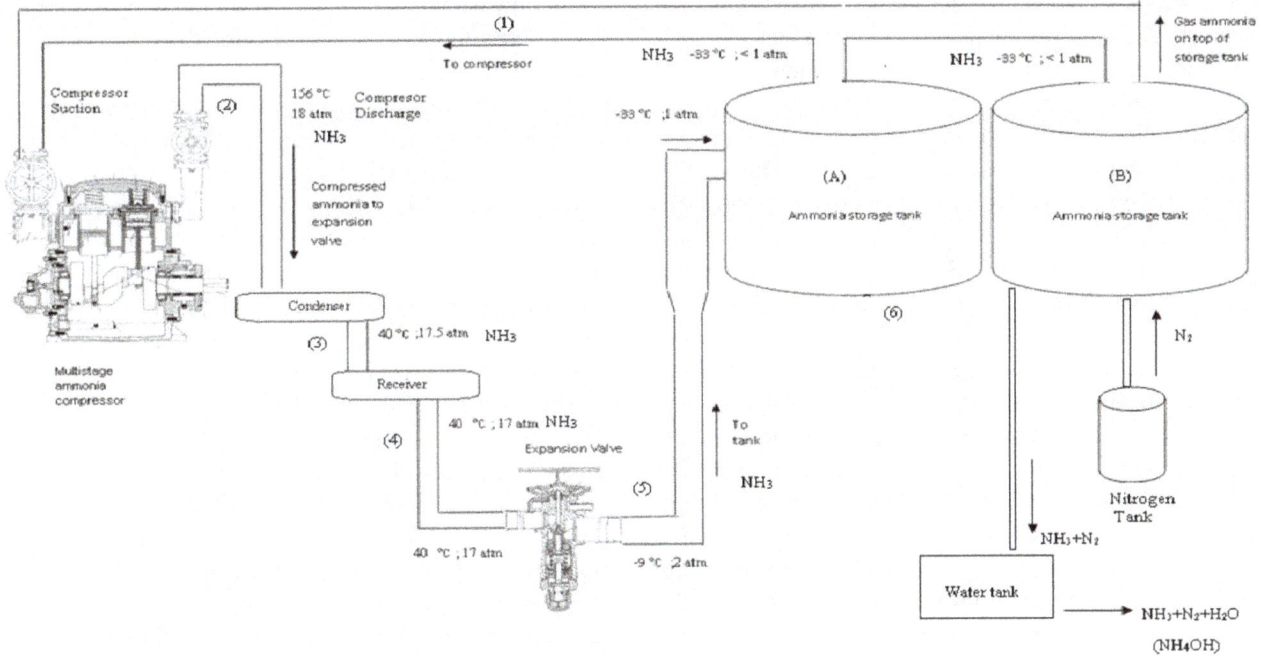

Fig. 1 Schematic representation of purging achieved using nitrogen during ammonia storage

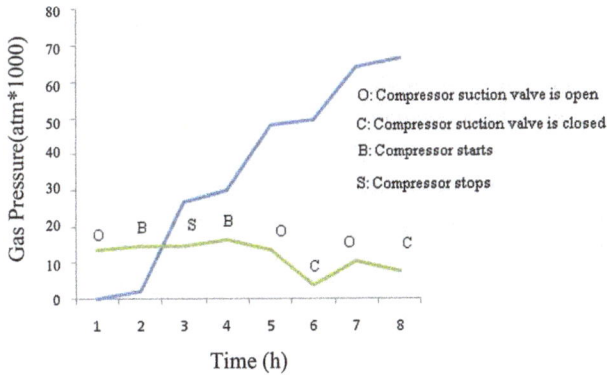

Fig. 2 Operational directions against time and pressure in tanks. The Y-axis shows the ammonia and nitrogen pressures; the X-axis expresses the time period of 8 h. The *green line* represents the pressure change with ammonia gas concentration during 8 h, and the *blue line* shows the pressure change of nitrogen gas during the purging process for 8 h. The *letters on the green line* show the changes made to purge ammonia gas

$$R = 0.082 \left( \frac{atm \times dm^3}{mol \times K} \right),$$

$T = 20\,°C$ (Temperature in tank B),

$$P \times V = n \times R \times T,$$

$$0.014 \times 2,990,000 = n \times 0.082 \times 293$$
$$= 1804.5 \ (mol \ h^{-1}) \times 28 \ (g \ mol^{-1})$$
$$\times (1 \ cm^3 / 0.001251 \ g)$$
$$= 38,995 \ dm^3 \ h^{-1}.$$

If $P$ increases ($P_A \gg P_B$), because of the pressure difference, the ammonia concentration in the empty tank B may increase.

Ammonia gas is transferred from tank A to tank B by the compressor suction lines. When the operation is complete and once the ammonia concentration is 100 %, the pressure difference between the two tanks, $P$, will be the lowest. Then, the ammonia concentration can be changed depending on pressure, as per the expression below [14]:

$$P \times V = n \times R \times T,$$

$$P = \left( \frac{n}{V} \right) \times R \times T.$$

By assuming inter-dependence between the variables concentration and pressure, the following correlation equation can be obtained [14]:

$$dP = dC \times R \times T,$$

$$P = C \times R \times T.$$

The temperature in the empty tank is thought to be equivalent to the air temperature because of the atmospheric pressure conditions maintained in the tank. Pressure ($P$, atm) and concentration ($C$, mol dm$^{-3}$) are the two variable parameters that must be considered in this equation.

If the pressures in tanks A and B are $P_A = 0.053$ atm and $P_B = 0.046$ atm, respectively, for increasing the pressure and ammonia concentration in tank B, the compressor suction valve in this tank is opened. However, the compressor does not start working. Vaporized ammonia from tank A flows into tank B because of the concentration difference, and thus, the ammonia gas pressure in tank B increases. When the ammonia gas pressure in tank A increases, the compressor begins to work; hence, the gas pressure in tank A is ensured to decrease. During this period, the compressor is running, and the compressor suction on tank B side is closed in order to prevent the flow of ammonia gas from tank B to the compressor suction line. When the pressure in tank A decreases to 0.043 atm, the compressor is stopped, and the opening compressor suction on tank B side is opened to ensure that ammonia gas flows to tank B.

It is noteworthy that the ammonia gas pressure in tank B increases when ammonia flows into the tank after the compressor suction valve on tank B side is opened. During the ammonia gas transfer from tank A to tank B, nitrogen gas is discharged from the ammonia discharge point at the bottom of tank B. This nitrogen gas results from the purging using nitrogen in tank B. In this manner, a large pressure increase in tank B is prevented by discharging nitrogen gas.

In the present operation, the amount of purged nitrogen gas must be calculated. This purged nitrogen gas is absorbed by water and will minimize the ammonia content in the nitrogen/ammonia gas mixture. Thus, precipitation of ammonia as ammonium hydroxide is an eco-friendly approach.

Another purpose of the operation is to ensure that the entire amount of ammonia is used to exert pressure on the nitrogen gas, for purging nitrogen in tank B. The entire amount of ammonia gas in the available system is transferred from tank A to tank B. The vaporized part of liquid ammonia in tank A is sent to the multistage compressor for liquefaction, and this liquefied ammonia is sent to tank A for storage. Most of the vaporized ammonia in tank A flows to tank B because of the concentration gradient and the difference between B and C.

In this manner, ammonia gas is sent to tank B, compressed, and made to discharge nitrogen into the atmosphere. According to the values of the operational parameters shown in Fig. 2, because of the increasing pressure difference between tanks A and B caused by the compressor in function, vaporized ammonia gas flows from tank A to tank B, which has lower ammonia gas pressure. Because of the low concentration of ammonia, the gas flows to tank B via the compressor suction valve, without flowing into the compressor (Fig. 1).

The ammonia gas entering tank B forces the nitrogen gas in this tank to be expelled from the discharge point. Similarly, when the pressure difference between tanks A and B decreases, the compressor suction valve is closed and the compressor is allowed to run for allowing nitrogen

gas to flow between the tanks and for discharging nitrogen gas by creating pressure in the tank. Thus, the vaporized ammonia in tank A is liquefied and stored in tank A. Once the pressure of ammonia gas in the tank is increased by the multistage compressor, the compressor stops, and thus, liquid ammonia in tank A is vaporized.

Depending on the increase in the pressure of ammonia gas in tank A, the compressor suction valve of tank B opens, and thus, the entire amount of ammonia gas enters tank B from tank A. When the compressor suction valve is open, the compressor does not run, because the ammonia gas, which is vaporized in tank A and passed to tank B, can decrease the ammonia gas concentration in tank B. The main principle in this process is the increase in the ammonia concentration in tank B.

One way for this increase is as follows: The ammonia gas is vaporized in tank A, but in the meantime, the ammonia gas pressure in tank A must be increased extensively within a short process time. This operation will continue until the ammonia gas concentration in tank B is 100 %. When the ammonia gas concentration is 100 %, liquid ammonia can enter tank B. Moreover, in the time frame shown in Fig. 2, the ammonia gas concentration is monitored in tank B, and accordingly, the amount of purged nitrogen gas is calculated.

The data in Fig. 2 show the gasification ability of ammonia with time and can be used to obtain important information on the compressor efficiency.

When the concentration of ammonia gas is 100 %, the compressor suction valve in tank B and the liquid ammonia entry line of the compressor to tank B are opened.

## Destruction and exergy analysis for the nitrogen cycle during purging

Exergy analysis is carried out to optimize the operational and geometrical parameters of the ammonia pump and ammonia storage tank in the nitrogen cycle of the limiting ammonia concentration system. This analysis facilitates assessment of the combined effect of heat transfer and pressure drop through simultaneous interactions with the pump. However, the pressure drop in the ammonia pump and tank affects exergy losses. Thus, exergy analysis for the nitrogen cycle of the limiting ammonia concentration system is necessary.

The operating parameters for the compression system are listed in Table 1.

For the ammonia pump component, the exergy losses are evaluated from the exergy balance. The exergy is defined by [15]:

$$e = T_a(s_{out} - s_{in}).$$

Table 1 Operating parameters of the ammonia compression system

| Parameters | Compression mode |
| --- | --- |
| $T_a$ (K) | 308 |
| $T_{cd}$ (K) | 313 |
| $T_{exp}$ (K) | 278 |
| $T_{cp}$ (K) | 278 |
| $T_{pump}$ (K) | 283 |
| $T_{vessel}$ (K) | 293 |

Using the exergy analysis method, the exergy losses occurring in all five components of the compression cycle (Fig. 1) are calculated as follows:

The exergy loss in the multistage compressor is: $e_{cp} = T_{cp}(s_2 - s_1)$.

The exergy loss in the condenser is: $e_{cd} = T_{cd}(s_3 - s_2)$. The exergy loss in the expansion valve is: $e_{exp} = T_{exp}(s_5 - s_4)$.

The exergy loss in the ammonia pump is: $e_{pump} = T_{pump}(s_7 - s_6)$.

The exergy loss in the vessel is: $e_{vessel} = T_{vessel}(s_1 - s_6)$. The total exergy loss in the compression cycle is: $e = e_{cp} + e_{cd} + e_{exp} + e_{pump} + e_{vessel}$.

The exergy efficiency of the compression system is: $\% \eta = 1 - (e/w)$.

The exergy analysis results of the ammonia compression system with operating modes are listed in Table 2.

### Chemical exergy analysis of purging system

When nitrogen purging is performed, nitrogen gas and ammonia gas are discharged to the waste water tank. During this time, ammonium hydroxide is formed by the reaction between water and ammonia. The chemical exergy equation for this reaction is:

$$\psi_r = v_i g_{f_i},$$

Table 2 Exergy analysis results of the ammonia compression system

| Parameters | Ammonia compression mode |
| --- | --- |
| $w$ (kJ kg$^{-1}$) | 41.258 |
| $e_{cp}$ (kJ kg$^{-1}$) | 9.251 |
| $e_{cd}$ (kJ kg$^{-1}$) | 4.442 |
| $e_{exp}$ (kJ kg$^{-1}$) | 2.369 |
| $e_{pump}$ (kJ kg$^{-1}$) | 2.254 |
| $e_{vessel}$ (kJ kg$^{-1}$) | 5.124 |
| $e$ (kJ kg$^{-1}$) | 23.44 |
| $(\%)\ \eta = 1 - \frac{e}{w}$ | 43.18 |

where $\psi_r$ is the molar exergy of the synthesis reaction at 1 atm and 25 °C [16].

For the purging process, the chemical exergy is determined using the molar exergy values for ammonia and water.

$$\psi_{NH_4OH} = g_{f,NH_3} + g_{f,H_2O}$$

The exergy for ammonium hydroxide formation from ammonia and water [16] is:

$$g_{f,NH_3} = 327.4 \text{ kJ mol}^{-1}, \quad g_{f,H_2O} = 1.3 \text{ kJ mol}^{-1},$$

$$\psi_{NH_4OH} = 328.7 \text{ kJ mol}^{-1}.$$

### Results and discussion

Purging of refrigerated materials such as ammonia requires special conditions for storage and necessitates the use of some special processes. The residual ammonia in the system after purging with nitrogen is released into the environment. Our process helps in mitigating the environmental hazards caused by ammonia. In this study, a mixture of ammonia and nitrogen, which is present in an empty tank made inert by using nitrogen, is used to discuss how the empty tank purged by ammonia is filled by liquid ammonia. During purging of nitrogen from the tank, environmental factors must be considered, despite the low concentration of ammonia (<10 %) resulting from the small amount of ammonia purged.

In the abovementioned process, each storage tank has a compressor suction line and a valve, and a liquid ammonia entry point for the compressor discharge line. According to the process conditions, this system can have liquid ammonia pumps and a gas return line to the tank (Fig. 3). Before liquid ammonia is taken to the empty tank, ammonia gas should be filled in the tank because ammonia is a refrigerated liquid, and hence, liquid ammonia conditions should be maintained in the tank (1 atm, −33 °C). On the other hand, before liquid ammonia is transferred to the empty storage tank, because of the process conditions, the ammonia gas concentration in the tank does not show a linear change. The change in the ammonia gas density with temperature is more active than the change in nitrogen gas density with temperature. Finally, with the increasing concentration of ammonia gas in the tank, a sudden increase in the nitrogen discharge rate is observed, which confirms the temperature–density correlation.

During purging of nitrogen gas in tank B, environmental factors should be considered, and dissolution in water would be the most economical solution to minimize environmental contamination by ammonia. Therefore, an eco-

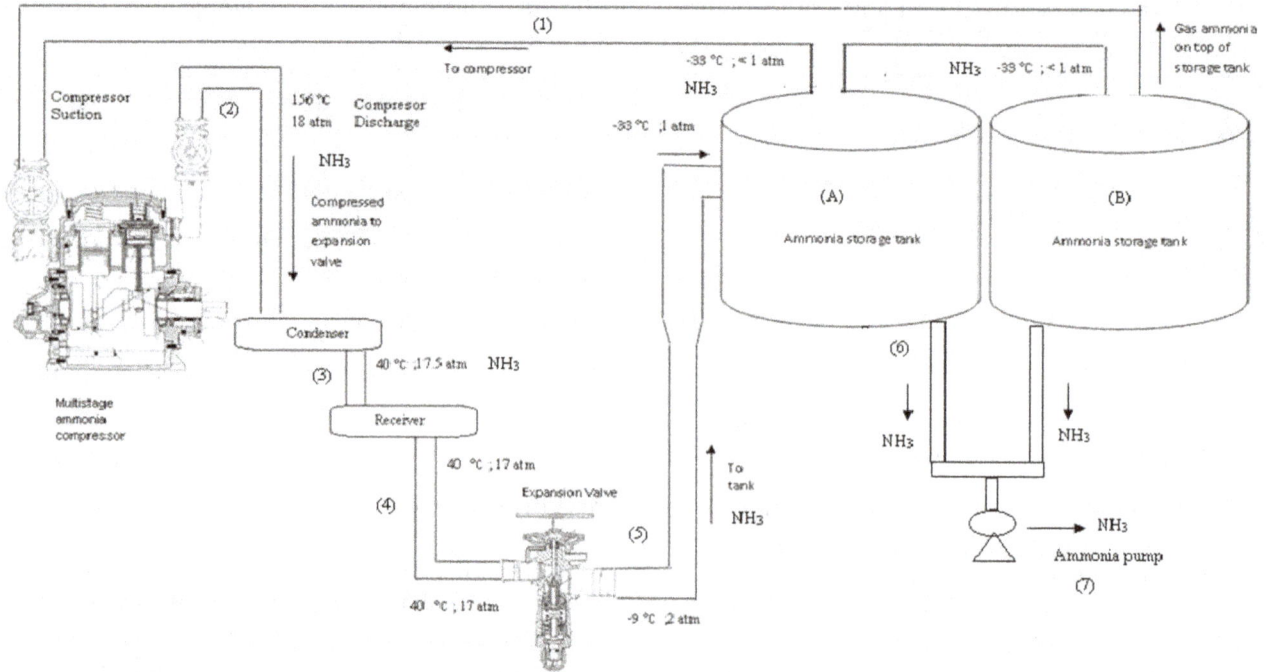

**Fig. 3** Operation of tank filled with liquid ammonia (vaporized ammonia from tank A to tank B goes to the compressor and is sent to the storage tank upon liquefaction by the compressor)

friendly method would be to precipitate ammonia as ammonium hydroxide.

Transfer of ammonia to tank B can be accomplished by various methods; however, the best option when using the present system will be to feed compressed and liquefied ammonia from the multistage compressor. In this method, because of the ammonia gas in the filled tank, the ammonia concentration in the empty tank is brought to 100 %, and then, some part of vaporized ammonia in the filled tank is compressed by the compressor; thus, liquefied ammonia is sent by the compressor discharge line to the filled tank. Exergy analysis performed for this method reveals that exergy losses for the present system result from losses in the ammonia tank and ammonia pumps.

According to Fig. 2, because of the pressure difference between tanks A and B, the ammonia gas in tank B forces the nitrogen gas in the tank to expel. Therefore, nitrogen gas can purge to the atmosphere from the discharge points in tank B. Another detail given in Fig. 4 is the nitrogen gas ratio for every pressure value of the nitrogen–ammonia mixture. According to Fig. 4, 350 min after the purging of ammonia gas, the pressure and concentration of the gas in tank B were measured as 600 $mmH_2O$ ($\sim 0.058$ atm) and 92.5 %, respectively. Then, the purging operation continued for further 140 min, during which time the pressure in this tank was 100 $mmH_2O$ ($\sim 0.0096$ atm) and the ammonia gas concentration was 87 %. The ammonia pressure in tank B showed a large alteration after the second measurement. In

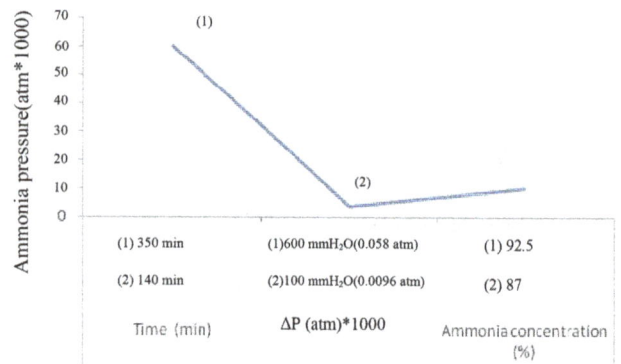

**Fig. 4** Ammonia concentration purged due to pressure difference (for tank B)

large-volume industrial ammonia storage systems (23,000 $m^3$), the uncompressed ammonia gas from the compressor line and leakages in other processes can induce slight differences in the pressure values measured during the purging operation. We expect the ammonia gas pressure to decrease in this process because we attempt to decrease the ammonia gas concentration. However, if the process is not isolated on the industrial scale, a sharp increase in ammonia gas pressure can occur when the compressor is started or in the event of any leakage. As illustrated in Fig. 2, after 6–7 h of operation, the ammonia gas concentration can increase; this is because when the compressor suction valve is first closed (C) and then opened (O), gasified ammonia cannot be discharged from the system, which results in increased ammonia gas pressure. However, when the compressor is started, it imparts a driving force to expel ammonia gas from the discharge points, thereby causing a decrease in the gas pressure.

With this ratio and the pressure difference between tanks A and B, the amount of nitrogen purged in tank B (Fig. 1) can be determined, and exergy analysis can be performed. Related data are given as nitrogen concentration in the nitrogen–ammonia gas mixture as a function of pressure and time. The aim here is to calculate the discharge rate of purged nitrogen in the mixture.

The measurement data for the ammonia storage vessel are acquired by an ammonia manometer pressure gauge with a measurement range 0–60″ $H_2O$ (accuracy 1.0 %, glycerin fillable).

The studies discussed here, in particular for industries, in which corrosive and cryogenic materials are used, explain how to remove corrosive materials, like ammonia, by purging the lines during periodic maintenance work. According to these studies, gas purging is most commonly used in industries that apply thermodynamic cycles, like petroleum refineries and energy production facilities.

The purpose of this study was to clean ammonia gas from an ammonia storage system and to purge the ammonia in an optimal way. The cheapest, detected method was the absorption of ammonia purged into water. Because the formation of ammonium hydroxide via reaction between ammonia and water is inevitable, it will be purged optimally. This ensures that there is no damage to the flora when ammonium hydroxide is discharged into empty agricultural fields.

## Conclusions

In industrial processes, various problems related to environmental pollution and equipment corrosion are encountered during purging with corrosive liquids such as ammonia. For this reason, proper maintenance of storage systems is important. A possible, inexpensive solution to this problem is the use of multistage compressors, so that the storage system is conditioned by liquefaction; that is, using nitrogen gas for purging ammonia gas. Ammonia is precipitated as ammonium hydroxide at the end of the process, and hence, there is minimum environmental load.

In this study, the operational parameters for purging of ammonia pipelines and storage tanks before liquid ammonia was transferred into the tank, was investigated by using real-time process data. Further, an exergy analysis was performed and the purge rate of compressed nitrogen gas was calculated. It is proposed that ammonia

in the ammonia–nitrogen gas mixture is purged by precipitating ammonia as ammonium hydroxide, which is one of the most important results. According to the data obtained for the present system, the amount of purged nitrogen is 38,995 $dm^3\ h^{-1}$. In addition, exergy analysis of the present system was performed using the thermodynamic parameters, and the exergy loss was determined to be 43.18 %.

The present system handles a very large industrial volume of around 23,000 $m^3$; hence, strict control of each system parameter is not easy. Therefore, the expected results were not obtained even after optimization of the process conditions. However, the proposed system is cost-effective and less time-consuming and it discharges relatively low amounts of environmental pollutants; thus, there are no notable concerns regarding the practical implementation of the system.

## References

1. Shi LL, Meng LC, Nandong X (2005) Researches on the factors affecting ammonia-purge efficiency of landfill leachate in purging tower. Ind Saf Dust Control 6:12
2. Oberski C, Cavataio G, Van Nieuwstadt MJ, Webb T, Ruona W (2010) U.S. Patent No. 7,770,384. U.S. Patent and Trademark Office, Washington, DC
3. Ishizaka T, Gunji I, Kannan H, Sawada I, Kojima Y (2003) U.S. Patent Application No. 10/516,311
4. Shaikh F, Lawrence D, Cooper S, Castleberry L Jr (2014) U.S. Patent No. 8,621,848. U.S. Patent and Trademark Office, Washington, DC
5. Sun J, Mupparapu S, Tarabulski TJ, Park PW (2013) U.S. Patent No. 8,459,012. U.S. Patent and Trademark Office, Washington, DC
6. Mayer F, Hornung M, Bürgi L, Gerner P (2014) U.S. Patent No. 8,771,598. U.S. Patent and Trademark Office, Washington, DC
7. Le Lostec B, Galanis N, Millette J (2012) Experimental study of an ammonia-water absorption chiller. Int J Refrig 35(8): 2275–2286
8. Wright CJ (2015) U.S. Patent No. 9,150,139. U.S. Patent and Trademark Office, Washington, DC
9. Ma D, Zhang Z, Li Y (2015) Investigation of gas purging process in pipeline by numerical method. Process Saf Environ Prot 94:274–284
10. Maekawa K (2010) U.S. Patent No. 7,691,208. U.S. Patent and Trademark Office, Washington, DC
11. Dincer I, Kanoglu M (2011) Refrigeration systems and applications, 2nd edn. Wiley, New York
12. Çengel YA, Boles M (1989) Thermodynamics: an engineering approach, 5th edn. McGraw-Hill, New York
13. Hogan JD (ed) (1996) Specialty gas analysis: a practical guidebook. Wiley
14. Maalouf S, Boulawzksayer E, Clodic D (2012) Orc finned—tube evaporator design and system performance optimization. In: International refrigeration and air conditioning conference, Purdue, US
15. Querol E, Borja G-R, Perez-Benedito JL (2013) Exergy concept and determination, practical approach to exergy and thermoeconomic analyses of industrial processes. Springer, London
16. Wall G (1998) Exergetics. Mölndal, Sweden. http://exergy.se

# 5

# Reactivity of naphtha fractions for light olefins production

Aaron Akah[1] · Musaed Al-Ghrami[1] · Mian Saeed[2] · M. Abdul Bari Siddiqui[2]

**Abstract** The catalytic cracking of naphtha fractions for propylene production was investigated under high severity catalytic cracking conditions (high temperatures and high catalyst to oil ratio). Straight run naphtha and cracked naphtha along with a with proprietary catalyst were used, and reaction was carried out using a catalyst to oil ratio (C/O) of 3–6 at 600–650 °C and 1 atm in a micro activity testing (MAT) unit. The results from this experiments show that light cracked naphtha (LCN) gave the highest propylene yield of 18% at 650 °C, and that propylene yield depends on the naphtha fraction being used as feed. The trend for reactivity and propylene yield was as follows: light cracked naphtha > heavy straight run naphtha > light straight run naphtha > heavy cracked naphtha.

**Keywords** Naphtha cracking · FCC catalysts · Light olefin production · High severity catalytic cracking

## Introduction

Light olefins such as ethylene and propylene are important building blocks for many end products like polyethylene and polypropylene. Recently, market analysis show that the demand for propylene is outpacing that of ethylene and the current supply cannot match the demand. A large proportion of propylene, about 65 wt%, is produced by steam

cracking and about 30 wt% during the fluid catalytic cracking (FCC) process as by product [1–3]. The propylene to ethylene ratio produced by steam cracking of naphtha is about 0.6, whereas the ethylene and propylene yields are about 2 and 6 wt% from conventional FCC process.

During catalytic cracking, the heavier and more complex hydrocarbon molecules are broken down into simpler and lighter molecules by the action of heat and catalyst. It is through this way that heavy oils can be upgraded into lighter and more valuable products (light olefin, gasoline and middle distillate components). The FCC is one of the most catalytic cracking technologies used widely in refinery for producing gasoline and diesel. However, current direction is to maximize olefins such as propylene and butylene by the addition of ZSM-5 to the catalyst formulation [4–14]. ZSM-5 shows high catalytic activity for the cracking of $C_7^+$ olefins into LPG range olefins and isomerization of n-olefins into i-olefins, while hydrogen transfer (a bimolecular reaction) is not allowed because of its small pore size [10, 15]. As a result, the entry of large branched hydrocarbons is restricted, thereby making the active sites accessible only to linear and monomethyl molecules [16].

Synergetic effects through mixing of conventional FCC catalyst (mostly USY zeolite) with ZSM-5 additive have been observed by several authors, and show that there is an increase in the yield of light olefins for the catalyst mixture, compared to product yield on the individual catalysts, suggesting that the reaction products are transferred between USY zeolite and ZSM-5 [6, 7, 9, 10, 17, 18]. Improvements in FCC catalyst, process design, hardware, and operation severity can boost high value light olefins yields, with propylene yield that can increase from 6 wt% up to 25 wt% or higher with VGO feed. However, additional efforts in the area of catalyst and process

✉ Aaron Akah
aaron.akah@aramco.com

1  Research and Development Center, Saudi Aramco, Dhahran 31311, Saudi Arabia

2  Center for Refining and Petrochemicals, Research Institute, KFUPM, P.O. Box 807, Dhahran 31261, Saudi Arabia

development are needed to be able to process light hydrocarbon such as naphtha.

Naphtha is the predominant feed for steam crackers, as more than half of the ethylene currently produced worldwide is derived from cracking naphtha feed. However, propylene production from steam crackers depends on the operating rates of the steam cracker and the type of feedstock. The propylene yield from steam cracking is directly proportional to the average molecular weight of the feed [19].

In the past, propylene was produced from steam cracking of naphtha and, as a result, propylene was available in substantial amounts. However, most modern steam crackers use ethane-based feed in place of heavy liquids leading to less propylene [19]. Consequently, it is expected that propylene production from steam crackers will further decrease as a result of the shift to ethane-based feeds. The knock-on effect is that more naphtha will be available as feedstock for the production of propylene via catalytic cracking and this can contribute to reducing the gap between the high demand and low supply of the propylene. The demand for propylene from FCC is growing at a faster rate than global FCC capacity and, therefore, propylene yields from FCC need to increase to keep up with demand.

This paper will discuss the evaluation of naphtha fractions as feedstock for the production of propylene and light olefins via catalytic cracking under high severity conditions which are high temperature and high catalyst/oil (C/O) ratio. The catalyst used in this study is made of ZSM-5 and USY in order to draw on the synergetic effect of mixing MFI and FAU zeolites to increase light olefin yield and boost the octane number of the gasoline produced.

## Materials and methods

### Naphtha feeds

Catalytic cracking experiments were carried out using naphtha fraction available at Saudi Aramco Refineries: light straight run naphtha (LSRN), heavy straight run naphtha (HSRN), light cracked naphtha (LCN) and heavy cracked naphtha (HCN). Detailed hydrocarbon analysis (PIONA) of the gasoline-range MAT liquid products was conducted using a Shimadzu PIONA GC equipped with an FID detector. The capillary column used was CP-Sil 5CB (50 m long, 0.32 mm ID).

### Catalyst

A proprietary catalyst was used for the evaluation of all feedstocks. Prior to testing, the fresh low-activity Y zeolite-based commercial catalyst and ZSM-5-based commercial additive were subjected to hydrothermal

deactivation treatment according to ASTM D4463. In a fluidized bed unit (Sakuragi Rikagaku, Japan), catalyst and additive for this study were treated separately in 100% steam environment at 810 °C for 6 h. The Y zeolite-based commercial catalyst and ZSM-5-based commercial additive were physically mixed in 3:1 ratio for all the cracking reactions presented in this paper.

## Experimental procedure

Naphtha catalytic cracking was carried out in a fixed-bed micro activity test (MAT) unit (Sakuragi Rikagaku, Japan), using a quartz tubular reactor (I.D. 22 mm, and 38 cm in length). A schematic representation of the experimental set up is shown in Fig. 1. A low-temperature circulating bath maintained at −10 °C was added to the unit instead of using conventional ice water. All experiments were conducted in the MAT unit at 30 s time-on-stream (TOS). The feed injector and reactor assembly were placed in the heating zone. Before feed injection, the system was purged with $N_2$ flow at 30 mL/min for about 15 min. Liquid receiver with the product vial was then connected to the bottom of the reactor. The other end of the receiver was connected to the burette for gas collection. A leak test was performed and a low-temperature bath was raised to cover the liquid receiver. The system was continuously purged with $N_2$ gas for further 15 min. The reactor was charged with a known amount of catalyst and about 1 g of naphtha was then fed to the reactor during 30 s along with 30 mL/min of $N_2$ flow. After the reaction, stripping of catalyst was carried out for 5 min using 30 mL/min of $N_2$ flow. The low-temperature bath was removed and stripping of liquid was continued for three more

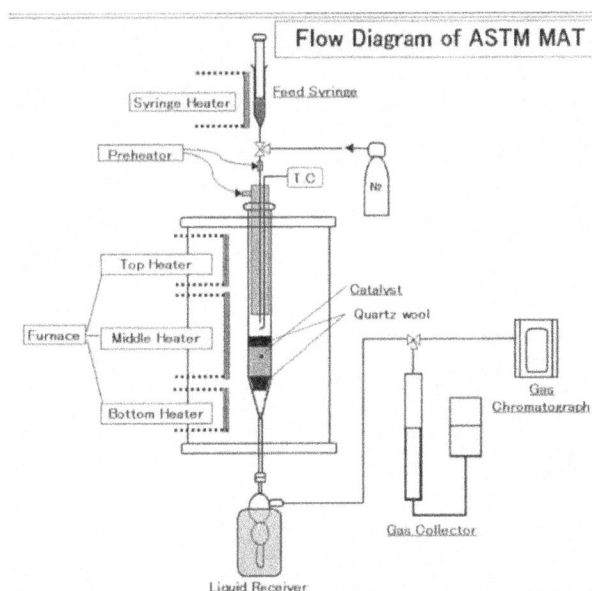

Fig. 1 Schematic diagram of the ASTM MAT Unit

minutes to remove the gas product dissolved in the liquid. During the reaction and stripping modes, gaseous products were collected in a gas burette by water displacement. Weight of the feed syringe was taken before and after experiments to obtain the exact weight of oil fed. Catalytic cracking experiments were performed at temperatures between 600 and 650 °C and the effect of catalyst/oil (C/O) ratio for each temperature was studied.

## Analysis of MAT products

MAT products comprised gas, liquid, and coke. Mass balance was considered acceptable within the limits of 95–103 wt%. A thorough gas chromatographic analysis of all MAT products was conducted to provide detailed yield patterns and information on the performance of the feed being tested. Gases were analyzed using two Varian GCs equipped with Flame Ionization Detector (FID) and Thermal Conductivity Detector (TCD). This allowed the quantitative determination of all light hydrocarbons up to $C_4$, $C_5$ paraffins, hydrogen and fixed gases. Hydrocarbons from $C_1$ to $C_4$, and $C_5$ paraffins, could be determined accurately. After gas analysis, the weight of each gas component was added and the weight of all components heavier than $C_4$ was added to gasoline fraction. The detailed composition of the product was obtained from the gas analysis which was normalized to account for the differences in mass.

Coke on spent catalyst was determined by Horiba Carbon–Sulfur Analyzer Model EMIA-220 V. About 1 g of spent catalyst (with tungsten and tin added as combustion promoters) was burnt in the high temperature furnace. The resulting combustion gas ($CO_2$) was passed through an Infra-Red Analyzer and carbon content was calculated as a percent of catalyst weight. All the results in this work are presented as weight percent (wt%) of the product. The conversion of naphtha feeds is defined as the total yield of the hydrocarbons from $C_1$ to $C_4$, hydrogen and coke.

Naphtha conversion (wt%) = Yield (wt%) of total gas

   + coke (wt%).

$$\tag{1}$$

The terms and definitions used in this work are summarized in Table 1.

## Results

### PIONA Analysis of Naphtha Feeds

Detailed hydrocarbon (PIONA) analysis of naphtha feeds is presented in Table 2. This table indicates that LSRN was mainly a mixture of pentanes and hexanes ($C_5$ and $C_6$ paraffins) which make up about 94 wt% of the feed. The remaining 6% was made up of naphthenes and there were no aromatics in LSRN. It also shows that about 69.9 wt% of the HSRN feed consisted of paraffins, while the remaining portion of HSRN was almost equally distributed between naphthenes (14.4%) and aromatics (15.7%). While LCN feed is mainly a mixture of iso-paraffins, olefins and aromatics, HCN feed consists predominantly of aromatics compounds.

### Screening of naphtha feeds

The naphtha fractions were initially screened at 650 °C, using a C/O ratio of 6 to determine the reactivity for light olefin production. The results are summarized in Table 3. It can be seen that the conversion of HCN was very low compared to the other three types of naphtha. Similar trends were also observed for total light olefin yield and LPG olefin. However, the coke yield on HCN was much higher than that on all the other naphtha fractions. The low reactivity of HCN and its tendency to produce more coke can be attributed to its high aromatics content.

Using the C/O of 6, the naphtha fractions were further screened at lower temperatures and the results are summarized in Fig. 2 below.

Figure 2 shows that while conversion increased with increasing temperatures for different naphtha fractions, conversion of HCN had insignificant increase. A similar trend was observed for propylene yield. Based on the screening results, which showed that HCN was the least reactive feed, it was decided not to conduct further study with HCN feed. The low reactivity and high coke yield of HCN are attributed to the fact that HCN is made up of predominantly aromatics which are highly stable and difficult to convert and they also act as coke precursors. The hydrocarbon composition of each feed affects it reactivity and it has been shown that a feed that is high in paraffin and aromatic content shows low reactivity during catalytic cracking, while a feed rich in olefins is very reactive.

For LSRN, HSRN and LCN, detailed cracking patterns were obtained by varying C/O ratios in the range of 3–6 at selected temperatures of 600, 625 and 650 °C.

### Catalytic cracking of LSRN and HSRN

Figure 3 shows the change in conversion with increasing C/O from 3–6 at different temperatures for both LSRN and HSRN. Both feeds showed an increasing trend in conversion with increase in temperature. HSRN showed higher conversion when compared to that of LSRN because it has less thermal stability than the LSRN. For LSRN, the increase in conversion with the increase in temperature

**Table 1** Definition of terms used in the estimation of performance

| Term | Description |
| --- | --- |
| Naphtha Conversion | All non-condensable components in the product + coke ($H_2$, $C_1$–$C_4$ hydrocarbons and coke) |
| Dry gas | $H_2$, $C_1$ and $C_2$ and $C_2$= |
| LPG | All $C_3$ and $C_4$ hydrocarbons (paraffins + olefins) |
| Light olefins | All olefins in $C_2$–$C_4$ range |
| LPG olefins | Propylene + total butylenes |
| %Yield | Percentage of respective product based on total feed |
| % Selectivity | Percentage of respective product in the converted feed only |

**Table 2** Composition of naphtha feeds

| | Component/wt% | N-Paraffins | iso Paraffins | Olefins | Naphthenes | Aromatics | Total |
| --- | --- | --- | --- | --- | --- | --- | --- |
| LSRN | C-5 | 29.8 | 8.2 | | 2.3 | | 40.3 |
| | C-6 | 26.9 | 28.5 | | 3.7 | | 59.1 |
| | C-7 | | 0.6 | | | | 0.6 |
| | Total | 57.3 | 36.7 | | 6.0 | | 100.0 |
| HSRN | C-6 | 4.8 | 1.5 | | 2.2 | 0.3 | 8.7 |
| | C-7 | 11.4 | 8.5 | | 5.5 | 2.5 | 27.9 |
| | C-8 | 10.0 | 9.4 | | 3.3 | 6.4 | 29.1 |
| | C-9 | 7.0 | 8.6 | | 2.7 | 5.0 | 23.2 |
| | C-10 | 2.8 | 3.6 | | 0.6 | 1.3 | 8.4 |
| | C-11 | 0.9 | 1.1 | | 0.1 | 0.1 | 2.3 |
| | C-12 | 0.2 | 0.1 | | | | 0.3 |
| | Total | 37.0 | 32.9 | | 14.4 | 15.7 | 100.0 |
| LCN | C-4 | | | 0.6 | | | 0.6 |
| | C-5 | 4.5 | 24.7 | 22.2 | 1.6 | | 53.1 |
| | C-6 | 1.5 | 9.4 | 4.2 | 3.5 | 11.8 | 30.4 |
| | C-7 | 0.8 | 2.9 | 0.9 | 3.7 | 6.7 | 14.9 |
| | C-8 | | 0.5 | 0.3 | 0.18 | | 1.0 |
| | Total | 6.8 | 37.5 | 28.2 | 9.0 | 18.5 | 100 |
| HCN | C-5 | 0.2 | 1.5 | 1.1 | 0.1 | | 2.9 |
| | C-6 | 0.1 | 0.4 | 0.5 | 0.2 | | 1.2 |
| | C-7 | | | | 0.1 | 20.7 | 20.8 |
| | C-8 | 0.3 | 1.0 | 1.3 | 0.6 | 33.0 | 36.2 |
| | C-9 | 0.3 | 1.5 | 0.3 | 0.6 | 23.4 | 26.0 |
| | C-10 | 0.3 | 2.1 | 0.1 | 0.2 | 6.0 | 8.6 |
| | C-11 | 0.2 | 2.1 | | 0.1 | 1.6 | 3.9 |
| | C-12 | | 0.2 | | | | 0.2 |
| | Total | 1.4 | 8.8 | 3.3 | 1.9 | 84.6 | 100.0 |

from 625 °C to 650 °C was more pronounced compared with the increase in conversion from 600 to 625 °C. This is probably because the hydrocarbon molecules undergo further cracking at 650 °C than when the temperature is 625 °C or 600 °C. Dry gas yield showed an increasing trend with conversion at all temperatures for both LSRN and HSRN. The rate of increase in dry gas with conversion is higher for the LSRN feed. For both LSRN and HSRN, total light olefins ($C_2$= + $C_3$= + $C_4$=) yield showed a linear trend with increasing conversion at all temperatures,

whereas HSRN showed higher total light olefins than LSRN.

Figure 4 shows the trend of each of the light olefins yields with conversion. Propylene yield showed a linear trend with increasing conversion at all temperatures for both LSRN and HSRN. HSRN showed higher propylene yields when compared to those of LSRN, because HSRN is made up of molecules with longer chains which are more reactive than those found in LSRN. Ethylene yields showed an increasing trend with increasing conversion at all

**Table 3** Comparison of naphtha cracking at 650 °C

| Feed | HCN | LCN | HSRN | LSRN |
|---|---|---|---|---|
| CAT/OIL | 5.98 | 6.04 | 5.71 | 5.95 |
| CONV.(%) | 14.71 | 40.07 | 34.79 | 26.55 |
| Yields (wt%) | | | | |
| Methane | 0.85 | 2.81 | 1.81 | 2.33 |
| Ethylene ($C_2$=) | 4.75 | 9.14 | 5.45 | 4.79 |
| Propylene ($C_3$=) | 4.78 | 17.75 | 12.94 | 10.12 |
| Coke | 0.79 | 0.07 | 0.66 | 0.70 |
| Groups | | | | |
| $H_2$–$C_2$ (Dry gas) incl $C_2$= | 6.60 | 13.48 | 9.04 | 9.78 |
| All $C_3$–$C_4$ (LPG) | 7.32 | 26.53 | 25.09 | 16.08 |
| $C_2$= – $C_4$= (Total light olefins) | 11.23 | 33.63 | 25.79 | 18.46 |
| $C_3$= + $C_4$= (LPG olefins) | 6.48 | 24.49 | 20.34 | 13.66 |
| $C_4$= (Butenes) | 1.70 | 6.74 | 7.40 | 3.54 |
| Selectivities | | | | |
| Dry gas | 44.86 | 33.63 | 25.98 | 36.82 |
| Propylene | 32.49 | 44.28 | 37.19 | 38.12 |
| Coke | 5.36 | 0.16 | 1.89 | 2.64 |

temperatures for both LSRN and HSRN. Butylenes yield showed a linear trend with increasing conversion at all temperatures for both LSRN and HSRN. HSRN showed higher butylenes yields when compared to those of LSRN. For HSRN, the difference in butylenes yields at 625 and 650 °C was less pronounced.

## Catalytic cracking of LCN

Figure 5 shows the change in conversion and yield pattern with increasing $C/O$ from 3 to 6 at different temperatures for LCN. At all temperatures, cracked naphtha feed showed an increasing trend in conversion. The conversion at the highest $C/O$ of 6 for temperature 650 °C was about 31%

more than that at 600 °C. Propylene and ethylene yields showed a linear trend with increasing conversion at all temperatures for LCN feed. Although butylenes yield showed an increasing trend with conversion, this trend was not very sharp at higher temperatures of 625 and 650 °C. Dry gas yield showed an almost linear increase with conversion at all temperatures. Total light olefins ($C_2$= + $C_3$= + $C_4$=) yield with conversion showed a linear trend with increasing conversion at all temperatures for LCN.

## Hydrogen transfer index (HTI) and cracking mechanism ratio (CMR)

The hydrogen transfer index (HTI) and the cracking mechanism ratio (CMR) for the cracking of the three naphtha fractions are shown in Fig. 6. The HTI describes the degree of hydrogen transfer reaction, which reduces the olefin yield in the products and, in this study, the HTI was measured using the ratio of $C_4$=/$C_4$. From Fig. 6, the general trend was that the HTI increased with increasing temperature and, for the same temperature, the HTI decreased with conversion. This is because hydrogen transfer is an exothermic reaction with a slower reaction rate and it is not favored by a high reaction temperature and shorter reaction time, but being a bimolecular reaction, it is promoted by a higher acid site density, which is provided by increasing the $C/O$ ratio [20, 21].

The cracking mechanism ratio (CMR), which is defined as the ratio of dry gases (methane, ethane, and ethylene) to isobutane in the gas products, is used to measure the ratio of monomolecular to bimolecular types of cracking, since $C_1$ and $C_2$ are typical products from protolytic cracking, while $iC_4$ is a typical product formed by β-scission of branched products [18, 22, 23].

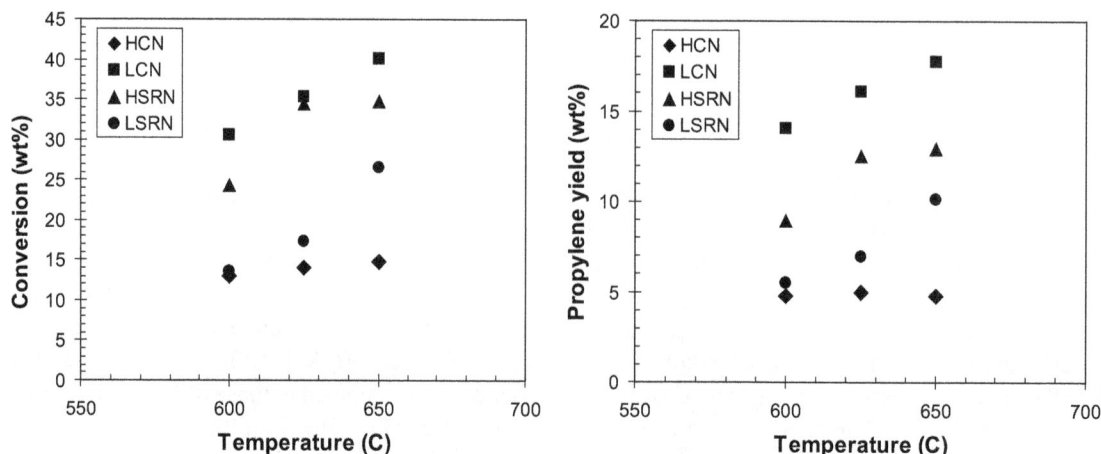

**Fig. 2** Impact of temperature on various naphtha feeds cracking at $C/O$ of about 6

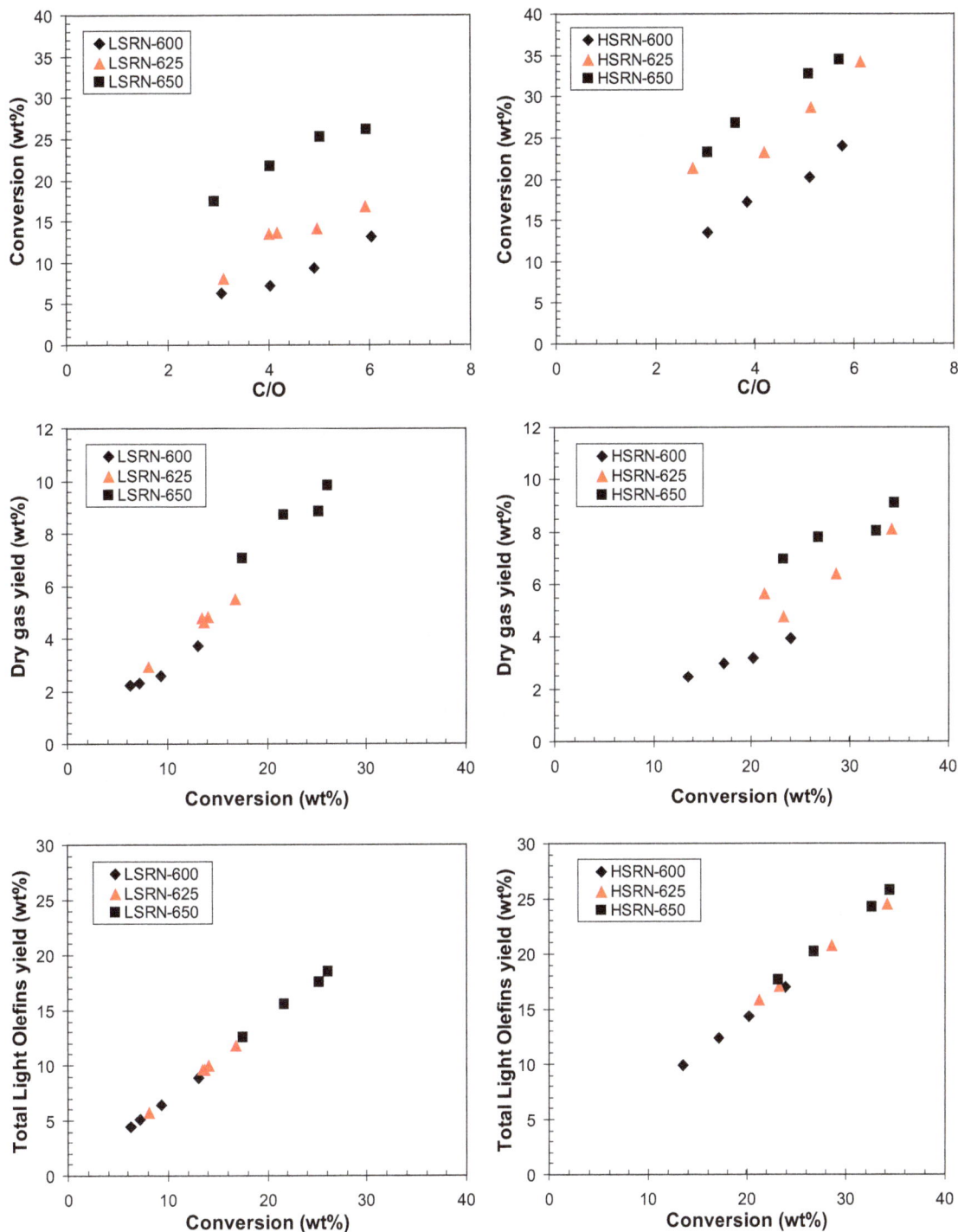

**Fig. 3** Conversion and yield data for LSRN and HSRN fractions

A qualitative estimation of the relative importance of the two cracking mechanisms can be made using the cracking mechanism ratio (CMR):

$$CMR = \frac{(C_1 + \sum C_2)}{iC_4},$$ (2)

where $C_1$, $C_2$, and $iC_4$ denote the selectivities to methane, ethane and ethylene, and i-butane, respectively.

A high CMR value (>1) reflects an important contribution of the protolytic cracking route, while a low value (<1) indicates the prevalence of the classical β-scission

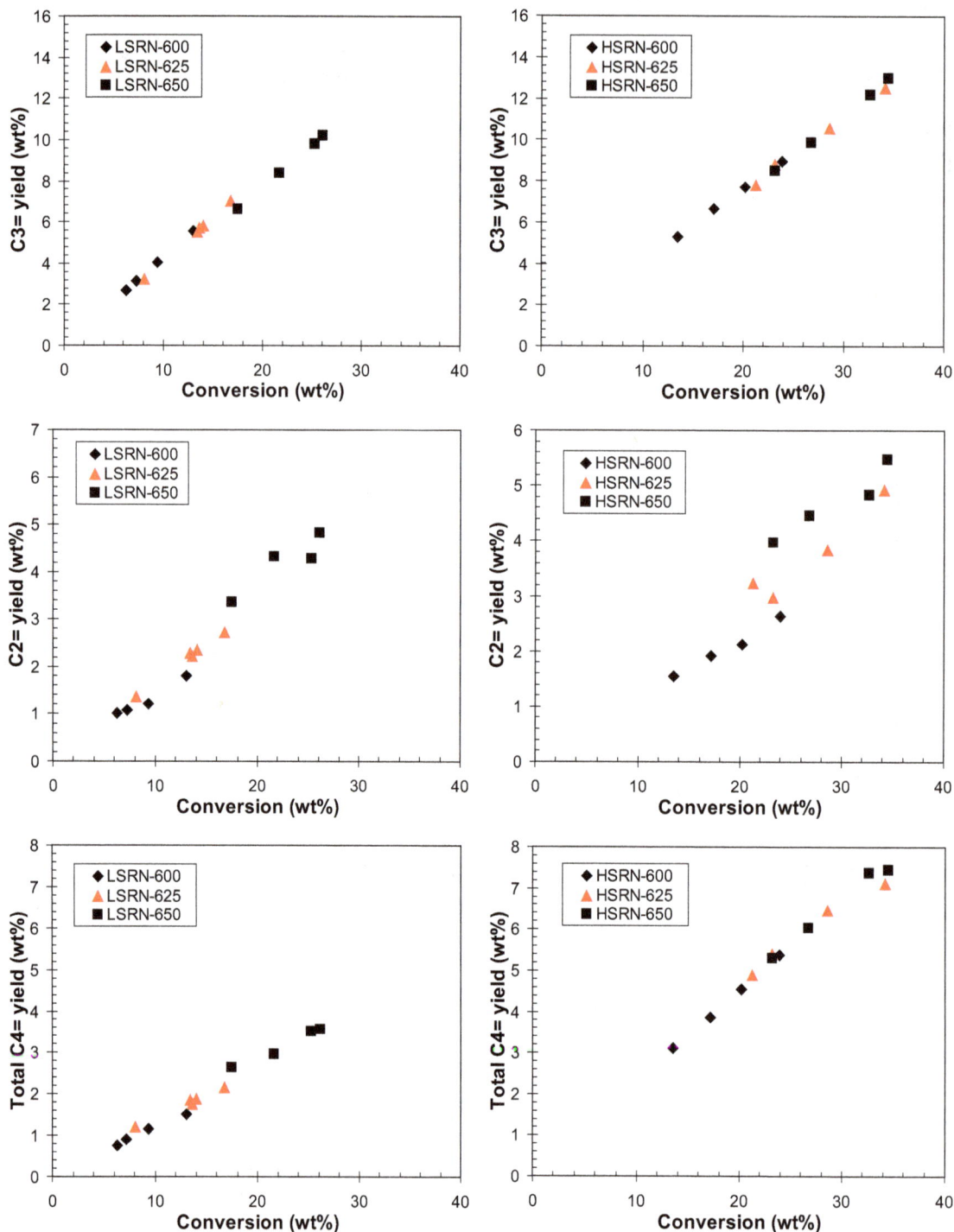

**Fig. 4** Light olefins yields of LSRN and HSRN cracking at different temperatures

cracking mechanism. From Fig. 6, it can be seen that the protolytic (monomolecular) cracking mechanism was more predominant than the beta scission (bimolecular) cracking mechanism for all the three naphtha fractions. It was also found that for each reaction temperature, as the conversion increased due to an increase in $C/O$, the CMR decreased as the contribution of the wide pore zeolite was increased. This shows that the production of light olefins is favored when protolytic cracking mechanism becomes dominant over classical bimolecular cracking reactions.

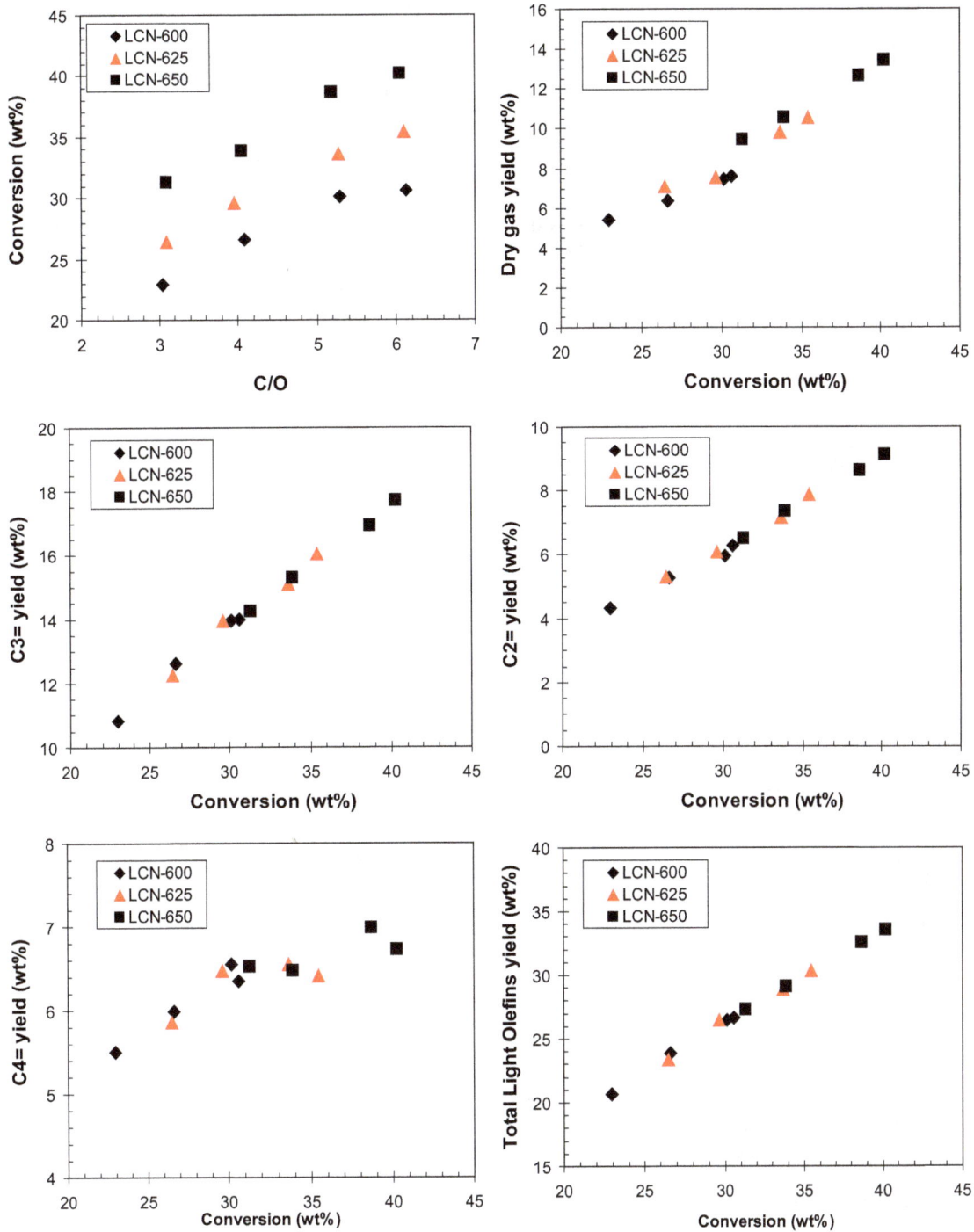

Fig. 5 Impact of increasing *C/O* on the conversion and yields of LCN cracking

## Thermal cracking index (TCI)

The contribution of thermal cracking to the cracking of naphtha is illustrated in Fig. 7. The contribution of thermal cracking was measured using the thermal cracking index (TCI), defined as the weight ratio of the sum of $C_1$ and $C_2$ yields to the sum of isobutane and isobutene yields [21].

$$TCI = \frac{C_1 + \sum C_2}{iC_4 + iC_4^=}. \tag{3}$$

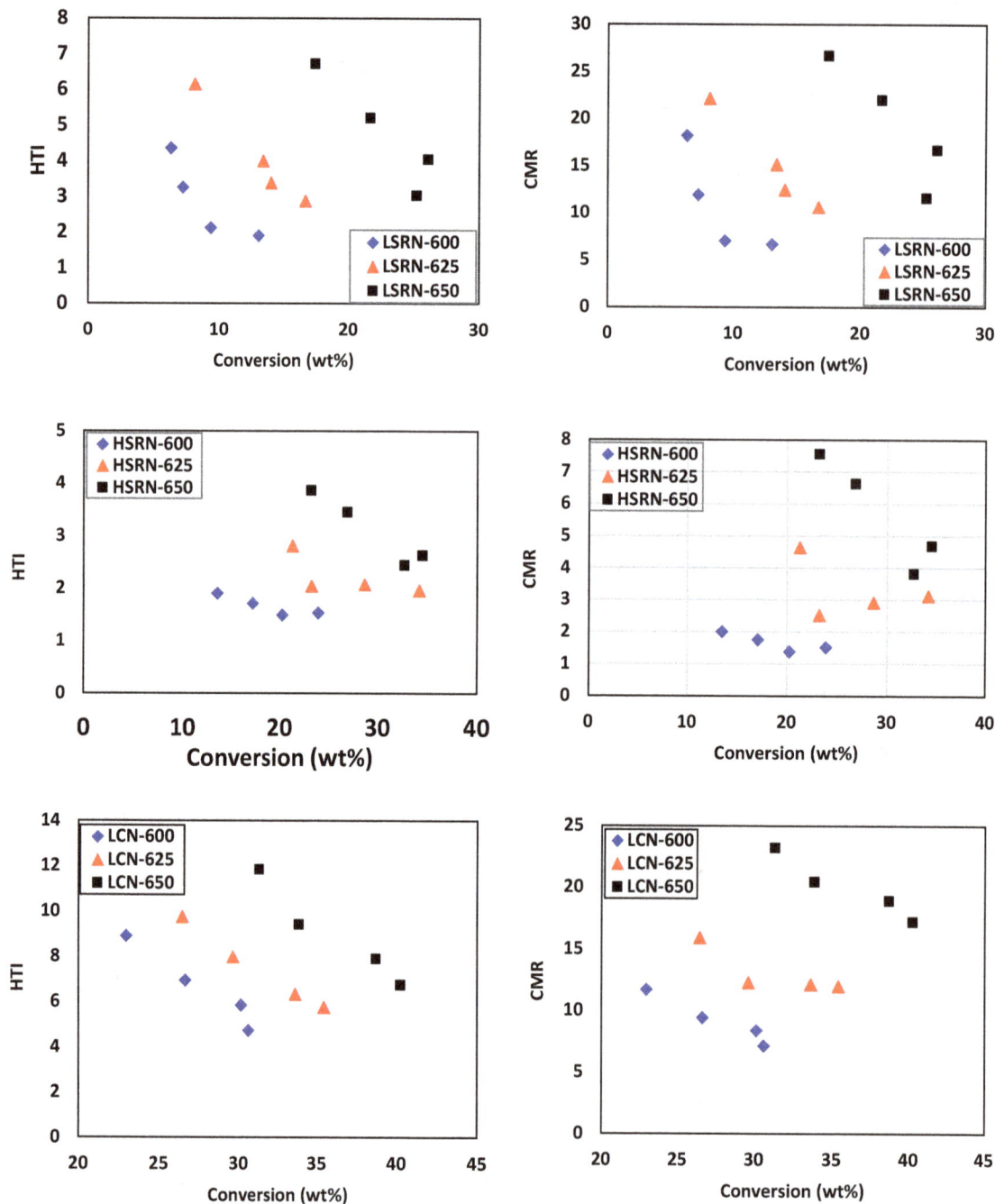

**Fig. 6** Impact of temperature on HTI and CMR

A value of TCI <0.6 means that catalytic cracking is the main reaction, while a value of TCI >1.2 means that thermal cracking is serious [21].

From Fig. 7, it can be seen that the TCI was much greater than 1 under most conditions indicating that at these temperatures, thermal cracking was serious and as such contributed to the product yield.

The general trend for all three fractions was that TCI increased as the reaction temperature increased. This is expected because an increase in temperature will lead to more thermal cracking contributions as less stable intermediates undergo further reaction.

**Discussion**

The cracking reactions can be categorized into two types, namely catalytic cracking and thermal cracking. Catalytic cracking is endothermic, occurs on the surface of the

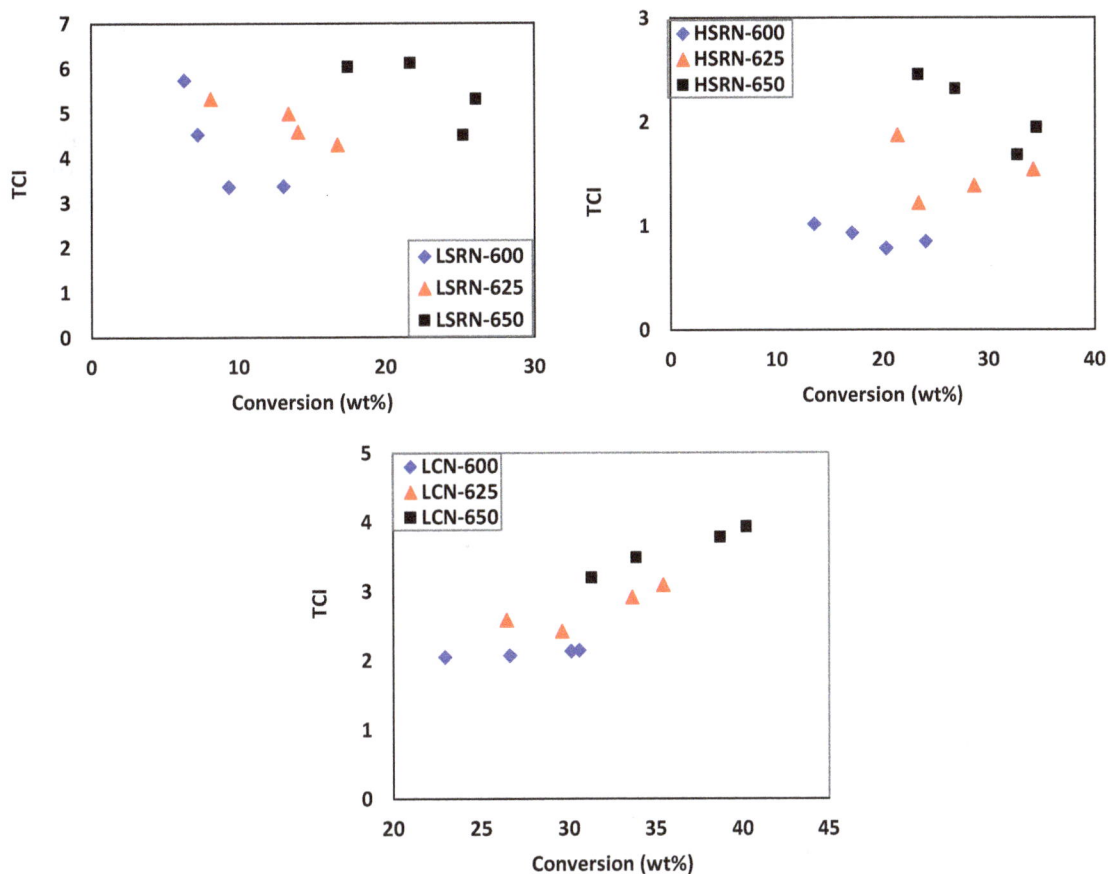

**Fig. 7** Effect of temperature on TCI

catalyst and yields high $C_3$ and $C_4$ olefins. So, the reaction temperature and acidic properties of the catalyst play a major role in the activity and selectivity of the reaction. The thermal cracking reaction is also endothermic and follows the free radical mechanism which generates ethylene and dry gas. For increasing propylene production, the reaction conditions have to be optimized to minimize thermal cracking.

The results from this study (Figs. 2, 3, 4, 5) show that reaction temperature, C/O ratio and feed characteristics have an effect on light olefin yield during catalytic cracking. The yield of light olefins and selectivity to propylene when using HCN feed are the lowest compared to the other feeds because of its high aromatic content. It has been shown in the literature that hydrocarbon feeds with a high C/H ratio (low hydrogen content) are difficult to convert under FCC conditions [3]. The production of propylene requires a disproportionate share of the hydrogen and co-products such as propane, and dry gas requires an even greater share of hydrogen. Therefore, the amount of hydrogen available from the feedstock can limit the potential to produce propylene. Subsequently, propylene production is highly dependent on feed properties.

In contrast to HCN, LCN is the most reactive and shows the highest conversion. It also shows the highest yield for light olefins as well as selectivity. This is because olefins make up a great portion of this feed (28.5 wt%), and with the right type of catalyst this can act as intermediate for high yield of light olefins.

LSRN is predominantly $C_5$–$C_6$ while HSRN is mostly $C_6$–$C_{10}$ with much smaller fractions of $C_{11}$–$C_{12}$. Their compositions should explain why HSRN is more reactive than LSRN, because HSRN is made up of longer chain molecules, which are easier to crack than the short length LSRN counterpart. This also has an effect on the product yield, as HSRN shows a higher light olefin yield.

The overall classification in terms of reactivity is as follows: LCN > HSRN > LSRN > HCN.

## Influence of catalyst-to-oil ratio on FCC naphtha cracking

In order for propylene to be produced, the cracking of FCC naphtha has to be carried out at high temperatures, but increasing temperature favors thermal cracking which leads to the formation of dry gas from hydrocarbon [24].

Raising the temperature is most effective in controlling the exothermic hydrogen transfer reaction and accelerating the catalytic cracking which is endothermic. In a commercial unit, the high reaction temperature is achieved by raising the catalyst circulation rate. This helps to improve the rate of reaction as well as the propylene yield. Since the hydrogen transfer reaction is a secondary reaction, it is better controlled using a short contact time to limit the hydrogenation of olefin over cracking reactions [3, 8].

By increasing $C/O$ ratio, contact opportunities between catalyst active centers and hydrocarbon molecules are improved leading to enhanced selectivity and yields of light olefins such as propylene and butylene during catalytic cracking of hydrocarbons.

As can be seen from Figs. 3 and 5, when there is an increase in $C/O$, the conversion, total light olefin yield and the yields of dry gas increase. The reactions of olefins take place more easily at high $C/O$ due to its high reactivity over catalyst active centers and this explains the high yield of light olefins with LCN as feed. Compared to olefins, paraffins have a lower reactivity towards cracking due to energetically more difficult formation of carbenium ions.

Increasing the catalyst to oil ratio increases the severity of the hydrogenation and cracking reaction leading to over cracking and more saturates. Subsequently, more dry gas is produced, which in this case is made up of mainly methane and ethane. For thermal cracking, free radicals are formed through hydrocarbons splitting their C–C bonds and H–C bonds, and then inclined to undergoing alpha scission, beta scission, and polymerization to produce $H_2$, methane, ethylene and coke [24]. However, at higher $C/O$, thermal cracking is effectively inhibited as hydrocarbons have more opportunities to contact with active centres of catalysts leading to catalytic cracking. Therefore, the yields of $C_1$ and $H_2$ decrease as the $C/O$ is enhanced. But because of more small pore-zeolite (ZSM-5) being available at higher $C/O$, which are more favorable for the formation of ethane, therefore, mostly the yields of ethylene and ethane in dry gas are enhanced. Also, a very high $C/O$ does not necessarily lead to a significant increase of propylene and butylene due to the enhanced hydrogen transfer reaction stemming from the presence of more Y zeolite-based catalyst. As Y zeolite-based catalyst are used in FCC units, the bimolecular mechanism competes with cracking to form light olefins.

For maximizing light olefin production, FCC units should be operated at high severity, which includes high $C/O$ ratios, high temperatures, and short contact time to minimize the hydrogen transfer reaction.

It has been well established that the conversion of light hydrocarbons occurs via carbenium ion chemistry [10, 25–27] involving the monomolecular and bimolecular mechanisms.

The steps for the catalytic cracking mechanism of alkanes [10, 25, 26, 28] are summarized below and illustrated in Fig. 8:

1. A carbenium ion is first generated by protonation of a paraffin molecule on a Brönsted acid site to form a carbonium ion (a pentacoordinated ion),
2. The carbonium ion then immediately cracks to give a carbenium ion and a paraffin.
3. Then, the carbenium ion cracks to give an olefin and an acidic proton on the surface of the catalyst.

For the Bimolecular mechanism, the carbenium ion in step 3 acts as an active center by reacting with a feed paraffin molecule to form a larger molecule that further cracks according to the following steps:

4. A paraffin reacts with carbenium ion to form a smaller paraffin and a bigger carbenium ion.
5. The bigger carbenium ion then cracks into an olefin and a carbenium ion through beta scission.

The bimolecular mechanism explain the hydrogen transfer reaction and the formation of heavier compounds than those found in the feed such as coke, light cycle oil and heavy cycle oil. Hydrogen transfer reaction usually occurs from an olefin to a carbenium ion on the catalyst surface giving rise to a paraffin and a hydrogen-deficient species (allylcarbenium ion) [29]. The hydrogen-deficient species can be further transformed into aromatics and coke via further dehydrogenation and cyclization reactions. Hydrogen transfer reactions are prevalent at high conversions and help to decrease the selectivity to olefins as shown in Fig. 8.

Thus, a bimolecular cracking mechanism will preferentially lead to the formation of aromatics from the naphthenes, since the carbenium may be directly formed from the feed molecule through a hydride transfer, while a monomolecular cracking mechanism will preferentially lead to short olefins. Then, the balance between hydrogen transfer reactions and cracking reactions will be critical in orienting the selectivity of olefins and naphthenes' conversion towards either propylene or gasoline-range aromatics or paraffins.

The balance between these two mechanisms will depend on the surface coverage by carbenium ions. The presence of carbenium ions is influenced by reaction temperature and catalyst acid site density. The higher number of acid sites favors the bimolecular mechanism. The acid site density may be decreased by dealumination or coking. Catalysts with small pore sizes such as ZSM-5 favor the monomolecular mechanism since there is limited space for the formation of the intermediate species for the bimolecular mechanism. Propylene production is hampered by hydrogen transfer reaction as it consumes the olefins generated by the catalytic cracking. This hydrogen transfer is

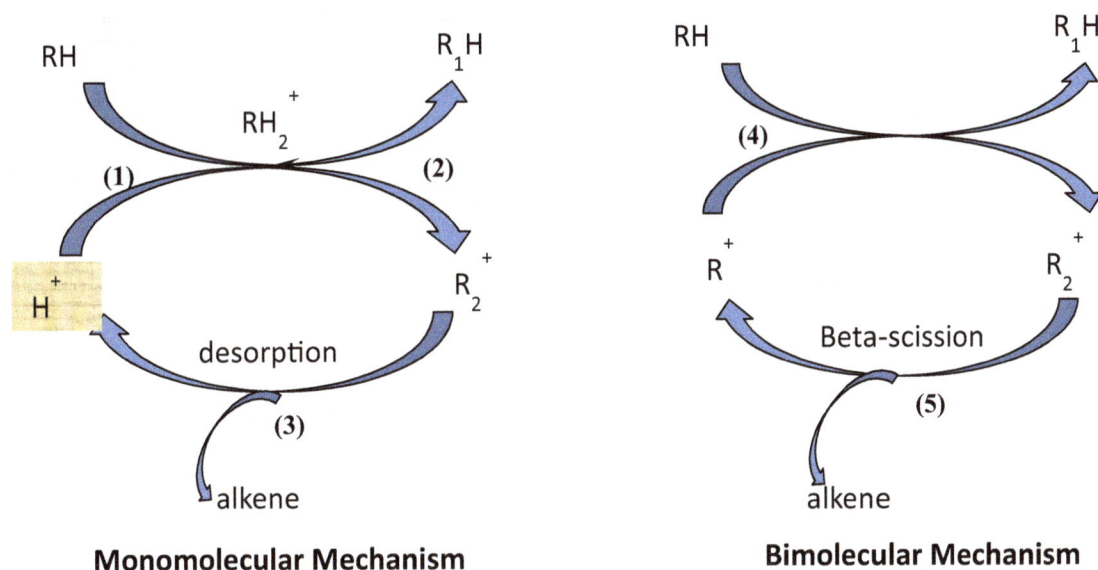

**Fig. 8** Illustration of cracking mechanism

generally known to have a bimolecular reaction and it depends on the acidity of the catalyst, contact time and reaction temperature.

## Conclusions

The results from these experiments show that propylene yield depends on the naphtha fraction being used as feed. LCN which contains a high percentage of olefins showed high reactivity and gave the highest propylene and light olefin yields, while HCN which is made up of mostly aromatics was the least reactive naphtha fraction. For straight run naphtha, HSRN showed a higher reactivity compared to LSRN. This is because, LSRN is made up of mainly $C_5$–$C_6$ molecules and are difficult to crack compared to HSRN which is made up of mainly $C_6$–$C_{11}$.

The results from this study further provide a guideline for processing naphtha fractions under high severity conditions for the production of light olefins. This will require the fine tuning of the catalyst system and reaction conditions to maximize the yields of light olefins.

## References

1. Brookes T (2012) New Technology developments in the petrochemical industry-refinery integration with petrochemicals to achieve higher value uplift in Egypt Petrochemicals Conference, Cairo-Egypt
2. Hyde B (2012) Light Olefins Market Review. In Foro Pemex Petroquimica, Mexico
3. Akah A, Al-Ghrami M (2015) Maximizing propylene production via FCC technology. Appl Petrochem Res 5(4):377–392
4. DOWNFLOW CONCURRENT CATALYTIC CRACKING. PHILLIPS PETROLEUM CO, US: United States
5. Triantafillidis CS et al (1999) Performance of ZSM-5 as a fluid catalytic cracking catalyst additive: effect of the total number of acid sites and particle size. Ind Eng Chem Res 38(3):916–927
6. Zhao X, Harding RH (1999) ZSM-5 additive in fluid catalytic cracking. 2. effect of hydrogen transfer characteristics of the base cracking catalysts and feedstocks. Ind Eng Chem Res 38(10):3854–3859
7. Zhao X, Roberie TG (1999) ZSM-5 additive in fluid catalytic cracking. 1. effect of additive level and temperature on light olefins and gasoline olefins. Ind Eng Chem Res 38(10):3847–3853
8. Aitani A, Yoshikawa T, Ino T (2000) Maximization of FCC light olefins by high severity operation and ZSM-5 addition. Catal Today 60(1–2):111–117
9. Degnan TF, Chitnis GK, Schipper PH (2000) History of ZSM-5 fluid catalytic cracking additive development at Mobil. Microporous Mesoporous Mater 35–36:245–252
10. den Hollander MA et al (2002) Gasoline conversion: reactivity towards cracking with equilibrated FCC and ZSM-5 catalysts. Appl Catal A 223(1–2):85–102
11. Corma A et al (2004) Different process schemes for converting light straight run and fluid catalytic cracking naphthas in a FCC unit for maximum propylene production. Appl Catal A 265(2):195–206
12. Abul-Hamayel MA, Aitani AM, Saeed MR (2005) Enhancement of propylene production in a downer FCC operation using a ZSM-5 additive. Chem Eng Technol 28(8):923–929
13. Arandes JM et al (2009) HZSM-5 zeolite as catalyst additive for residue cracking under FCC conditions. Energy Fuels 23(9):4215–4223
14. Lee J et al (2013) Catalytic cracking of C5 raffinate to light olefins over lanthanum-containing phosphorous-modified porous ZSM-5: effect of lanthanum content. Fuel Process Technol 109:189–195
15. Shimada I et al (2015) Increasing Octane value in catalytic cracking of n-hexadecane with addition of *BEA type zeolite. Catalysts 5(2):703
16. Haag WO, Lago RM, Weisz PB (1981) Transport and reactivity of hydrocarbon molecules in a shape-selective zeolite. Faraday Discuss Chem Soc 72:317–330
17. Wallenstein D, Harding RH (2001) The dependence of ZSM-5 additive performance on the hydrogen-transfer activity of the REUSY base catalyst in fluid catalytic cracking. Appl Catal A 214(1):11–29

18. Zhongqing L et al (2003) Cracking behavior of MCM-22, ZSM-5 and Beta as FCC catalyst additives. Prepr Pap Am Chem Soc Div Fuel Chem 48(2):714

19. Brooks R (2013) Modeling the North American market for natural gas liquids. In 32nd US Association of Energy and Economics (USAEE) Conference, Anchorage, 28–31 July

20. Cheng WC, Rajagopalan K (1989) Conversion of cyclohexene over Y-zeolites: a model reaction for hydrogen transfer. J Catal 119(2):354–358

21. Zhang J et al (2013) Synergistic process for coker gas oil catalytic cracking and gasoline reformation. Energy Fuels 27(2):654–665

22. Bastiani R et al (2013) Application of ferrierite zeolite in high-olefin catalytic cracking. Fuel 107:680–687

23. Wielers AFH, Vaarkamp M, Post MFM (1991) Relation between properties and performance of zeolites in paraffin cracking. J Catal 127(1):51–66

24. Wang G, Xu C, Gao J (2008) Study of cracking FCC naphtha in a secondary riser of the FCC unit for maximum propylene production. Fuel Process Technol 89(9):864–873

25. Corma A et al (2005) Light cracked naphtha processing: controlling chemistry for maximum propylene production. Catal Today 107–108:699–706

26. Kotrel S, Knözinger H, Gates BC (2000) The Haag-Dessau mechanism of protolytic cracking of alkanes. Microporous Mesoporous Mater 35–36:11–20

27. Buchanan JS, Santiesteban JG, Haag WO (1996) Mechanistic considerations in acid-catalyzed cracking of olefins. J Catal 158(1):279–287

28. Rahimi N, Karimzadeh R (2011) Catalytic cracking of hydrocarbons over modified ZSM-5 zeolites to produce light olefins: a review. Appl Catal A 398:1–17

29. Komatsu T (2010) Catalytic cracking of paraffins on zeolite catalysts for the production of light olefins, in 20th annual Saudi-Japan symposium catalysts in petroleum refining & petrochemicals. Dhahran, Saudi Arabia

# 6

# Inquiring the photocatalytic activity of cuprous oxide nanoparticles synthesized by a green route on methylene blue dye

Manoranjan Behera[1] · Gitisudha Giri[1]

**Abstract** We synthesized cuprous oxide ($Cu_2O$) nanoparticles (NPs) with an average crystallite size of 8.8 nm in presence of Arka (*Calotropis gigantea*) leaves extract. The photo-bleaching activity of $Cu_2O$ NPs on the aqueous methylene blue (MB) dye was studied by illumination of visible light. In the absorption spectra, a decrease in the absorption peak intensity at 665 nm of MB was observed in presence of $Cu_2O$ NPs. A red shift in its peak position as a function of irradiation time is suggesting that oxide particles are degrading the organic dye in an aqueous medium. In the vibration spectra, red shift in the C–H stretching band (2954, 2926, and 2855 $cm^{-1}$) of methylene group and C–N stretching band (1343 and 1226 $cm^{-1}$) of MB in presence of $Cu_2O$ NPs proposes a surface adsorption of MB over NP's surfaces. Quenching in the emission band intensity and red shifts in the peak maxima of MB in presence of $Cu_2O$ NPs is ascribed to the charge transfer interaction between MB and oxide NPs. A linear Stern–Volmer plot reveals that decrease in the emission intensity of MB dye occurs via the dynamic quenching mechanism. Synthesis of $Cu_2O$ NPs of various architectures using a green route could be use as an approach towards the cost-effective treatment of water pollutants.

**Keywords** Photodegradation · Methylene blue dye · Surface adsorption · Charge transfer interaction · Photoluminescence quenching

## Introduction

Organic dyes are the major threats to our environments. It is not an easy task to remove them from dye-bearing waste waters owing to their stability towards oxidizing agents [1–3]. Amongst the various organic dyes, methylene blue (MB)—a cationic dye mostly used in paper, rubber, and textile industry as colorants—is found in waste water [4–6]. It is reported that acute exposure to MB dye might cause tissue narcosis, heart stroke, jaundice, etc., in humans [7, 8]. Presently a variety of physical, chemical, and biological methods were available for the treatment of dye contaminated water. From an extensive literature study, we concluded that a chemical method such as adsorption process is an economical and efficient route for elimination of toxic dyes from polluted water [9–12]. In this regards, various low-cost adsorbents such as fly ash, metal sulfides ($Ag_2S$), and metal oxides (titania $TiO_2$, cuprous oxide $Cu_2O$) have already been tested to treat polluted water [4, 9–13]. In recent years, semiconductor nanoparticles (NPs) were extensively used by scientists and academicians around the globe for the removal of various organic dyes via adsorption route. Pourahmad [4] reported that $Ag_2S$ NPs encapsulated in a mesoporous material can efficiently degrade MB dye in aqueous solution. In another work, Srinivasan and White [13] reported an accelerated photodegradation of MB over three-dimensionally ordered macroporous $TiO_2$ pore sizes lies between 0.5 and 1 μm. However, owing to high band gap of 3.2 eV, $TiO_2$ cannot perform in the visible region.

Among the various semiconductor metal oxides as photocatalyst, $Cu_2O$ NPs find a special place in photocatalysis under visible light [10–12, 14–16]. Wide applications of this oxide NPs are mainly owing to its non-toxicity, easy availability of cheap and up-scalable

✉ Manoranjan Behera
mano.silicon@gmail.com

[1] Silicon Institute of Technology, Bhubaneswar, Odisha, India

synthetic routes, lying of the band gap (i.e., 2.17 eV) in the visible range, tunability of band gap, and strong tendency to adsorb molecular oxygen which helps in scavenging the photogenerated electrons so that electron–hole pairs recombination can be restrained easily at the interface [10–12, 14–16]. It is reported that size and shape are of paramount importance in tailoring the various properties and applications of the $Cu_2O$ NPs. A variety of $Cu_2O$ nanostructures such as wires, boxes, cubes, truncated cubes, octahedra, nanocages, nanomultipods, spheres, and a variety of hollow structures have already been synthesized and tested for their photocatalytic activity on various organic compounds [14, 15]. To develop diverse architectures of $Cu_2O$, the various methods widely used includes hydrothermal method, microemulsion method, surfactant-assisted route, and wet chemical method [10–12, 14–16]. Sun et al. [9] have reported an enhanced photocatalytic activity for $Cu_2O$–graphene oxide (GO) nanocomposite synthesized via solvothermal route towards Rhodamine B (RhB) dye. They reported that more than 65 % RhB was degraded within 80 min of visible light irradiation. Cai et al. [10] have reported synthesis of $Cu_2O$-reduced GO (rGO) composite by a one-pot hydrothermal method using glucose as reducing and cross-linking agent. They reported that as rGO promotes the charge carrier separation, it increases the aqueous photocatalytic efficiency. But, nearly 70 % methyl orange (MO) degradation was reported for this nanocomposite after a long irradiation time of 300 min. In another article, Zhang et al. [11] have reported 80 % MO degradation after 30 min irradiation by graphene/defected $Cu_2O$ nanocomposite synthesized via a chemical vapor deposition method. They stressed on the importance of O-atoms towards the charge carrier separation. Zou et al. have reported synthesis of $Cu_2O$–rGO nanocomposites of various $Cu_2O$ crystal facets. They have reported only 72, 60 and 28 % MB degradation after 120 min for octahedral, dodecahedral and cubic faceted $Cu_2O$–rGO NPs.

A few reports are available on photocatalytic activity study of $Cu_2O$ NPs synthesized by green chemical route. $Cu_2O$ microcrystals with well-formed facets were synthesized by a simple hydrothermal method by Zheng et al. [14] and investigated the surface stabilities and photocatalytic properties of the synthesized $Cu_2O$ microcrystals. It is reported that $Cu_2O$ {100} and {110} facets gradually disappear and transform into nanosheets during the photodegradation of MO dye. With the increase of irradiation time, $Cu_2O$ microcrystals completely transform into nanosheets with {111} facets and the finally formed nanosheets exhibit stable photocatalytic activities. Zhu et al. [17] have synthesized $Cu_2O$ micro-/nanocrystals using a simple liquid phase reduction process under microwave irradiation. In particular, for the dandelion

morphology, the photocatalytic degradation rates of RhB dye is reported to be highest, i.e., 56.37 %. From above literature studies we concluded that photodegradation efficiency of $Cu_2O$ towards a dye depends on various parameters such as nature of the dye, synthesis route, shape and size of NPs, attachment of charge carrier separating agent to NP, adsorption capability, surface area, etc.

As we have not found any report on MB degradation by $Cu_2O$ NPS obtained via a green route, in this manuscript we report on visible light photodegradation of MB dye by biosurfactant-capped $Cu_2O$ NPs synthesized via a green route using leaves extract of Arka plant (i.e., *Calotropis gigantea*). The saponin molecules present in the plant extract not only acts as stabilizing agent [18] but is also believed to act as charge carrier separator. The synthesized nano-powders were characterized using UV–visible spectroscopy, Fourier transform infrared (FTIR) spectroscopy, X-ray photoelectron spectroscopy (XPS), X-ray diffraction (XRD), zeta potential, dynamic light scattering (DLS), photoluminescence (PL) spectroscopy, field emission scanning electron microscope (FESEM) and transmission electron microscope (TEM).

## Experimental

### Synthesis of $Cu_2O$ NPs

We synthesized $Cu_2O$ NPs by a green synthetic route using copper sulfate ($CuSO_4 \cdot 5H_2O$) crystals as precursor salt, Arka leaves extract as encapsulating agent, and hydrazine hydrate as reducing agent in an aqueous medium. In an aim to synthesize $Cu_2O$ NPs, first we prepared a precursor salt solution (1 M) by dissolving $CuSO_4 \cdot 5H_2O$ crystals in double-distilled water. After that, an aqueous plant extract was made by adding about 10 g of Arka leaves to 100 mL water in a conical flask and was boiled for 1 h in a hot plate. Now, the two solutions were mixed in a proper ratio and stirred in a magnetic stirrer for half an hour at 70 °C. Under the stirring condition, reducing agent was added dropwise till the blue color (i.e., due to $Cu^{2+}$ ions) changed to a permanent reddish-brown suspension consisting of cuprous oxide ($Cu_2O$) NPs. The aqueous suspensions thus obtained was kept standstill overnight to settle down $Cu_2O$ NPs. After careful decantation of upper supernatant liquid, those solid particles which were settled at the bottom of the container were kept in an electrical oven maintained at 80 °C for 2 h for complete removal of solvent so that samples can be collected in powder form. To confirm the formation of $Cu_2O$ NPs, we added 2 mL of concentrated ammonia solution to 10 mg $Cu_2O$ powders taken in a 25-mL beaker. A clear blue solution results from formation of $[Cu(NH_3)_4(H_2O)_2]^{2+}$ complex after shaking the beaker for a few seconds which indicates formation of copper(I) oxide [19]. The dried nano-powders were stored in desiccators for characterizations.

## Characterization techniques

We characterized the as-synthesized $Cu_2O$ powders using UV–visible spectroscopy, FTIR spectroscopy, XPS, XRD, zeta potential, DLS, PL spectroscopy, FESEM and TEM. UV–visible absorption spectra were acquired on a Perkin-Elmer spectrophotometer. The crystal phase of synthesized NPs was identified using an X-ray diffractometer (Rigaku D/Max 2000). Vibration spectra have been studied for the powder samples with a Thermo Nicolet Corporation FTIR Spectrometer (Model NEXUS–870). X-ray photoelectron spectrum (XPS) was collected on a VG ESCALAB MK-II spectrometer with a monochromatic Al Kα source ($hv = 1486.6$ eV) operated at 10 kV and 20 mA at $10^{-9}$ Pa. Sample for XPS was prepared by drop-casting method. A drop of aqueous dispersion of $Cu_2O$ NPs was placed on the silicon substrate and allowed to dry in desiccators by keeping overnight at room temperature. Zeta potential was measured using a Malvern Nano ZS instrument using phase analysis light scattering technique. Solid samples were sonicated for 20 min to well disperse oxide NPs in water prior to measuring zeta potential ($\xi$). The Malvern Nano ZS instrument was also used to find hydrodynamic diameter ($D_h$) of the aqueous suspension before and after adding $Cu_2O$ NPs to MB dye. The dispersed samples were analyzed at least for thrice at 25 °C to get accurate results. The PL spectra have been recorded with a computer-controlled Perkin-Elmer (Model-LS 55) luminescence spectrometer in conjugation with a red-sensitive photomultiplier tube detector (RS928) and a high-energy pulsed Xe-discharge lamp as an excitation source (average power 7.3 W at 50 Hz). Image of a biosurfactant-modified $Cu_2O$ nanocomposite powder was studied using FESEM of *ZEISS SUPRA-40* and TEM of JEM-2100 (JEOL, Japan) machines. TEM sample was prepared by placing one drop of diluted solution on a carbon-coated 600-mesh copper grid and allowing the sample to dry in desiccators at room temperature. In the FESEM studies, the sample was spin-coated on a (110) silicon plate and then a thin gold coating was sputtered to make a conducting surface.

## Photocatalytic degradation

To study the photodegradation of MB dye in an aqueous medium, first we prepared a suspension of $Cu_2O$ particles (100 mg) by dispersing it in a 500 mL solution of MB (20 mg/L) in water. The aqueous suspension was then sonicated for 20 min in dark to make the powders disperse

well in the solution before irradiation with visible light. The photocatalytic reaction was conducted in a 500-mL cylindrical glass reactor. We used 500 W Xenon lamp with a UV cutoff filter as visible light source. The dispersion was bubbled with 100 mL/min $O_2$ and stirred magnetically at 30 °C. At regular intervals, 15 mL of the suspension was sampled and separated by centrifugation at 4000 rpm for 10 min to remove the powder. The concentration of residual MB (i.e., supernatant liquid) was measured by its absorbance ($A$) at 665 nm with a Perkin-Elmer spectrophotometer. The degradation efficiency ($D$) of MB was calculated using the equation, $D = (A_0 - A)/A_0 \times 100$ %, where $A_0$ is the absorbance of 20 mg/L MB at 665 nm, $A$ is the absorbance at the same wavelength of the extracted solution. As a control, another sample was analyzed without irradiating to test the adsorption ability of the oxide catalyst; other test methods were the same, as mentioned above.

## Results and discussion

### UV–visible spectra

Figure 1 shows evolution of absorption spectra (400–900 nm) of MB in presence of $Cu_2O$ NPs upon irradiation of visible light; (a) 0, (b) 10, (c) 20, (d) 30, (e) 60, (f) 90, and (g) 120 min at room temperature. Before irradiating, MB solution with $Cu_2O$ NPs was kept in the dark for nearly 3 h to eliminate possibility of other effects such as adsorption and reaction without light to affect the result. As shown in Fig. 1, MB displays two absorption bands, one near 605 nm and other one near 665 nm. The decrease in the intensity of absorption peaks with time elapsed is ascribed to degradation of MB by $Cu_2O$ NPs. For

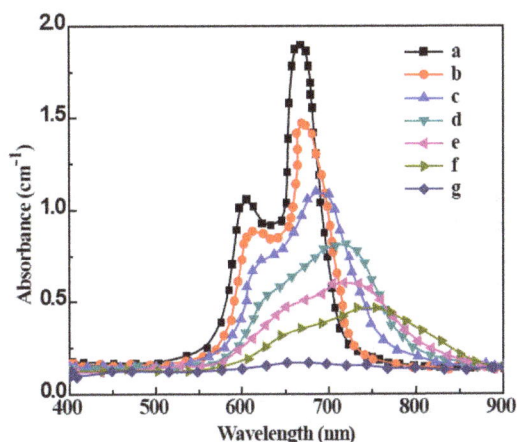

**Fig. 1** Absorption spectra of MB in presence of $Cu_2O$ NPs upon irradiation of visible light; *a* 0, *b* 10, *c* 20, *d* 30, *e* 60, *f* 90, and *g* 120 min

example, at 665 nm, the absorbance has decreased from 1.9 to 1.46 $cm^{-1}$ after irradiation for only 10 min and this value has decreased to as small as 0.14 $cm^{-1}$ after 120 min irradiation of visible light. This reveals that rate of decomposition of MB increases very fast with time. The extremely low intensity of 665 nm band after 120 min irradiation of light suggests that concentration of residual MB is very low.

We studied the degradation of MB to verify the photocatalytic activity of aqueous suspensions of $Cu_2O$ NPs by irradiation with visible light. For this, we monitored the absorbance of MB solution at the absorption maximum wavelength 665 nm. The % degradation was calculated from the change in absorbance value during photodegradation with time of light irradiation. Figure 2 depicts the % degradation of MB before and after light irradiation against time of exposure. As can be seen from the Fig. 2, in the dark, the effect of photolysis on MB is not apparent. The concentration of MB solution slightly decreases due to the adsorption of MB on the surfaces of the catalyst. On the other hand, in presence of light, the %decomposition of MB dye in presence of $Cu_2O$ NPs decreases non-linearly and reaches a value of 90 % after 120 min of light exposure. The observed photodegradation efficiency of biosurfactant-capped $Cu_2O$ NPs on MB dye is due to presence of biosurfactant on its surface. The role of biosurfactant here is to promote charge carrier separation to enhance the photocatalytic efficiency. Cai et al. [10] have reported that the efficiency of $Cu_2O$ NPs has increased owing to introduction of rGO which act as charge carrier separator. In comparison to the work by Zou et al. [12], our sample has performed well and may be ascribed to small size of the $Cu_2O$ cluster as seen from TEM and FESEM images. In an article, Zho et al. [20] reported a similar decrease in the intensity of 665 nm band of MB by halloysite nanotube-

supported silver NPs. They have reported nearly 90 % of MB degradation after 60 min irradiation. They ascribed superior efficiency to larger surface area of halloysite nanotubes.

## XRD pattern

Figure 3 depicts the XRD pattern of $Cu_2O$ nanopowder (a) before and (b) after degrading MB dye. According to the JCPDS card no. 78-2076, both the patterns can be indexed as face-centered cubic structure [9–12]. Before adsorption of MB dye solution, the characteristic peaks (Fig. 3a) as observed at $2\theta$ values of 29.3, 36.8, 42.3, 61.4, 73.5, and 77.6 are assigned to the crystal planes of (110), (111), (200), (220), (311), and (222) of crystalline $Cu_2O$, respectively [9–12]. Form the most intense (111) peak, the crystallite size of $Cu_2O$ NPs before and after MB adsorption is found to be 8.8 and 9.3 nm, respectively. The result implies that adsorption of MB dye on the surface of oxide particles hardly alters the crystallite size and the reclaimed $Cu_2O$ NPs retain their catalytic activity.

## FTIR spectra and DLS

Figure 4 which depicts vibration spectra in the ranges 3200–800 $cm^{-1}$ for MB (a) before and (b) after adsorption on the $Cu_2O$ NPs surface reveals photodegradation of the dye. As shown in Fig. 4a, the FTIR spectrum of MB shows three C–H stretching bands (2807, 2866, and 2924 $cm^{-1}$), one C–H bending vibration (920 $cm^{-1}$) and three C–N stretching bands (1131, 1253 and 1352 $cm^{-1}$) [21–23]. After 60 min of light irradiation, a significant decrease in the intensities of these bands occurs when the MB dye gets adsorbed via S-atom on the surface of oxide NPs. Yu et al. [23] reported that the intensity of vibration bands between

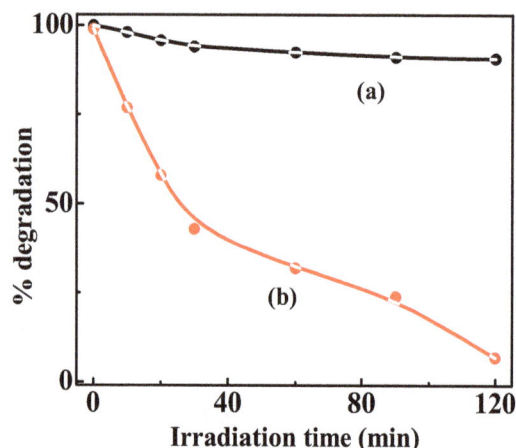

Fig. 2 Degradation of 20 mg/L MB in presence of $Cu_2O$ NPs a before and b after visible light irradiation

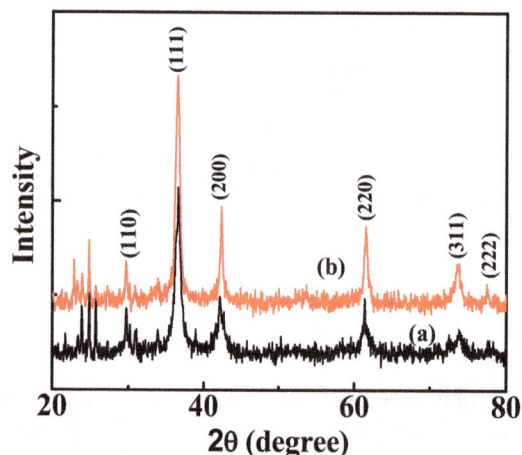

Fig. 3 XRD patterns of $Cu_2O$ NPs a before and b after MB adsorption

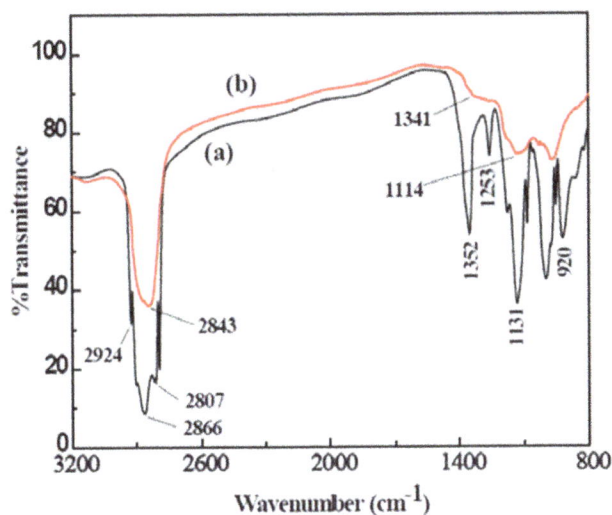

Fig. 4 FTIR spectra of MB *a* before and *b* after adsorption on $Cu_2O$ NPs surface

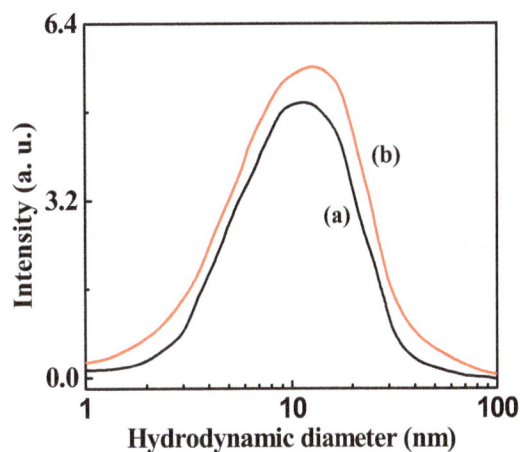

Fig. 5 Size distributions of $Cu_2O$ NPs *a* before and *b* after MB adsorption in an aqueous dispersion

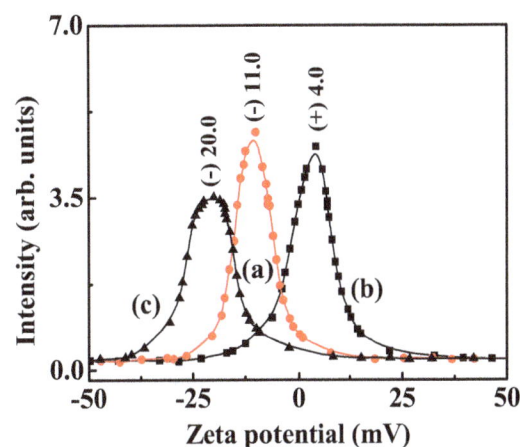

Fig. 6 Zeta potential distribution of $Cu_2O$ NPs *a* before, *b* immediately after MB addition and *c* after 30 min of MB adsorption

1700 and 1100 $cm^{-1}$ of MB dye has decreased in presence of $TiO_2$ NPs when illuminated with visible light. In another report, Amini et al. [24] showed that upon lighting with visible light, the intensity of C–N stretching band in MB dye has decreased drastically in presence of $MnO_x/WO_3$ catalyst and $H_2O_2$. A profound blue shift in these modes too indicates photodegradation of MB in presence of $Cu_2O$ NPs via surface adsorption process. A large blue shift of nearly 23 $cm^{-1}$ that is from 2866 to 2843 $cm^{-1}$ in the C–H stretching band and nearly 3 $cm^{-1}$ shift in C–N stretching bands suggests interfacial interaction between MB and oxide NPs. Disappearance of C–H bending vibration of heterocycle ring of MB in presence of oxide NPs also suggests that surface interaction arises between these two entities. In an experiment, Behera and Ram [25] have ascribed a small red shift of 5 $cm^{-1}$ in the C=O stretching vibration in presence of Au NPs to a weak interaction between poly(vinyl pyrrolidone) polymer and Au surface via the O-atom of the pyrrolidone ring.

We also studied the size distributions of $Cu_2O$ NPs (a) before and (b) after MB adsorption which are shown in Fig. 5. As it is seen from the DLS profile, the average hydrodynamic diameter of $Cu_2O$ NPs has increased from 11.7 to 12.7 nm. This clearly hints that the average size of $Cu_2O$ NPs has hardly increased when MB dye gets adsorbed on the oxide surface.

## Zeta potential and XPS study

Figure 6 displays the distribution of zeta potential ($\xi$) of $Cu_2O$ NPs (a) before and (b) after MB adsorption at 7.5 pH. As it is seen from the Fig. 6a, the zeta distribution profile of $Cu_2O$ NPs before MB addition gives an average $\xi$-value

(−) 11.0 mV. Negative zeta potential of $Cu_2O$ NPs represents accumulation of negative charges on the surface of oxide NPs. When MB dye is added to the aqueous suspension of $Cu_2O$ NPs and the $\xi$-value determined immediately, the sign has changed from negative to positive with an average $\xi$-value (+) 4.0 mV. The positive $\xi$-value could be ascribed to accumulation of positively charged S-atom of $MB^+$ dye (see Scheme 1) on the surface of oxide NPs. After 30 min of illumination, the same sample again acquires a negatively charged surface with an average $\xi$-value (−) 20.0 mV. This may be ascribed to accumulation of negatively charged electrons on the oxide surface when donation of non-bonding electrons of S-atom of MBH (see Scheme 2) occurs to the surface of $Cu_2O$ NPs [26]. Also it is observed from Table 1 that, when MB interacts with $Cu_2O$ NPs, the full width at half maximum (FWHM), i.e., width of zeta band has increased from 9.8 to 11.0 mV. This value has further increased to a value 13.7 mV when it gets

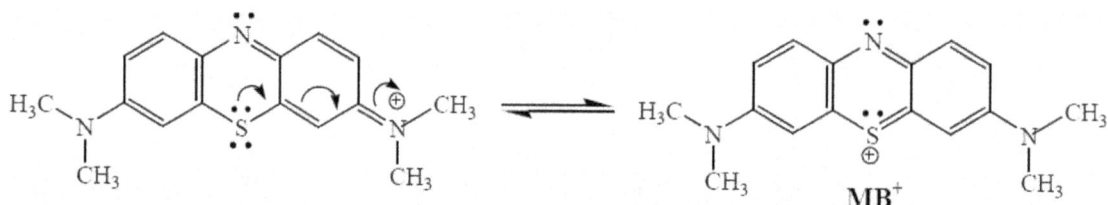

**Scheme 1**  Structure of methylene blue

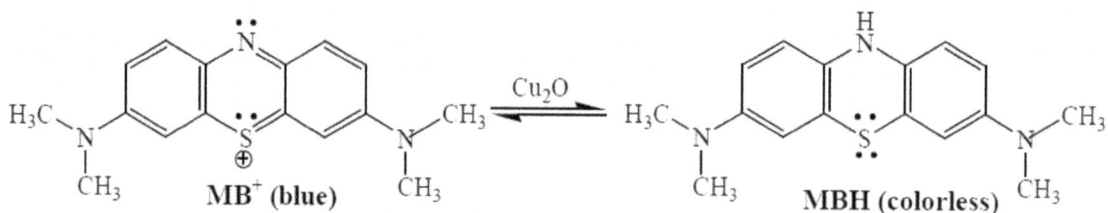

**Scheme 2**  Decolorization of methylene blue dye in presence of Cu$_2$O NPs

**Table 1**  Zeta potentials and full width at half maximum (FWHM) values for Cu$_2$O NPs before and after addition of MB dye

| Samples | $\xi$-value (mV) | FWHM (mV) |
|---|---|---|
| (a) Cu$_2$O NPs before MB addition | (−) 11.0 | 9.8 |
| (b) Cu$_2$O NPs immediately after MB addition | (+) 4.0 | 11.0 |
| (c) Cu$_2$O NPs after 30 min of MB adsorption | (−) 20.0 | 13.7 |

adsorbed on the surface of oxide NPs. These results clearly hint that an interfacial interaction exists between MB and oxide NPs and extent of surface interaction increases with time. Yao and Wang [27] reported that the solution pH is an important parameter which controls the photodegradation efficiency of TiO$_2$ NPs on MB dye as it influences the sign of surface electrical charge of the oxide NPs. At pH 2, the average $\xi$-value of TiO$_2$ NP is reported to be (+) 19.04 mV whereas at pH 9, the average $\xi$-value of TiO$_2$ NP is (−) 25.49 mV.

XPS measurement was done to study the surface interaction between oxide NP and MB molecules and to finding the chemical state of copper (Cu) atom in synthesized NPs. Figure 6a depicts the XPS spectrum recorded for $2p$ photoelectrons of Cu (A) before and (B) after adsorption of MB molecules. In Fig. 6a, the XPS spectrum for Cu$_2$O NPs before MB adsorption detects a doublet for Cu$2p$ band, i.e., Cu $^2p_{3/2}$ and Cu $^2p_{1/2}$ at 932 and 952 eV, respectively [28]. In sample-B, i.e., after MB adsorption, the Cu $^2p_{3/2}$ band appears at binding energy (BE) of 932.9 eV. The band at 932.9 eV is assigned for Cu $^2p_{3/2}$ peak of Cu$^0$. It is suggested that when Cu$_2$O NPs comes in contact with MB and irradiated with visible light, it oxidizes the dye and possibly itself gets reduced to Cu$^0$ from Cu$^+$ state. The O1$s$ peak of Cu$_2$O NPs before and after MB adsorption is shown in Fig. 6b. The broad peak which centered at 530.7 eV is for the metallic oxide (O$^{2-}$) [29]. Zhang et al. [11] have

reported O1$s$ peak of Cu$_2$O NPs at 531.94 eV owing to adsorbed O-atom. It is seen from the spectra that the intensity of this broad peak has drastically decreased in presence of MB. It may be ascribed to change in the chemical state of metallic oxide (O$^{2-}$) when MB molecules get adsorbed on the surface of oxide NPs. Lu et al. [30] reported that a clean Ti sample does not give an O1$s$ peak, but form a titanium oxide layer (O$^{2-}$) on the surface of Ti sample when exposed to excess oxygen, it results in a peak near 531.2 eV (Figs. 7, 8).

## Photoluminescence study

Figure 9 shows the emission spectra (620–680 nm) of MB–Cu$_2$O dispersions consisting of (a) 0, (b) 1.0, (c) 2.0, (d) 3.0, (e) 5.0, and (f) 10.0 μM Cu$_2$O NPS in water, excited at 630 nm. As it is seen from Fig. 9, addition of 1 μM Cu$_2$O NPs caused only 18 % quenching of emission intensity of MB molecules and it has reached about 78 % after adding as large as 10 μM Cu$_2$O NPs. Such a noticeable decay in the emission intensity is ascribed to existence of a surface interaction between oxide NP and MB molecules as non-bonding electron (n) transfer occurs from S-atom of MBH molecules (see Scheme 2) to the surface of oxide NPs. Similar quenching in the emission intensity of various fluorophores was already reported in presence of an efficient light quencher like Au NP and ascribed to

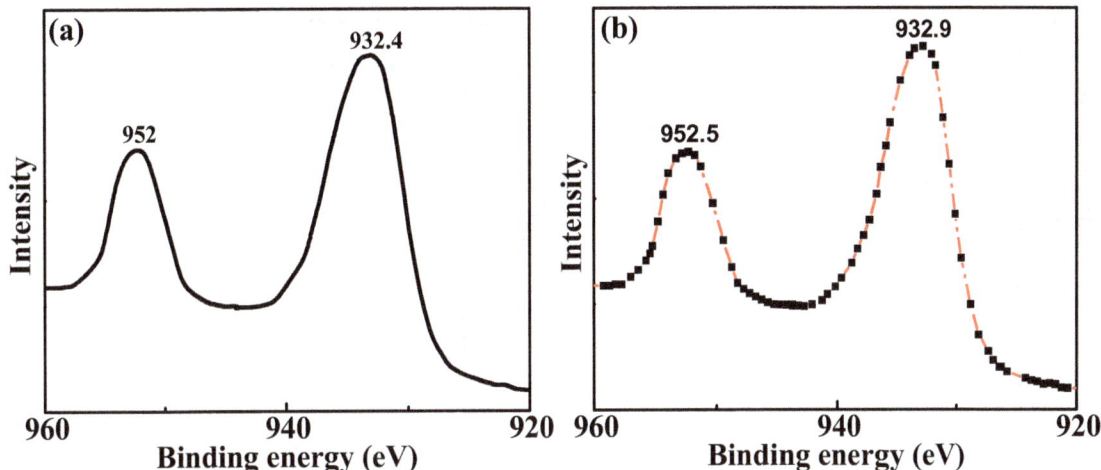

**Fig. 7** XPS spectra (Cu2$p$ bands) of Cu$_2$O nanocrystals: **a** without and **b** with MB dye

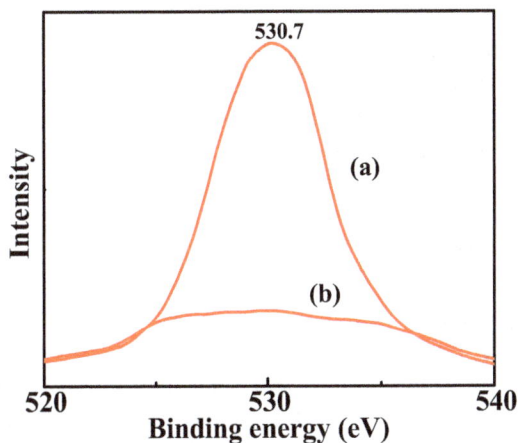

**Fig. 8** XPS spectra (O1$s$ band) of Cu$_2$O nanocrystals: $a$ before and $b$ after MB dye adsorption

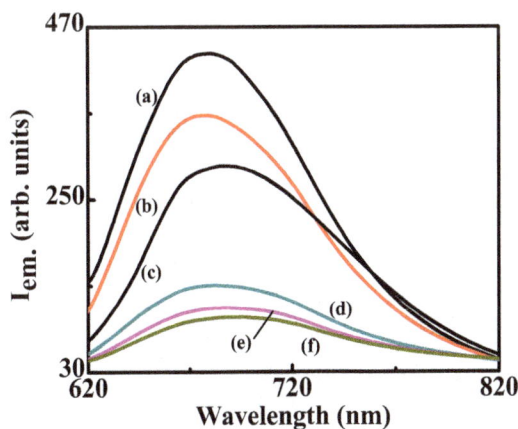

**Fig. 9** Emission spectra of MB with different Cu$_2$O NPs content in water; $a$ 0, $b$ 1.0, $c$ 2.0, $d$ 3.0, $e$ 5.0, and $f$ 10.0 μM, excited at 630 nm

charge transfer interaction between these two moieties [31–34]. Zou et al. [12] have reported interfacial charge transfer process between different Cu$_2$O crystal facets and rGO sheets. They ascribed the strong PL quenching in case of octahedral Cu$_2$O/rGO crystal to a strong interfacial interaction between Cu$_2$O NPs and rGO sheets.

We also studied the variation of integrated emission intensity of MB molecules against Cu$_2$O contents and it is found that emission intensity of MB molecules have quenched intensely by addition of Cu$_2$O NPs (Fig. 10a).

Further we studied PL quenching of MB molecules in presence of an oxide quencher using the well-known Stern–Volmer equation [25, 34] which can be written as:

$$F_0/F = 1 + K_{SV}[Q] \tag{1}$$

where '$F_0$' and '$F$' are the emission intensities of MB molecules in absence and presence of quencher (i.e., here it is Cu$_2$O NPs); $[Q]$ is the concentration of quencher (i.e., Cu$_2$O NP) and $K_{SV}$ is quenching constant which tells about efficiency of a quencher. Figure 10b demonstrates the dependence of $F_0/F$ on the concentration of Cu$_2$O NP. A linear relationship between $F_0/F$ and Cu$_2$O NPs content as observed from the plot reveals that only one type of quenching occurs in the system [33, 34]. We obtained a $K_{SV}$ value of $3.5 \times 10^5$ M$^{-1}$ which can be obtained from the slope of the linear line. A linear Stern–Volmer plot reveals dynamic quenching mechanism in this system [32, 34]. Further, such a large $K_{SV}$ value suggests existence of strong surface interaction between MB molecules and Cu$_2$O NP via S-atom of MBH molecule [32, 34, 35]. Tong et al. [35] have studied the interaction between MB dye and calf thymus deoxyribonucleic acid (ct-dRNA) by PL spectroscopic technique. In their study, they reported that

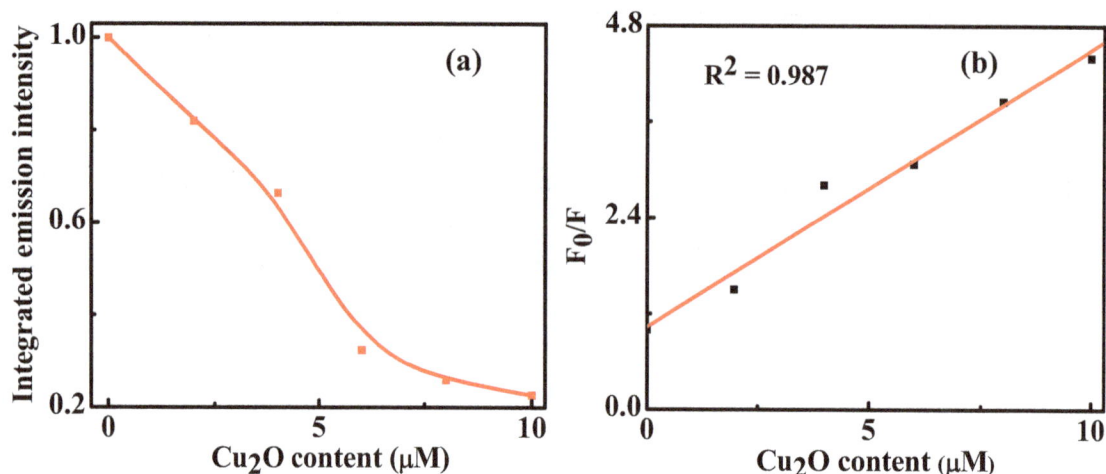

**Fig. 10 a** Variation of integrated emission intensity with $Cu_2O$ content and **b** Stern–Volmer plot of $F_0/F$ versus $Cu_2O$ content. The linear fit gives $F_0/F = 1.04 + 0.35 [Cu_2O]$

upon addition of 55.6 μM ct-dRNA, the fluorescence intensity of MB can be quenched up to 80 % with no shifts in the emission wavelength owing to intercalation binding between MB and the ct-dRNA.

## Microstructures

Morphologies of oxide NPs were studied from the FESEM and TEM images. In Fig. 11a, the SEM image shows nanoclusters of various architectures whose average diameters lies in the range of 5–15 nm. The TEM image in Fig. 11b from the same sample also illustrates nanoclusters of various shapes and sizes. An average size of 10 nm is estimated from the TEM image. Data are well supported by the DLS data (discussed in "FTIR spectra and DLS" section). Uniform distributions of $Cu_2O$ NPs are observed from microscopic images. Zou et al. [12] have reported that

relatively larger size particle lie in the range of 200–400 nm for $Cu_2O$ NPs of various shapes such as octahedral, dodecahedral and cubic crystals which were deposited on the surface of rGO.

## Conclusions

We synthesized $Cu_2O$ NPs using a green chemistry route and studied the photodegradation of MB dye in the visible region in presence of $Cu_2O$ NPs in aqueous medium. Absorption spectra revealed that %decomposition of MB dye decreases non-linearly with $Cu_2O$ contents and it reaches a value of 90 % after light exposure for 120 min. XRD result implies that adsorption of MB dye on the surface of oxide particles hardly alters the crystallite size. This implies that reclaimed $Cu_2O$ NPs will maintain their

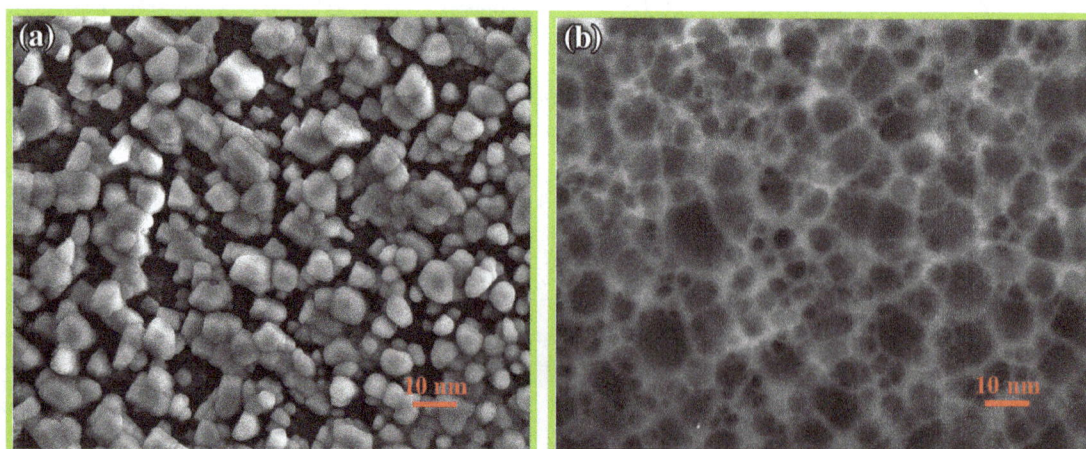

**Fig. 11 a** FESM image and **b** TEM image of surfactant-capped $Cu_2O$ NPs

original catalytic activity. XPS, FTIR, and zeta potential measurements suggest that an interfacial interaction which exists between MB dye and $Cu_2O$ NPs is responsible for photodegradation of MB dye in presence of oxide NPs. Extensive quenching in the emission intensity of MB molecules in presence of $Cu_2O$ NPs is ascribed to existence of a charge transfer-type interaction between oxide NP and MB molecules as transference of non-bonding electron ($n$) occurs from S-atom of MBH molecules to the surface of oxide NPs. A linear Stern–Volmer plot reveals existence of dynamic quenching mechanism in this system.

**Acknowledgments** Financial support of Silicon Institute of Technology is highly acknowledged.

**Authors' contributions** MB conceived the study, carried out all the experiments, and drafted the manuscript. GG helped MB in writing the manuscript. Both authors have read and approved the final manuscript.

**Compliance with ethical standards**

**Conflict of interest** The authors declare that they have no competing interests.

# References

1. Kar A, Smith YR, Subramaniam V (2009) Improved photocatalytic degradation of textile dye using titanium dioxide nanotubes formed over titanium wires. Environ Sci Technol 43:3260
2. Ma W, Li J, Tao X, He J, Xu Y, Yu JC, Zhao J (2003) efficient degradation of organic pollutants by using dioxygen activated by resin-exchanged iron(II) bipyridine under visible irradiation. Angew Chem Int Ed 42:1029
3. Pal J, Ganguly M, Mondal C, Roy A, Negishi Y, Pal T (2013) Crystal-plane-dependent etching of cuprous oxide nanoparticles of varied shapes and their application in visible light photocatalysis. J Phys Chem C 117:24640
4. Pourahmad A (2012) $Ag_2S$ nanoparticle encapsulated in mesoporous material nanoparticles and its application for photocatalytic degradation of dye in aqueous solution. Supperlattices Microstruct 52:276
5. Turhan K, Ozturkcan SA (2012) Decolorization and degradation of reactive dye in aqueous solution by ozonation in a semi-batch bubble column reactor. Water Air Soil Pollut 224:1353
6. Salleh MAM, Mahmoud DK, Karim WAWA, Idris A (2011) Cationic and anionic dye adsorption by agricultural solid wastes: A comprehensive review. Desalination 280:1
7. Nassar MM, Magdy YH (1997) Removal of different basic dyes from aqueous solutions by adsorption on palm-fruit bunch particles. Chem Eng J 66:223
8. Vasanthkumar K, Ramamurthi V, Sivanesan SJ (2005) Modeling the mechanism involved during the sorption of methylene blue onto fly ash. Colloid Interface Sci 284:14
9. Sun L, Wang G, Hao R, Han D, Cao S (2015) Solvothermal fabrication and enhanced visible light photocatalytic activity of $Cu_2O$-reduced graphene oxide composite microspheres for photodegradation of rhodamine B. Appl Surf Sci 358:91
10. Cai J, Liu W, Li Z (2015) One-pot self-assembly of $Cu_2O$/RGO composite aerogel for aqueous photocatalysis. Appl Surf Sci 358:146
11. Zhang D, Hu B, Guan D, Luo Z (2016) Essential roles of defects in pure graphene/$Cu_2O$ photocatalyst. Catal Commun 76:7
12. Zou W, Zhang L, Liu L, Wang X, Sun J, Wu S, Deng Y, Tang C, Gao F, Dong L (2016) Engineering the $Cu_2O$-reduced graphene oxide interface to enhance photocatalytic degradation of organic pollutants under visible light. Appl Catal B Environ 181:495
13. Srinivasan M, White T (2007) Degradation of methylene blue by three-dimensionally ordered macroporous titania. Environ Sci Technol 41:4405
14. Zheng Z, Huang B, Wang Z, Guo M, Qin X, Zhang X, Wang P, Dai Y (2009) Crystal faces of $Cu_2O$ and their stabilities in photocatalytic reactions. J Phys Chem C 113:14448
15. Zhang Y, Deneg B, Zhang T, Gao D, Xu AW (2010) Shape effects of $Cu_2O$ polyhedral microcrystals on photocatalytic activity. J Phys Chem C 114:5073
16. Abboud Y, Saffaz T, Chagraoui A, Bouari LL, Brouzi K, Tanane O, Ihssane B (2014) Biosynthesis, characterization and antimicrobial activity of copper oxide nanoparticles (CONPs) produced using brown alga extract (*Bifurcaria bifurcata*). Appl Nanosci 4:571
17. Zhu Q, Zhang Y, Wang J, Zhou F, Chu P (2011) Microwave synthesis of cuprous oxide micro-/nanocrystals with different morphologies and photocatalytic activities. J Mater Sci Technol 27:289
18. Behera M, Giri G (2014) Green synthesis and characterization of cuprous oxide nanoparticles in presence of a bio-surfactant. Mater Sci Pol 32:702
19. Donnan FG, Thomas JS (1911) The solubility of cuprous oxide in aqueous ammonia solutions, and the composition of the cuprous-ammonia complex. J Chem Soc Trans 99:1788
20. Zou M, Du M, Zhu H, Xu CS, Fu YQ (2012) Green synthesis of halloysite nanotubes supported Ag nanoparticles for photocatalytic decomposition of methylene blue. J Phys D Appl Phys 45:325302
21. Ovchinnikov OV, Chernykh SV, Smirnov MS, Alpatova DV, Vorob'eva RP, Latyshev AN, Evlev AB, Utekhin AN, Lukin AN (2007) Analysis of interaction between the organic dye methylene blue and the surface of AgCl(I) microcrystals. J Appl Spectrosc 74:809
22. Colthup NB, Daly LH, Wiberley SE (eds) (1990) Introduction to infrared and raman spectroscopy. Academic Press, Boston
23. Yu Z, Chuang SSC (2007) Probing methylene blue photocatalytic degradation by adsorbed ethanol with in situ IR. J Phys Chem C 111:13813
24. Amini M, Pourbadiei B, Purnima T, Ruberub A, Woob LK (2014) Catalytic activity of $MnO_x$/$WO_3$ nanoparticles: synthesis, structure characterization and oxidative degradation of methylene blue. New J Chem 38:1250
25. Behera M, Ram S (2013) Spectroscopy-based study on the interaction between gold nanoparticle and poly (vinylpyrrolidone) molecules in a non-hydrocolloid. Int Nano Lett 3:17
26. Anderson L, Wittkopp SM, Painter CJ, Lligel JJ, Schreiner R, Bell JA, Shakhasiri BZ (2012) What is happening when the blue bottle bleaches: An investigation of the methylene blue-catalyzed air oxidation of glucose. J Chem Edu 89:1425
27. Yao J, Wang, C (2010) Decolorization of methylene blue with sol via UV irradiation photocatalytic degradation. Int J Photoenergy. Article ID 643182, 6

28. Yin M, Wu C, Lou Y, Burda C, Koberstein J, Zhu Y, O'briens S (2005) Copper oxide nanocrystals. J Am Chem Soc 127:9506

29. Heng B, Xiao OT, Tao W, Hu X, Chen X, Wang B, Sun D, Tang Y (2012) Zn doping-induced shape evolution of microcrystals: the case of cuprous oxide. Cryst Growth Des 12:3998

30. Lu G, Stevan LB, Jeffery S (2000) Oxidation of a polycrystalline titanium surface by oxygen and water. Surf Sci 458:80

31. Alexandridis P (2011) Gold nanoparticle synthesis, morphology control, and stabilization facilitated by functional polymers. Chem Eng Technol 34:15

32. Lakowicz JR (1999) Principles of fluorescence spectroscopy. Plenum Press, New York

33. Pramanik S, Banerjee P, Sarkar A, Bhattacharya SC (2008) Size-dependent interaction of gold nanoparticles with transport protein: a spectroscopic study. J Lumin 128:1969

34. Behera M, Ram S (2013) Intense quenching of fluorescence intensity of poly (vinyl pyrrolidone) molecules in presence of gold nanoparticles. Appl Nanosci 3:543

35. Tong C, Hu Z, Wu J (2010) Interaction between methylene blue and calfthymus deoxyribonucleic acid by spectroscopic technologies. J Fluoresc 20:261

# 7

# Adsorption and desorption kinetics of toxic organic and inorganic ions using an indigenous biomass: *Terminalia ivorensis* seed waste

Jonathan O. Babalola[3] · Funmilayo T. Olayiwola[3] · Joshua O. Olowoyo[3] ·
Alimoh H. Alabi[3] · Emmanuel I. Unuabonah[2] · Augustine E. Ofomaja[1] ·
Martins O. Omorogie[1,2]

**Abstract** Environmental remediation has been a strategy employed by scientists to combat water pollution problems that have led to the scarcity of potable water. Hence, in this study, *Terminalia ivorensis* seed waste (TISW) was explored for the removal of Congo Red, Methylene Blue, Cadmium and Lead from aqueous solutions. Some experimental variables such as pH, biosorbent dose, initial solute ion concentration, agitation time and temperature were optimised. The surface microstructures of TISW were studied using proximate analysis, bulk density, specific surface area, pH of Point of Zero Charge, Fourier Transform Infra Red Spectroscopy, Thermogravimetric/Differential Thermal Analysis, Scanning Electron Microscopy and Energy Dispersive Analysis of X-ray. The maximum Langmuir monolayer saturation adsorption capacity, $q_{max_L}$, was obtained as 175.44 mg/g for the removal of Methylene Blue by TISW. Also, the $q_{max_L}$ for CR, Cd(II) ion and Pb(II) ion were 85.47, 12.58 and 52.97 mg/g, respectively. Also, the pseudo-first-order constant, $k_1$, and pseudo-second-order rate constant, $k_2$, are 0.008–0.026 min$^{-1}$ and 0.012–0.417 mg g$^{-1}$ min$^{-1}$, respectively. Hence, TISW is recommended as a good adsorbent for the removal of both toxic industrial dyes and toxic metal ions from polluted water.

**Keywords** *Terminalia ivorensis* · Thermodynamics · Kinetics · Mass transfer · Desorption

✉ Jonathan O. Babalola
bamijibabalola@yahoo.co.uk

✉ Martins O. Omorogie
omorogiem@run.edu.ng;
osaigbovooohireimen@gmail.com; martinso@vut.ac.za

[1]  Adsorption and Catalysis Research Laboratory, Department of Chemistry, Vaal University of Technology, Private Bag X021, Andries Potgieter Boulevard, Vanderbijlpark 1900, South Africa

[2]  Environmental and Chemical Processes Research Laboratory, Department of Chemical Sciences, Redeemer's University, P.M.B. 230, Ede, Osun State, Nigeria

[3]  Department of Chemistry, University of Ibadan, 200284 Ibadan, Nigeria

## Introduction

The quest for potable water has been a serious challenge to the global ecological balance, especially in developed countries. The recent interest in water shortage and water pollution has driven researchers very hard, so as to look for a lasting, effective and efficient solution to this menace combating the survival of human lives and the ecology. Today, 800 million people still lack access to potable water. For instance, only 46% of the Oceania population and 39% of the Sub-Sahara Africa population have access to potable water [1–4]. The sporadic rise in human activities over the decades has led to the consistent release of various recalcitrant anthropogenic pollutants into water bodies, thereby causing a sharp drop in the quality of the ecosystem. Eradication of these deleterious pollutants is difficult using classical water purification systems. Report from the World Health Organisation (WHO) showed that 1.8 million people die from polluted water-related diseases [2]. In Nigeria, this problem aggravates the sustainability of its ecology, leading to the precarious survival of flora and fauna. Hence, this has heralded an urgent need for researchers to proffer a sustainable solution to this life quagmire [5–8]. Toxic metals are known to be the most problematic pollutants, due to their non-biodegradable nature. Irrespective of the huge effort that has been put together by researchers to minimise their environmental impact, toxic metals in the environment still have adverse effects on human, fauna and flora. The toxicological effects of Cd(II) and Pb(II) in humans include bone lesions, cancers (kidney and lung), hypertension, inhibition of haemoglobin formation, sterility, infants brain impairment, learning disabilities, abortion, and kidney damage. To date, various remediation technologies have been used to treat polluted water containing toxic synthetic dyes and metals. These are coagulation, chemical oxidation, photodegradation, aerobic or anaerobic oxidation, precipitation, membrane filtration, dialysis, ion exchange, solvent extraction, reverse osmosis, among others [9, 10]. These technologies have some drawbacks, such as low selectivity, incomplete removal, cost ineffectiveness and production of large amount of secondary sludge. Adsorption is known as a green, sustainable and cost-effective technology, which is an alternative to other remediation technologies mentioned above. In recent times, natural or modified biomaterials (biosorbents and agricultural byproducts), which comprise lignin, cellulose, hemicelluloses and other organic compounds that have various functional moieties, have been found to be ubiquitous, easily sourced, cheap, green, sustainable and good adsorbents for clean recovery of toxic synthetic dyes and metals [9]. Some researchers in their previous treatises have used eucalyptus seeds [9], peanut hulls [11], corn cob [12], almond shell, hazelnut shell [13], olive cake [14], mungbean husk [15], mango peel waste [16], orange waste [17], *Scolymus hispanicus L.* [18], chemically modified orange peel [19], *Nauclea diderrichii* seed biomass [20, 21], *Zea mays* seed chaff [22], Mesoporous $SiO_2$/-graphene oxide nanoparticles-modified *Nauclea diderrichii* seed biomass [23, 24], *Parkia biglobosa* biomass [25], *Papaya*-clay combo [26], $MnO_2$ nanoparticles-modified *Nauclea diderrichii* biomass waste [27], $TiO_2$ nanoparticles-modified *Nauclea diderrichii* seed biomass [28], *Pentaclethra macrophylla* and *Malacantha alnifolia* barks [29]. Other researchers in the literature used other adsorbent and techniques to treat wastewaters in addition to adsorption [30–53].

*Terminalia ivorensis*, a timber tree, is the sole member of the genus that occurs naturally in West Africa. It is an indigenous plant (family *Combretaceae*) which is recognised in the South-western rain forest of Nigeria and Ghana. The bark is used as a lotion for the treatment of wounds, sores and cuts. *Terminalia ivorensis* is a large deciduous forest tree. On plantations, weeding up to the second year and line cleaning or creeper cutting from the third to sixth year may be necessary. With the diverse native uses of this species, its utility can be further enhanced for agroforestry development by the present research effort on it [54]. *Terminalia ivorensis* is found in the rainforest but is predominantly a tree of seasonal forest

zones. *Terminalia ivorensis* has numerous uses in the building, construction, carpentry industries and household equipment. Also, this plant can also use as ethnomedicine, a lotion for the treatment of wounds, sores and cuts because it healed without scar [54].

The seed epicarp of *Terminalia ivorensis* is initially removed before the seed is planted. This seed epicarp constitutes a colossal amount of obnoxious waste. This has informed our choice of exploring this waste as a biosorbent for this research. Conventional remediation technologies like coagulation, membrane filtration, dialysis, reverse osmosis, etc., utilized to treat water polluted with toxic metal ions and dyes are very expensive and result in secondary pollution. The use of *Terminalia ivorensis* seed waste will avail researchers and industries the opportunity to substitute the aforementioned cocktails of conventional technologies above with very cheap, convenient, suitable, eco-friendly and secondary pollution-free technology to treat polluted water.

This work explores the remediation potential of green, sustainable, cost-effective, ubiquitous, and locally sourced biomass of *Terminalia ivorensis* seed waste (TISW) for some toxic pollutants studied. To the best of the authors' knowledge, TISW is a new biological adsorbent which is reported to have good uptake capacities for MB, CR, Cd(II) and Pb(II).

## Experimental

### Preparation of *Terminalia ivorensis* seed waste (TISW)

*Terminalia ivorensis* seed waste (TISW) was obtained from the Forest Research Institute of Nigeria (FRIN), in Ibadan (7°23′16″ North, 3°53′47″ East), Nigeria. After collection, it was air dried for four weeks and later oven dried for 22 h at 70 °C. The dried bark was cut into small pieces and pulverised. Further drying was carried out for 6 h. The pulverised TISW was sieved to 250 μm, put into plastic containers and used for various adsorption experiments.

### Surface microstructures of TISW

*Proximate analysis, bulk density, pH of point of zero charge (PZC), specific surface area (SSA), Fourier transform infra red spectroscopy (FTIR), thermogravimetric/differential thermal analysis (TG/DTA), scanning electron microscopy (SEM) and energy dispersive analysis of X-ray (EDAX) of TISW*

Proximate analysis and bulk density of TISW were determined according to the published protocol of Ofomaja and

Naidoo [55]. The pH of point zero charge of TISW was determined according to the published protocol of Stumm and Morgan [56]. The specific surface area of TISW was determined according to the published protocol of Sears [57]. The Fourier Transform Infra Red spectroscopic analyses of unloaded, and toxic metals/dyes loaded TISW were carried out by Perkin Elmer FTIR Spectrophotometer (Spectrum Version 2) at the scanning frequencies of 400–4000 cm$^{-1}$. The Perkin Elmer thermogravimetric/differential thermal analysis (TG/DTA) was carried out by Perkin Elmer TG/DTA thermal analyser. Furthermore, the surface microstructures of TISW were carried out by Scanning Electron Microscopy (SEM) instrument, coupled with Energy Dispersive Analysis of X-ray (EDAX), JEOL JSM-6390 LV Model.

All the experimental details and mathematical equations used for these adsorption and desorption studies are shown in the Electronic Supporting Information (ESI) of this manuscript.

## Results and discussion

### Surface characterization of TISW microstructures

*Proximate analysis, bulk density, PZC, SSA, FTIR, TG/ DTA, SEM and EDAX of TISW*

The percentage proximate compositions of TISW on dry weight basis are depicted in Table 1. The determined bulk density of TISW was 0.05 g cm$^{-3}$. The SSA of TISW was 240.6 m$^2$ g$^{-1}$. The PZC of the TISW was 6.44.

The PZC or (isoelectric pH) of biosorbent is the pH at which the density of anionic moieties is equal to that of the cationic moieties on a biosorbent surface. At pH < PZC, the surface of a biosorbent is predominantly cationic, while at pH > PZC, its surface is predominantly anionic. The functional moieties on TISW surface probably acquire negative or positive charges depending on the solution pH [55]. The PZC result of TISW suggests that it possesses a broad solution pH range to retain its positive charges (see Fig. 1).

The FTIR spectra of TISW, with its loaded CR, MB, Pb$^{2+}$ and Cd$^{2+}$ analogues are shown in figures S1a-S1e of the ESI. The free O–H stretch for TISW appeared at 3871 and 3748 cm$^{-1}$. After loading TISW with CR, MB and Pb$^{2+}$, there was no shift in the vibration frequency of the free O–H stretch. For Cd$^{2+}$-loaded TISW, the vibration

frequency shifted to 3759 cm$^{-1}$. The N–H stretch of TISW was observed at 3456 cm$^{-1}$. The N–H stretch for CR, MB, Pb$^{2+}$ and Cd$^{2+}$-loaded TISW shifted to 3445, 3438, 3447 and 3443 cm$^{-1}$, respectively. These shifts in the vibration frequencies might be due to binding of these toxic ions onto the functional moieties in the cell walls of TISW [29]. The C–H of CH$_2$ and CH$_3$ for TISW was observed at 2935 cm$^{-1}$. No significant shift was observed for MB, Pb$^{2+}$ and Cd$^{2+}$-loaded TISW. But for CR-loaded TISW, the C-H vibration frequency decreased to 2924 cm$^{-1}$. The –C=O stretch for TISW was found at 1648 cm$^{-1}$. For MB-loaded TISW, the vibration frequency decreased 1623 cm$^{-1}$. There were negligible shifts in the vibration frequency of –C=O stretch for TISW loaded with CR, Pb$^{2+}$ and Cd$^{2+}$. The aromatic –C=C– stretch for TISW appeared at 1544 cm$^{-1}$. Also, there was no significant change in the –C=C– vibration frequency when TISW was loaded with CR, MB, Pb$^{2+}$ and Cd$^{2+}$. The C–H bend for TISW appeared at 1359 cm$^{-1}$. This vibration frequency increased to 1376 and 1384 cm$^{-1}$ for CR-loaded TISW, and Pb$^{2+}$ and Cd$^{2+}$-loaded TISW. For MB-loaded TISW, this vibration frequency decreased to 1348 cm$^{-1}$. The –C–O bend for TISW was observed at 1057 cm$^{-1}$. However, there were negligible shifts in the vibration frequency of –C–O bend for TISW loaded with MB, CR, Pb$^{2+}$ and Cd$^{2+}$. For TISW, the –C–O out of plane deformation bend was observed at 582 cm$^{-1}$. This vibration frequency shifted to 571 and 568 cm$^{-1}$ for CR and MB-loaded TISW, and Pb$^{2+}$ and Cd$^{2+}$-loaded TISW, respectively. The various shifts in the stretch and bend vibration frequencies after loading TISW with the studied pollutant ions/molecules might result from chelation, precipitation and ion exchange reactions that was involved in the biosorption process (see Table 2) [29].

The TG of TISW (see Fig. 2) showed that <220 °C, ca. 11% weight loss was observed due to the loss of surface water from TISW. At temperature >220 and <550 °C, ca. 55% weight loss was observed, which was assigned to the loss of the volatile components of lignocelluloses and hemicelluloses of TISW [20, 21]. Above 550 °C, TISW was thermally stable. For DTA (also see Fig. 2) of TISW, exothermic heat flow assigned to 1% decrease in derivative weight was observed from 33 to 150 °C, endothermic heat flow assigned to 1.5% increase in derivative weight was observed from >150 to 180 °C, another exothermic heat flow assigned to 5% decrease in derivative weight was observed from 180 to 350 °C and two endothermic heat flows assigned to 3.5 and 0.5% increases in derivative

| | % Moisture content | % Crude protein | % Crude fat | % Crude fibre | % Ash content |
|---|---|---|---|---|---|
| **Table 1** Proximate analysis of *Terminalia ivorensis* seed waste | 3.09 | 10.51 | 1.84 | 24.36 | 6.45 |

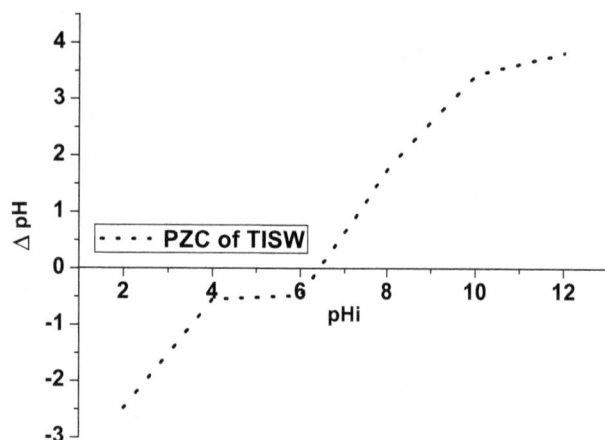

**Fig. 1** The pH of point of zero charge plot for *Terminalia ivorensis* seed waste

weight were observed from >350 to 400 °C and 400 to 580 °C, respectively. The exothermic heat flows led to the dissociation of some chemical bonds of lignocelluloses and hemicelluloses of TISW. Also, the endothermic heat flows led to the association of some chemical bonds to form small molecules such as CO, $CO_2$, $NH_3$, etc. [20, 21, 25, 58].

The SEM micrographs of TISW show tiny irregular, scattered and spaced fibrils of the lignocelluloses that are contained in the cell walls that comprise lumped particles of unconnected and broken walls. The EDAX of TISW showed the elemental compositions of TISW, which was an indication of the constituents of the functional moieties in TISW (see Fig. 3a–d).

## Influence of pH on the removal of sorbate ions onto TISW

The influence of solution pH on the biosorption capacity of TISW largely depends on the speciation of the sorbate or solute ions/molecules at different pHs of the aqueous solutions. The removals of CR and MB onto TISW were highest at pH 2.0 and pH 12.0, respectively. These were obtained as 3.34 and 12.36 mg/g, respectively. The biosorption of CR was optimum at pH 2.0 due to increase in the proton cloud densities of TISW (see Fig. 4). This enabled negatively charged CR molecules to bind easily to the surface of TISW. On the other hand, the removal of MB onto TISW was found to be optimum at pH 12.0 due to the low proton cloud densities on the surface of TISW. This accounted for the increase in the removal of positively charged MB molecules onto TISW [21, 29].

Furthermore, the removal of $Cd^{2+}$ onto TISW was highest at pH 6.0. This was obtained at 10.60 mg/g. Also, >pH 7.0, there was a very slight increase in the removal of $Cd^{2+}$ to 10.64 mg/g. The negatively charged

**Table 2** Fourier transform infrared spectra characteristics of *Terminalia ivorensis* seed waste before and after biosorption of Methylene Blue, Congo Red, Cadmium and Lead

| Dyes/metals | Absorption band peak ($cm^{-1}$) | | | Functional groups |
|---|---|---|---|---|
| | Before | After | Difference | |
| CR | 3871.14 | 3871.14 | – | Free –OH |
| MB | | 3871.14 | – | Free –OH |
| Pb(II) | | 3865.54 | 5.600 | Free –OH |
| Cd(II) | | 3871.14 | – | Free –OH |
| CR | 3747.89 | 3742.29 | 5.600 | Free –OH |
| MB | | 3747.89 | – | Free –OH |
| Pb(II) | | 3742.29 | 5.600 | Free –OH |
| Cd(II) | | 3759.10 | 11.21 | Free –OH |
| CR | 3456.00 | 3445.00 | 11.00 | N–H Stretch |
| MB | | 3438.00 | 18.00 | N–H Stretch |
| Pb(II) | | 3447.00 | 9.000 | N–H Stretch |
| Cd(II) | | 3443.00 | 13.00 | N–H Stretch |
| CR | 2935.57 | 2924.36 | 11.21 | C–H Stretch |
| MB | | 2929.97 | 5.600 | C–H Stretch |
| Pb(II) | | 2929.97 | 5.600 | C–H Stretch |
| Cd (II) | | 2935.57 | – | C–H Stretch |
| CR | 1648.20 | 1647.56 | 0.640 | C=O—Stretch |
| MB | | 1623.00 | 25.20 | C=O—Stretch |
| Pb(II) | | 1650.20 | 1.800 | C=O—Stretch |
| Cd(II) | | 1645.00 | 3.200 | C=O—Stretch |
| CR | 1544.05 | 1538.46 | 5.590 | Aromatic C=C |
| MB | | 1538.46 | 5.590 | Aromatic C=C |
| Pb(II) | | 1538.46 | 5.590 | Aromatic C=C |
| Cd(II) | | 1541.25 | 2.800 | Aromatic C=C |
| CR | 1359.44 | 1376.22 | 16.78 | S=O—Stretch |
| MB | | 1348.25 | 11.19 | S=O—Stretch |
| Pb(II) | | 1384.61 | 25.17 | S=O—Stretch |
| Cd(II) | | 1384.61 | 25.17 | S=O—Stretch |
| CR | 1057.34 | 1054.54 | 2.800 | C–O– Stretch |
| MB | | 1048.95 | 8.390 | C–O—Stretch |
| Pb(II) | | 1051.74 | 5.600 | C–O—Stretch |
| Cd(II) | | 1057.34 | – | C–O—Stretch |
| CR | 581.81 | 570.62 | 11.19 | C–S—Stretch |
| MB | | 570.62 | 11.19 | C–S—Stretch |
| Pb(II) | | 567.83 | 23.98 | C–S—Stretch |
| Cd(II) | | 565.03 | 16.78 | C–S—Stretch |

TISW surface at weak acidic and neutral pH favoured the high uptake of $Cd^{2+}$ due to deprotonation (low proton density). The very negligible increase for the removal of $Cd^{2+}$ by TISW observed at >pH 6.0 might be due to the likelihood of the precipitation of $Cd^{2+}$ on the functional sites on the cell walls of TISW. On the other hand, the removal of $Pb^{2+}$ by TISW was highest at pH 6.0. This was obtained at 12.13 mg/g. At pH >6.0–7.0, $Pb^{2+}$ no longer exists in the solution. The speciation of $Pb^{2+}$ at this pH led

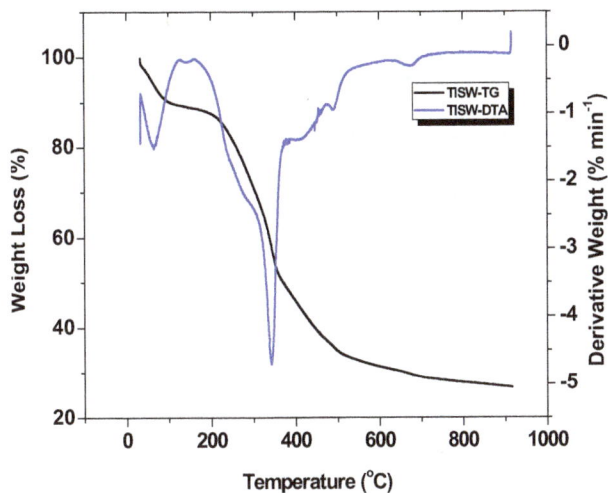

**Fig. 2** Thermogravimetric/differential thermal analysis for *Terminalia ivorensis* seed waste

to the formation of Pb(OH)$^+$, Pb(OH)$_2$, Pb(OH)$_3^-$ and Pb(OH)$_4^{2-}$. From the PZC value of TISW, which was 6.44, it suggested that TISW retained positive charges on its surface over a pH range <6.44. Due to this, TISW would biosorbed high amount of CR molecules than the positive pollutant ions/molecules through ion exchange mechanism and electrostatic attraction. The adsorption of Pb$^{2+}$ continued until precipitation occurred at pH >6.0. Hence, at pH >6.0–7.0, the binding of Pb(OH)$^+$ species to the TISW biosorbent surface occurred by complexation mechanism and, at this point, further biosorption of Pb$^{2+}$ ceased. Hence, the main mechanisms by which MB and CR molecules interact with TISW were by electrostatic attraction and π–π hydrophobic interactions. These π–π hydrophobic interactions are higher in CR molecule than MB molecule due to the many aromatic rings in CR thus it has higher propensity to bind to the negatively charged

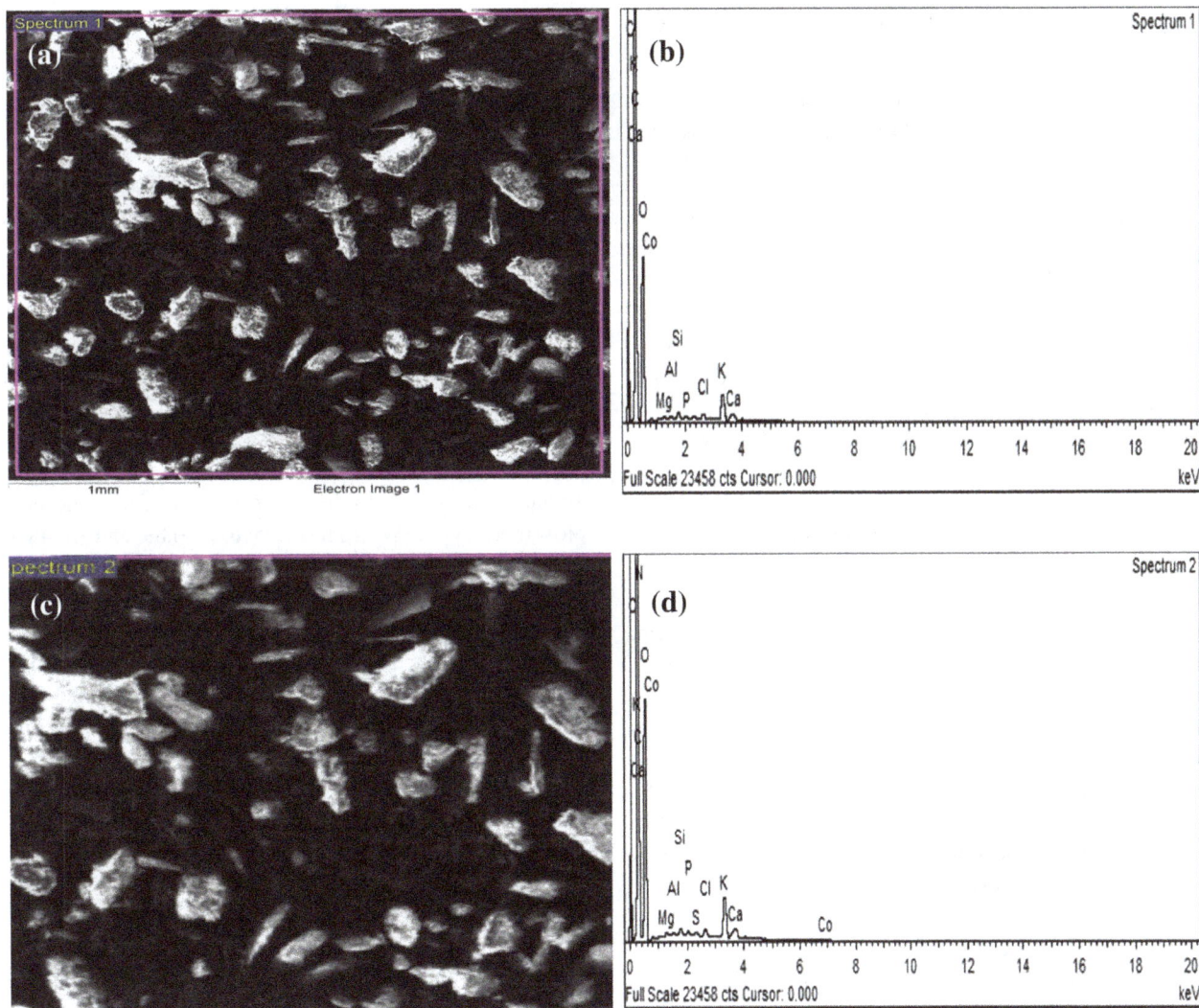

**Fig. 3 a–d** Scanning electron micrographs and energy dispersive analysis of X-ray for *Terminalia ivorensis* seed waste

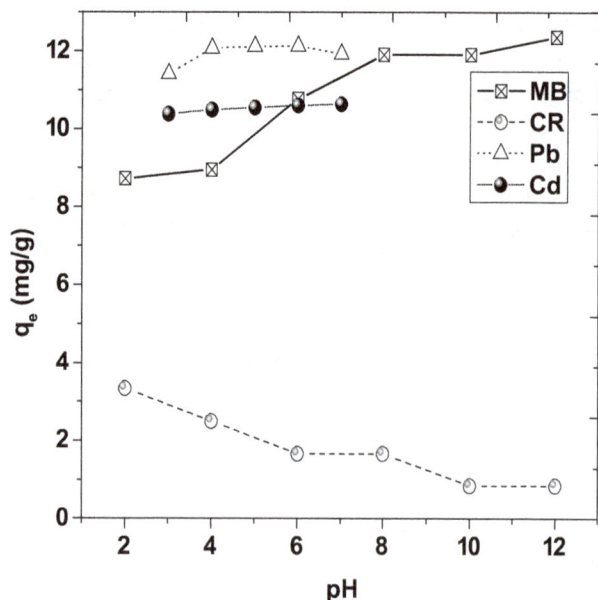

**Fig. 4** The plots of the amounts of Methylene Blue, Congo Red, Cadmium and Lead biosorbed by *Terminalia ivorensis* seed waste, $q_e$ (mg/g) against pH (pH 2.0–7.0 for Cadmium and Lead; pH 2.0–12.0 for Methylene Blue and Congo Red); adsorbent dose = 25 mg; agitation speed = 200 rpm; initial metal ion concentration = 100 mg/L; agitation time = 180 min; temperature = 300 K)

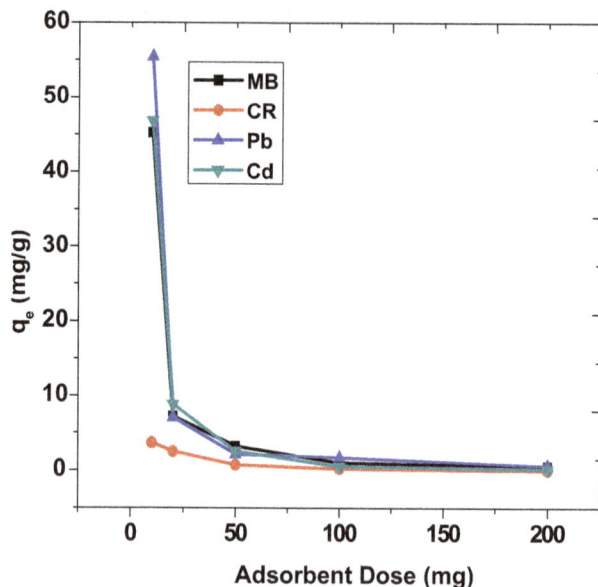

**Fig. 5** The plots of the amounts of Methylene Blue, Congo Red, Cadmium and Lead biosorbed by *Terminalia ivorensis* seed waste, $q_e$ (mg/g) against adsorbent dose (adsorbent dose = 10–200 mg; agitation speed = 200 rpm; initial metal ion concentration = 100 mg/L; agitation time = 180 min; temperature = 300 K)

functional moieties on the surface of TISW by $\pi$–$\pi$ hydrophobic interactions than MB. Electrostatic attraction occurred when MB or CR molecules bound to the negatively charged functional moieties on the surface of TISW depending on the pH of the aqueous solutions. On the other hand, the main mechanisms for the uptake of $Cd^{2+}$ and $Pb^{2+}$ by TISW were by electrostatic attraction and complexation, which are dependent on the pH of the working solutions as expounded above [29].

**Influence of adsorbent dose on the removal of sorbate ions onto TISW**

The amount of sorbate ions biosorbed decreased with increase in biosorbent dose from 10 to 200 mg. This decrease became more pronounced as the weight of the biosorbent increased, resulting in the agglomeration or aggregation of biosorbent particles, probably blockage of biosorption sites, reduced surface area for sorbate–sorbent interaction and the lengthening of diffusion path [55]. Figure 5 shows the plots of the amount of solute ions adsorbed at equilibrium, $q_e$ (mg/g) against adsorbent dose (mg).

**Influence of initial sorbate ions concentration on the biosorption capacity of TISW**

The initial sorbate ions concentration is a cogent factor that determines the mass transfer and diffusion dynamics of the

biosorption of sorbate or solute ions onto biomasses. Also, increase in the initial sorbate ions concentration is the driving force that is needed to overcome the mass transfer resistance of the uptake of solute ions onto the surface of the biomass [59].

As the initial concentrations of CR and MB increased from 25 to 750 mg/L, there was a commensurate rise in their removal from 6.30 to 362.80 and 10.17 to 362.04 mg/g respectively. Also, the increase in the initial concentrations of $Cd^{2+}$ and $Pb^{2+}$ from 25 to 750 mg/L also amounted in the proportionate increase in $Cd^{2+}$ and $Pb^{2+}$ biosorbed by TISW from 11.55 to 146.89 and 10.48 to 176.22 mg/g, respectively. It is noteworthy to mention here that the increase in the initial concentration of sorbate ions led to the corresponding increase in their uptakes, as they occupied the functional moieties on the surface of TISW. The binding of these solute ions to these functional moieties in the active sites occurred in quick successions until they were occupied [25, 29].

**Influence of agitation time on the removal of sorbate ions onto TISW**

The residence time of solute ions plays a salient role in the solute ions uptake onto the biosorbent surface, due to its influence on the sorbent–solute interface. Ho [60] and Onal et al. [61] proposed that biosorption consists of three steps, which are: (a) the diffusion of adsorbate through the solution to the external surface of the adsorbent or the boundary layer

**Table 3** Linear equilibrium parameters for the removal of Methylene Blue, Congo Red, Cadmium and Lead onto *Terminalia ivorensis* seed waste

| | Langmuir model | | | Freundlich model | | | Temkin model | | |
|---|---|---|---|---|---|---|---|---|---|
| | $q_{max_L}$ (mg/g) | $k_L$ (L/mg) | $R^2$ | $k_f$ (L/mg)$^{\frac{1}{n}}$ (mg/g) | $1/n$ | $R^2$ | $k_T$ (L/mg) | $b_T$ | $R^2$ |
| MB | 175.44 | 0.05 | 0.99 | 0.15 | 2.41 | 0.95 | 6.23 | 12.46 | 0.83 |
| CR | 85.47 | 0.02 | 1.00 | 2.63 | 0.93 | 0.99 | 3.74 | 40.60 | 0.93 |
| Cd(II) | 12.58 | 0.04 | 0.99 | 14.79 | 0.43 | 0.78 | 3.03 | 70.60 | 0.66 |
| Pb(II) | 52.91 | 0.04 | 0.96 | 3.89 | 0.53 | 0.87 | 3.53 | 123.54 | 0.94 |

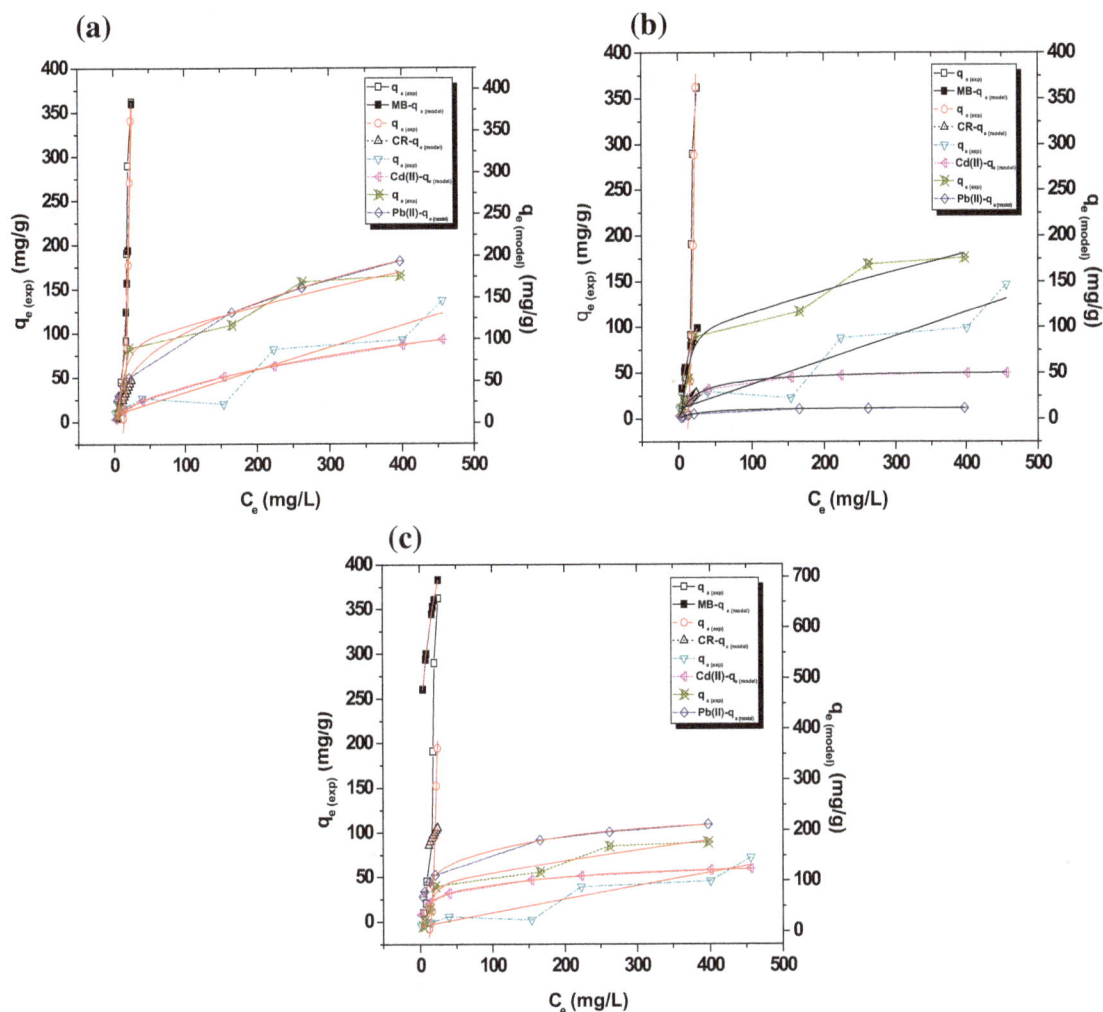

**Fig. 6 a** Nonlinear Freundlich isotherm plots of Methylene Blue, Congo Red, Cadmium and Lead biosorbed by *Terminalia ivorensis* seed waste (temperature = 300 K; adsorbent dose = 25 mg; agitation speed = 200 rpm; initial metal ion concentration = 25–500 mg/L; agitation time = 180 min). **b** Nonlinear Langmuir isotherm plots of Methylene Blue, Congo Red, Cadmium and Lead biosorbed by *Terminalia ivorensis* seed waste (temperature = 300 K; adsorbent dose = 25 mg; agitation speed = 200 rpm; initial metal ion concentration = 25–500 mg/L; agitation time = 180 min). **c** Nonlinear Temkin isotherm plots of Methylene Blue, Congo Red, Cadmium and Lead biosorbed by *Terminalia ivorensis* seed waste (temperature = 300 K; adsorbent dose = 25 mg; agitation speed = 200 rpm; initial metal ion concentration = 25–500 mg/L; agitation time = 180 min)

diffusion of the solute molecules, (b) the gradual adsorption step, in which intraparticle diffusion may be rate limiting and (c) the diffusion of adsorbate particles to adsorption sites either by pore diffusion through the liquid-filled pores or by a solid diffusion mechanism.

The biosorption of CR by TISW at temperature of 300 K increased up to 15 min. Thereafter, equilibrium was attained. Also, the biosorption of CR by TISW increased up to 30 and 60 min at 320 and 340 K respectively, after which equilibrium was reached. It could be said that increase in temperature led to an increase in the time equilibrium was attained for the biosorption of CR onto TISW. At 300 K, CR molecules rapidly occupied the available functional moieties on the active sites in 15 min of biosorption process. At 320 and 340 K, the available functional moieties on the active sites were occupied at 30 and 60 min, respectively. Although, increase in the temperature from 300 to 340 K was meant to increase in the driving force (mass transfer) of CR molecules onto the functional moieties in the pores of TISW due to the increase in the collision frequency and mobility of CR molecules [29, 62]. This temperature increase might have led to the rupture of some chemical bonds of the functional moieties of TISW, distorting the $\pi$–$\pi$ stacking from hydrophobic interactions of CR aromatic rings. Hence, this phenomenon decreased the propensity of CR molecules to bind to TISW surface [29]. The biosorption of MB onto TISW at 300 and 320 K increased up to 60 min. Thereafter, a slight increase in the uptake of MB was observed up to 120 min, after which equilibrium was attained. At 340 K, the biosorption of MB increased up to 120 min, after which equilibrium was reached. Increment in the temperature made MB to exhibit a dissociated form that decreased the rate at which it bound onto the functional moieties in the pores of TISW [29, 62].

The biosorption of $Cd^{2+}$ onto TISW at 300 K increased up to 30 min, respectively, and >30 min of $Cd^{2+}$ uptake culminated into equilibrium. For the biosorption of $Pb^{2+}$ by TISW at 320 and 340 K, the uptake of $Pb^{2+}$ increased up to 15 min, after which equilibrium was observed. The increase in the mass transfer of $Pb^{2+}$ onto the functional moieties in the TISW pores was due to the fast diffusion of $Pb^{2+}$ from the solute phase onto the sorbent [61, 62].

## Equilibrium, kinetics and thermodynamics of CR, MB, $Cd^{2+}$ and $Pb^{2+}$ removal by TISW

To know the biosorption capacity of TISW for CR, MB, $Cd^{2+}$ and $Pb^{2+}$, the experimental data were modelled with some equilibrium models. Langmuir isotherm fits the experimental data for the removal of CR, MB, $Cd^{2+}$ and $Pb^{2+}$ by TISW better than the Freundlich and Temkin isotherms. The correlation co-efficient values for the Langmuir isotherm were higher than those for Freundlich and Temkin isotherms. The saturation monolayer adsorption capacities, $q_{max_L}$, of TISW for CR, MB, $Cd^{2+}$ and $Pb^{2+}$ were 85.47, 175.44, 12.58 and 52.91 mg g$^{-1}$, respectively (see Table 3). Lucidly, it is noteworthy to say that the homogeneous nature of TISW surface favoured the uptake of MB molecules more than the other pollutant ions of interest in this work. From Freundlich isotherm, it can be deduced that the values for the biosorption affinity, $1/n$, for the uptake of CR, MB, $Cd^{2+}$ and $Pb^{2+}$ onto TISW were

**Table 4** Nonlinear equilibrium parameters for the removal of Methylene Blue, Congo Red, Cadmium and Lead onto *Terminalia ivorensis* seed waste

| | MB − $q_e$ (exp) | MB − $q_e$ (model) | CR − $q_e$ (exp) | CR − $q_e$ (model) | $Cd^{2+}$ − $q_e$ (exp) | $Cd^{2+}$ − $q_e$ (model) | $Pb^{2+}$ − $q_e$ (exp) | $Pb^{2+}$ − $q_e$ (model) |
|---|---|---|---|---|---|---|---|---|
| **Nonlinear Freundlich model** | | | | | | | | |
| Reduced $\chi^2$ | 974.71 | 36.14 | 894.71 | 3.57E−5 | 403.34 | 0.34 | 32.15 | 0.68 |
| Residual sum of squares | 1949.41 | 72.27 | 1789.42 | 7.15E−5 | 806.68 | 0.68 | 111.87 | 1.36 |
| Adjusted $R^2$ | 0.95 | 0.99 | 0.70 | 1.00 | 0.91 | 1.00 | 0.97 | 1.00 |
| **Nonlinear Langmuir model** | | | | | | | | |
| Reduced $\chi^2$ | 479.73 | 0.21 | 948.07 | 2.99E−5 | 403.35 | 0.05 | 2251.11 | 0.01 |
| Residual sum of squares | 1939.41 | 0.43 | 9814.22 | 5.97E−5 | 818.20 | 0.11 | 651.93 | 0.01 |
| Adjusted $R^2$ | 0.95 | 1.00 | 0.67 | 1.00 | 0.93 | 1.00 | 0.99 | 1.00 |
| **Nonlinear Temkin model** | | | | | | | | |
| Reduced $\chi^2$ | 974.72 | 7.84E−5 | 849.17 | 0.01 | 403.34 | 0.53 | 757.62 | 1.60 |
| Residual sum of squares | 1949.41 | 1.57E−4 | 1798.14 | 0.01 | 608.86 | 1.06 | 223.50 | 3.21 |
| Adjusted $R^2$ | 0.96 | 1.00 | 0.67 | 1.00 | 0.91 | 1.00 | 0.98 | 1.00 |

0.93, 2.41, 0.43 and 0.53, respectively. This suggests that the binding of MB onto TISW would be least favourable by heterogeneous surface, while the binding of $Cd^{2+}$ onto TISW would be most favourable by heterogeneous surface $(0 < 1/n < 1$-favourable, $1/n > 1$-non favourable).

The nonlinear plots (Langmuir, Freundlich and Temkin isotherms) for the removal of CR, MB, $Pb^{2+}$ and $Cd^{2+}$ by TISW are shown in Fig. 6a–c. From these nonlinear isotherm plots, it was deduced that the equilibrium data better fit the nonlinear isotherms than the linear isotherms due to the higher correlation coefficient values ($R^2$ values) obtained for these plots, when compared with those obtained for linear isotherm plots. For the nonlinear isotherm plots, the reduced Chi square values ($\chi^2$ values) and the residual sum of squares values indicated that Langmuir isotherm better fits the equilibrium data than Freundlich and Temkin isotherms (see Table 4).

The kinetics of biosorption gives a good idea of how the experimental timing controls the mechanism, chemical reaction and the bulk transfer of the solute ions onto the biosorbent surface. Table 5 indicates that pseudo-second-order kinetic equation gave better fits to the experimental data than pseudo-first-order kinetic equation for the removal of CR, MB, $Cd^{2+}$ and $Pb^{2+}$ by TISW. The correlation co-efficient values for pseudo-second-order kinetic equation were higher than those obtained for pseudo-first-order kinetic equation at the experimental temperatures. This suggests that the mechanism for the biosorption process is chemisorption. The pseudo-second-order rate constants, $k_2$, for the removal of CR, MB, $Cd^{2+}$ and $Pb^{2+}$ by TISW at 300–340 K were $k_2 \leq 0.417$ g $mg^{-1}$ $min^{-1}$, while the pseudo-first-order rate constants, $k_1$, for the removal of CR, MB, $Cd^{2+}$ and $Pb^{2+}$ by TISW at 300–340 K were $k_1 \leq 0.026$ $min^{-1}$. The Weber–Morris

intraparticle diffusion equation indicated that the plots did not pass through the origin, due to the fact that the values of the boundary film or the thickness of the boundary layer were $c \geq 0$ in all cases. This implies that intraparticle diffusion process was not the only rate controlling or limiting step for the biosorption process, but pore diffusion and film diffusion might have taken place in the rate controlling or limiting step [29, 63, 64].

The nonlinear plots (pseudo-first-order, pseudo-second-order and intraparticle diffusion equations) for the removal of CR, MB, $Pb^{2+}$ and $Cd^{2+}$ onto TISW are shown in Tables 6 and 7. It was deduced that the kinetic data better fit the nonlinear kinetic equation than the linear kinetic equations. This resulted from the fact that the correlation coefficient values ($R^2$ values) obtained for these plots were higher than those obtained for the linear kinetic equations. For the nonlinear kinetic plots, the reduced Chi square values ($\chi^2$ values) and the residual sum of squares values suggested that pseudo-second-order equations gave better fit to the kinetic data when compared to pseudo-first-order and intraparticle diffusion equations. Figure 7a–c depicts the nonlinear kinetic plots for pseudo-first-order, pseudo-second-order and intraparticle diffusion equations for biosorbed CR, MB, $Cd^{2+}$ and $Pb^{2+}$ by TISW. But this work has shown that low-cost TISW demonstrated remediation potential for both toxic industrial dyestuffs and toxic metal ions. Hence, TISW has a wide array of applications for environmental remediation.

From thermodynamic standpoint (see Table 8), $+\Delta G^\circ$ (kJ/mol) values increased from 6.53 to 7.40, 7.67 to 8.69, 23.87 to 27.05 and 13.31 to 15.09 as TISW biosorbed CR, MB, $Cd^{2+}$ and $Pb^{2+}$ from 300 to 340 K, respectively. The $\Delta H^\circ$ (kJ/mol) values were $-0.27$, $-0.40$, $+0.09$ and $+0.22$ for the biosorption of CR, MB, $Cd^{2+}$ and $Pb^{2+}$ by TISW,

**Table 5** Linear kinetic parameters for the removal of Methylene Blue, Congo Red, Cadmium and Lead onto *Terminalia ivorensis* seed waste

|  | Temperature (K) | PFOM | | | PSOM | | | WMID | | |
|---|---|---|---|---|---|---|---|---|---|---|
|  |  | $k_1$ (/min) | $q_e$ (mg/g) | $R^2$ | $k_2$(g/mg min) | $q_e$ (mg/g) | $R^2$ | $k_{id}$ (g/mg $min^{1/2}$) | $R^2$ | $c$ |
| MB | 300 | 0.026 | 0.974 | 0.806 | 0.011 | 12.920 | 0.997 | 0.081 | 0.880 | 11.283 |
|  | 320 | 0.023 | 1.458 | 0.805 | 0.057 | 12.390 | 0.999 | 0.130 | 0.687 | 10.738 |
|  | 340 | 0.019 | 1.734 | 0.694 | 0.085 | 11.420 | 0.999 | 0.106 | 0.849 | 10.076 |
| CR | 300 | 0.020 | 4.575 | 0.929 | 0.012 | 7.949 | 0.994 | 0.307 | 0.923 | 3.951 |
|  | 320 | 0.022 | 2.557 | 0.764 | 0.046 | 7.868 | 0.999 | 0.021 | 0.511 | 5.003 |
|  | 340 | 0.022 | 2.177 | 0.904 | 0.051 | 4.235 | 0.995 | 0.014 | 0.606 | 2.180 |
| Cd(II) | 300 | 0.011 | 0.679 | 0.952 | 0.417 | 11.710 | 1.000 | 0.029 | 0.964 | 11.915 |
|  | 320 | 0.008 | 1.332 | 0.755 | 0.339 | 12.300 | 1.000 | 0.023 | 0.927 | 11.417 |
|  | 340 | 0.021 | 0.556 | 0.712 | 0.303 | 12.240 | 1.000 | 0.005 | 0.508 | 11.560 |
| Pb(II) | 300 | 0.004 | 0.771 | 0.062 | 0.224 | 12.250 | 1.000 | 0.034 | 0.694 | 11.831 |
|  | 320 | 0.011 | 1.178 | 0.805 | 0.142 | 11.600 | 1.000 | 0.087 | 0.951 | 10.976 |
|  | 340 | 0.008 | 1.270 | 0.694 | 0.085 | 12.150 | 1.000 | 0.013 | 0.379 | 10.298 |

*PFOM* nonlinear pseudo-first-order model, *PSOM* nonlinear pseudo-second-order model, *WMID* Weber–Morris intraparticle diffusion model

respectively. The $-\Delta S°$ (J mol$^{-1}$ K$^{-1}$) values for CR, MB, Cd$^{2+}$ and Pb$^{2+}$ biosorbed by TISW were 217.72, 255.71, 79.56 and 44.38, respectively. Table 8 depicts that the values of Gibb's free energy, ($\Delta G°$) for CR, MB, Cd$^{2+}$ and Pb$^{2+}$ biosorbed by TISW were non-spontaneous at all temperatures. The values of enthalpy change, ($\Delta H°$) for CR and MB biosorbed by TISW were exothermic, while those for Cd$^{2+}$ and Pb$^{2+}$ biosorbed by TISW were endothermic. The values of entropy change, ($\Delta S°$) for CR, MB, Cd$^{2+}$ and Pb$^{2+}$ biosorbed by TISW showed decreasing disorderliness or chaos of the biosorption process. Table 9 shows the adsorption capacities, $q_{max_L}$ (mg/g) of some adsorbents for

CR, MB, Cd$^{2+}$ and Pb$^{2+}$ used by researchers as reported in literature.

**Desorption kinetics**

Figure 8 shows that desorption of MB from MB-loaded TISW was maximum at 86.37% for 0.1 M HCl desorbent and also maximum at 79.77% for 0.1 M HNO$_3$ desorbent after 30 min. Similarly, desorption of CR from CR-loaded TISW was maximum at 92.73% for 0.1 M HCl desorbent and likewise maximum at 55.65% for 0.1 M HNO$_3$ desorbent after 30 min. The desorption kinetics of Cd$^{2+}$

**Table 6** Nonlinear kinetic parameters for the removal of Methylene Blue and Congo Red onto *Terminalia ivorensis* seed waste

| | MB | | | | | | CR | | | | | |
| --- | --- | --- | --- | --- | --- | --- | --- | --- | --- | --- | --- | --- |
| | 300 K | | 320 K | | 340 K | | 300 K | | 320 K | | 340 K | |
| | $q_{t(exp)}$ | $q_{t(model)}$ | $q_{t(exp)}$ | $q_{t(model)}$ | $q_{t(exp)}$ | $q_{t(model)}$ | $q_{t(exp)}$ | $q_{t(model)}$ | $q_{t(exp)}$ | $q_{t(model)}$ | $q_{t(exp)}$ | $q_{t(model)}$ |
| PFOM | | | | | | | | | | | | |
| Reduced $\chi^2$ | 0.09 | 0.05 | 0.05 | 0.02 | 0.02 | 0.01 | 0.31 | 0.08 | 0.09 | 0.03 | 0.10 | 0.01 |
| Residual sum of squares | 0.56 | 0.31 | 0.31 | 0.12 | 0.10 | 0.01 | 1.87 | 0.46 | 1.12 | 0.21 | 0.62 | 0.08 |
| Adjusted $R^2$ | 0.98 | 0.99 | 0.98 | 1.00 | 0.98 | 1.00 | 0.94 | 0.99 | 0.94 | 1.00 | 0.91 | 0.99 |
| PSOM | | | | | | | | | | | | |
| Reduced $\chi^2$ | 1.25 | 0.76 | 0.05 | 0.04 | 0.03 | 0.02 | 0.31 | 0.19 | 0.19 | 0.11 | 0.10 | 0.05 |
| Residual sum of squares | 10.88 | 4.59 | 0.31 | 0.28 | 0.21 | 0.10 | 1.87 | 1.18 | 1.12 | 0.65 | 0.62 | 0.32 |
| Adjusted $R^2$ | 0.93 | 0.97 | 0.98 | 1.00 | 0.98 | 1.00 | 0.94 | 0.98 | 0.94 | 0.98 | 0.91 | 0.98 |
| WMID | | | | | | | | | | | | |
| Reduced $\chi^2$ | 20.83 | 4.70E−5 | 0.05 | 1.21E−4 | 0.02 | 8.08E−5 | 0.31 | 6.74E−4 | 0.19 | 3.06E−6 | 0.10 | 0.01 |
| Residual sum of squares | 1.22 | 2.82E−4 | 0.31 | 7.27E−4 | 0.10 | 4.85E−4 | 1.87 | 0.01 | 1.12 | 1.84E−5 | 0.62 | 0.05 |
| Adjusted $R^2$ | 0.93 | 1.00 | 0.98 | 1.00 | 0.88 | 1.00 | 0.94 | 1.00 | 0.94 | 1.00 | 0.91 | 0.99 |

*PFOM* nonlinear pseudo-first-order model, *PSOM* nonlinear pseudo-second-order model, *WMID* Weber–Morris intraparticle diffusion model

**Table 7** Nonlinear kinetic parameters for the removal of Cadmium and Lead onto *Terminalia ivorensis* seed waste

| | Cd(II) | | | | | | Pb(II) | | | | | |
| --- | --- | --- | --- | --- | --- | --- | --- | --- | --- | --- | --- | --- |
| | 300 K | | 320 K | | 340 K | | 300 K | | 320 K | | 340 K | |
| | $q_{t(exp)}$ | $q_{t(model)}$ | $q_{t(exp)}$ | $q_{t(model)}$ | $q_{t(exp)}$ | $q_{t(model)}$ | $q_{t(exp)}$ | $q_{t(model)}$ | $q_{t(exp)}$ | $q_{t(model)}$ | $q_{t(exp)}$ | $q_{t(model)}$ |
| PFOM | | | | | | | | | | | | |
| Reduced $\chi^2$ | 0.02 | 1.10E−4 | 0.01 | 1.25E−4 | 1.77 | 0.01 | 2.51 | 0.05 | 2.51 | 0.11 | 2.88 | 0.01 |
| Residual sum of squares | 0.01 | 6.60E−4 | 0.01 | 1.68E−4 | 2.81 | 0.02 | 3.23 | 0.09 | 6.55 | 0.53 | 6.79 | 0.03 |
| Adjusted $R^2$ | 0.97 | 1.00 | 0.97 | 1.00 | 0.97 | 1.00 | 0.92 | 1.00 | 0.93 | 1.00 | 0.91 | 1.00 |
| PSOM | | | | | | | | | | | | |
| Reduced $\chi^2$ | 0.02 | 6.22E−4 | 2.88 | 0.01 | 3.95 | 0.01 | 3.33 | 0.03 | 3.67 | 0.02 | 1.22 | 0.05 |
| Residual sum of squares | 0.25 | 0.03 | 31.65 | 0.02 | 5.99 | 0.03 | 12.56 | 0.07 | 6.48 | 0.35 | 8.15 | 0.07 |
| Adjusted $R^2$ | 0.97 | 1.00 | 0.95 | 1.00 | 0.91 | 1.00 | 0.95 | 1.00 | 0.95 | 1.00 | 0.97 | 1.00 |
| WMID | | | | | | | | | | | | |
| Reduced $\chi^2$ | 1.40 | 1.51E−6 | 0.02 | 6.13E−6 | 1.33 | 1.33 | 13.18 | 0.03 | 8.31 | 0.05 | 1.01 | 0.11 |
| Residual sum of squares | 8.40 | 4.25E−6 | 0.01 | 3.88E−5 | 7.82 | 0.03 | 5.15 | 0.05 | 7.07 | 1.03 | 7.24 | 0.85 |
| Adjusted $R^2$ | 0.97 | 1.00 | 0.97 | 1.00 | 0.96 | 1.00 | 0.96 | 1.00 | 0.94 | 1.00 | 0.97 | 1.00 |

*PFOM* nonlinear pseudo-first-order model, *PSOM* nonlinear pseudo-second-order model, *WMID* Weber–Morris intraparticle diffusion model

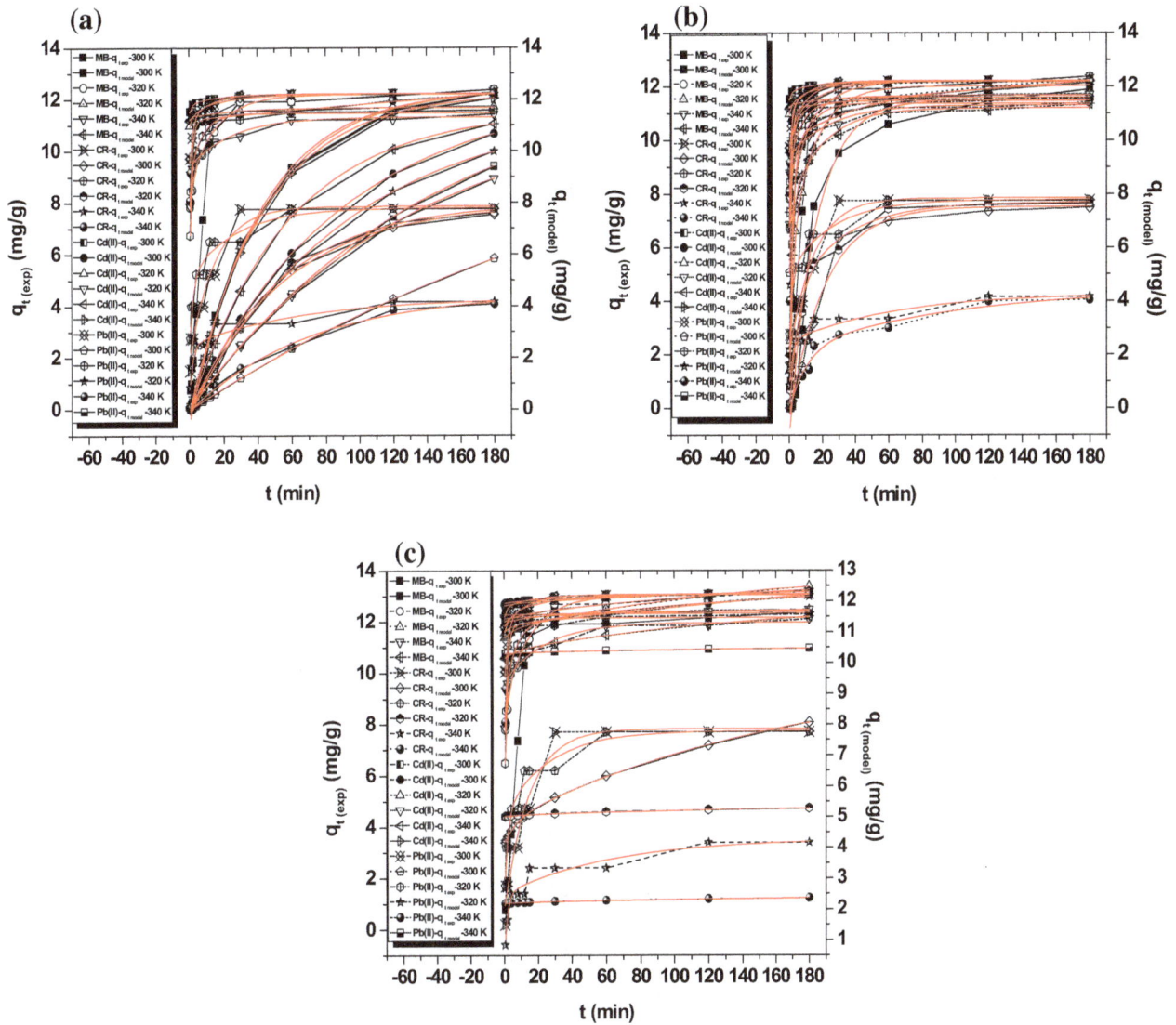

**Fig. 7** **a** Nonlinear pseudo-first-order kinetic plots of Methylene Blue, Congo Red, Cadmium and Lead biosorbed by *Terminalia ivorensis* seed waste (temperature = 300–340 K; adsorbent dose = 25 mg; agitation speed = 200 rpm; initial metal ion concentration = 100 mg/L; agitation time = 0.5–180 min). **b** Nonlinear pseudo-second-order kinetic plots of Methylene Blue, Congo Red, Cadmium and Lead biosorbed by *Terminalia ivorensis* seed waste (temperature = 300–340 K; adsorbent dose = 25 mg; agitation

speed = 200 rpm; initial metal ion concentration = 100 mg/L; agitation time = 0.5–180 min). **c** Nonlinear Weber–Morris intraparticle diffusion kinetic plots of Methylene Blue, Congo Red, Cadmium and Lead biosorbed by *Terminalia ivorensis* seed waste (temperature = 300–340 K; adsorbent dose = 25 mg; agitation speed = 200 rpm; initial metal ion concentration = 100 mg/L; agitation time = 0.5–180 min)

| Table 8 Thermodynamic parameters for the removal of Methylene Blue, Congo Red, Cadmium and Lead onto *Terminalia ivorensis* seed waste at different temperatures | $+\Delta G^{\circ}$ (kJ/mol) | | | $\Delta H^{\circ}$ (kJ/mol) | $-\Delta S^{\circ}$ (J/mol/K) |
|---|---|---|---|---|---|
| | 300 K | 320 K | 340 K | | |
| TISW-CR | 6.53 | 6.97 | 7.40 | -0.27 | 217.72 |
| TISW-MB | 7.67 | 8.18 | 8.69 | -0.40 | 255.71 |
| TISW-Pb$^{2+}$ | 13.31 | 14.20 | 15.09 | +0.22 | 44.38 |
| TISW-Cd$^{2+}$ | 23.87 | 25.46 | 27.05 | +0.09 | 79.56 |

**Table 9** The adsorption capacities, $q_{max_L}$ (mg/g) of some adsorbents used by researchers

| Adsorbents | References | $q_{max_L}$ (mg/g) | | | |
|---|---|---|---|---|---|
| | | Cd$^{2+}$ | Pb$^{2+}$ | MB | CR |
| Eucalyptus seed waste | [9] | 71.15 | | | |
| *Solanum melongena* | [10] | | 71.42 | | |
| Peanut hull | [11] | 6.00 | 30.00 | | |
| Corn cob | [12] | 5.38 | | | |
| Hazelnut shell | [13] | 5.47 | 16.46 | | |
| Olive cake | [14] | 65.40 | | | |
| Mungbean husk | [15] | 35.41 | | | |
| Mango peel waste | [16] | 68.92 | 99.05 | | |
| Orange waste biomass | [17] | 41.58 | | | |
| *Scolymus hispanicus* L. | [18] | 54.05 | | | |
| Chemically modified orange peel | [19] | 13.70 | 73.53 | | |
| *Nauclea diderrichii* seed waste | [20] | 6.30 | | | |
| *Zea mays* seed chaff | [22] | 121.95 | 384.62 | | |
| Graphene oxide/*Nauclea diderrichii* seed waste | [23] | 7.54 | 7.94 | | |
| *Parkia biglobosa* biomass | [24] | 157.98 | 94.25 | | |
| *Papaya*-clay combo | [25] | | | 35.46 | |
| Nano-titania/*Nauclea diderrichii* seed waste | [26] | | 7.49 | | |
| *Pentaclethra macrophylla* bark | [29] | 43.76 | 348.43 | 251.26 | 157.23 |
| *Malacantha alnifolia* bark | [29] | 255.75 | 133.87 | 89.00 | 800.00 |
| *Cedrela odorata* seed waste | [39] | | | 111.88 | 128.84 |
| *Parkia biglobosa* cellulosic extract | [40] | | | 1498.42 | 266.67 |
| TISW | This study | 12.58 | 52.91 | 175.44 | 85.47 |

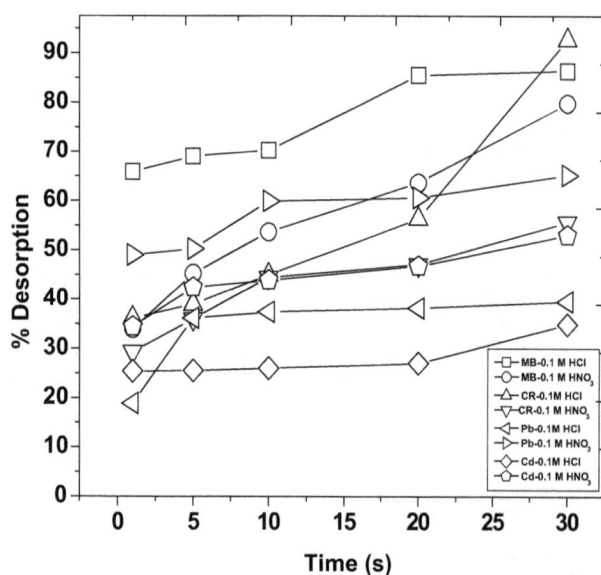

**Fig. 8** Desorption kinetics of Methylene Blue, Congo Red, Cadmium and Lead loaded-*Terminalia ivorensis* seed waste using 0.1 M HCl and 0.1 M HNO$_3$ (adsorbent dose = 25 mg; agitation speed = 200 rpm; agitation time = 1–30 min; temperature = 300 K)

showed that the maximum of 34.89 and 53.01% of Cd$^{2+}$ was desorbed from Cd$^{2+}$ loaded TISW using 0.1 M HCl and 0.1 M HNO$_3$ desorbents, respectively, after 30 min.

Also, the maximum of 39.5 and 65.16% of Pb$^{2+}$ was desorbed from Pb$^{2+}$ loaded TISW using 0.1 M HCl and 0.1 M HNO$_3$ desorbents, respectively, after 30 min. Conversely, desorption of CR and MB from TISW was kinetically faster than that of Cd$^{2+}$ and Pb$^{2+}$ from TISW. This desorption kinetics result reveals that TISW can be regenerated and reused for another cycle of biosorption.

## Conclusion

For the first time, *Terminalia ivorensis* seed waste (TISW) demonstrated a good potential for the removal of CR, MB, Cd$^{2+}$ and Pb$^{2+}$. Proximate analysis, bulk density, specific surface area, pH of point of zero charge, Fourier transform infra red spectroscopy, scanning electron microscopy, thermogravimetric/differential thermal analysis and energy dispersive analysis of X-ray were used to study the surface texture or morphology of TISW. The equilibrium data best fit the Langmuir isotherm, with a maximum Langmuir monolayer saturation adsorption capacity, $q_{max_L}$ = 175.44 mg MB per g of TISW. The kinetic data best fit the pseudo-second-order equation for the removal of CR, MB, Cd$^{2+}$ and Pb$^{2+}$ onto TISW. This study revealed that TISW, a benign agricultural seed waste, could be recommended as ubiquitous and cheap biological adsorbent for environmental remediation of toxic industrial dyestuffs and toxic metals.

**Acknowledgements** Dr. Martins O. Omorogie, who is currently a postdoctoral research fellow at Department of Chemistry, Vaal University of Technology (VUT), Vanderbijlpark, South Africa, appreciates the Department for providing TG/DTA (Perkin Elmer) for the surface textural characterisations of TISW. Dr. Martins O. Omorogie and Prof. Jonathan O. Babalola acknowledge the Department of Chemistry, University of Ibadan, Nigeria for the provision of FTIR (Perkin Elmer), FAAS (Buck Scientific 205) and UV/Visible Spectrophotometer (Cecil), which were used for various analyses.

# References

1. Cabral JPS (2010) Water microbiology: Bacterial pathogens and water. Int J Environ Res Public Health 7:3657–3703
2. United Nations (2012) The millennium development goals report. United Nations, New York
3. Onda K, LoBuglio J, Bartram J (2012) Global access to safe water: accounting for water quality and the resulting impact on MDG progress. Int J Environ Res Public Health 9:880–894
4. Kandile NG, Mohamed HM, Mohamed MI (2015) New heterocycle modified chitosan adsorbent for metal ions (II) removal from aqueous systems. Int J Biol Macromol 72:110–116
5. Hameed BH, Ahmad AA (2009) Batch adsorption of methylene blue from aqueous solution by garlic peel, an agricultural waste biomass. J Hazard Mater 164:3115–3121
6. Saha B, Das S, Saikia J, Das G (2011) Preferential and enhanced adsorption of different dyes on iron oxide nanoparticles: a comparative study. J Phys Chem C 115:8024–8033
7. Pang H, Wang WQ, Yan ZZ, Zhang H, Li XX, Chen J, Zhang JS, Zhang B (2012) Porous $Mn_3[Co(CN)_6]_2.nH_2O$ nanocubes as a rapid organic dyes adsorption material. RSC Adv 2:9614–9618
8. Chen ZH, Zhang JN, Fu JW, Wang MH, Wang XZ, Han RP, Xu Q (2014) Adsorption of methylene blue onto poly(cyclotriphosphazene-co-4,4′-sulfonyldiphenol) nanotubes: kinetics, isotherm and thermodynamics analysis. J Hazard Mater 273:263–271
9. Kiruba UP, Senthil Kumar P, Prabhakaran C, Aditya V (2014) Characteristics of thermodynamic, isotherm, kinetic, mechanism and design equations for the analysis of adsorption in Cd(II) ions-surface modified Eucalyptus seeds system. J Taiwan Inst Chem 45:2957–2968
10. Yuvaraja G, Krishnaiah N, Subbaiah MV, Krishnaiah A (2014) Biosorption of Pb(II) from aqueous solution by *Solanum melongena* leaf powder as a low-cost biosorbent prepared from agricultural waste. Colloids Surf B 114:75–81
11. Brown P, Jefcoat IA, Parrish D, Gill S, Graham E (2002) Evaluation of the adsorptive capacity of peanut hull pellets for heavy metals in solution. Adv Environ Res 4:19–29
12. Leyva-Ramos CR, Bernal-Jacome LA, Acosta-Rodriguez I (2005) Adsorption of cadmium(II) from aqueous solution on natural and oxidized corn cob. Sep Purif Technol 45:41–49
13. Bulut Y, Tez Z (2007) Adsorption studies on ground shells of hazelnut and almond. J Hazard Mater 149:35–41
14. Anber ZA, Matouq MAD (2008) Batch adsorption of cadmium ions from aqueous solution by means of olive cake. J Hazard Mater 151:194–201
15. Saeed A, Iqbal M, Holl WH (2009) Kinetics, equilibrium and mechanism of $Cd^{2+}$ removal from aqueous solution by mungbean husk. J Hazard Mater 168:1467–1475
16. Iqbal M, Saeed A, Zafar SI (2009) FTIR spectrophotometry,
17. Marin ABP, Ortuno JF, Aguilar MI, Meseguer VF, Saez J, Llorens M (2010) Use of chemical modification to determine the binding of Cd(II), Zn(II) and Cr(III) ions by orange waste. Biochem Eng J 53:2–6
18. Barka N, Abdennouri M, Boussaoud A, Makhfouk M-EI (2010) Biosorption characteristics of cadmium(II) onto *Scolymus hispanicus L.* as low-cost natural biosorbent. Desalination 258:66–71
19. Lashee MR, Ammar NS, Ibrahim HS (2012) Adsorption/desorption of Cd(II), Cu(II) and Pb(II) using chemically modified orange peel: equilibrium and kinetic studies. Solid State Sci 14:202–210
20. Omorogie MO, Babalola JO, Unuabonah EI, Gong JR (2012) Kinetics and thermodynamics of heavy metal ions sequestration onto novel *Nauclea diderrichii* seed biomass. Bioresour Technol 118:576–579
21. Omorogie MO, Babalola JO, Unuabonah EI, Song W, Gong JR (2016) Efficient chromium abstraction from aqueous solution using a low-cost biosorbent: *Nauclea diderrichii* seed waste. J Saudi Chem Soc 20(1):49–57
22. Babalola JO, Omorogie MO, Babarinde AA, Unuabonah EI, Oninla VO (2016) Optimization of the biosorption of $Cr^{3+}$, $Cd^{2+}$ and $Pb^{2+}$ using new biowaste: *Zea mays* seed chaff. Environ Eng Manage J 15(7):1571–1580
23. Omorogie MO, Babalola JO, Unuabonah EI, Gong JR (2014) Solid phase extraction of hazardous metals from aqua system by nanoparticle-modified agrowaste composite adsorbents. J Environ Chem Eng 2(1):675–684
24. Omorogie MO, Babalola JO, Unuabonah EI, Gong JR (2014) Hybrid materials from agrowaste and nanoparticles: implications on the kinetics of the adsorption of inorganic pollutants. Environ Technol 35(5):611–619
25. Ogbodu RO, Omorogie MO, Unuabonah EI, Babalola JO (2015) Biosorption of heavy metals from aqueous solutions by *Parkia biglobosa* biomass: equilibrium, kinetics, and thermodynamic studies. Environ Prog Sustain Energy 34(6):1694–1704
26. Unuabonah EI, Adedapo AO, Nnamdi CO, Adewuyi A, Omorogie MO, Adebowale KO, Olu-Owolabi BI, Ofomaja AE, Taubert A (2015) Successful scale-up performance of a novel Papaya-clay combo adsorbent: up-flow adsorption of a basic dye. Desalin Water Treatment 56(2):536–551
27. Omorogie MO, Babalola JO, Unuabonah EI, Gong JR (2016) Clean technology approach for the competitive binding of toxic metal ions onto $MnO_2$ nano-bioextractant. Clean Techn Environ Policy 18(1):171–184
28. Omorogie MO, Babalola JO, Unuabonah EI, Gong JR (2015) New facile benign agrogenic-nanoscale titania material; Remediation potential for toxic inorganic cations. J Water Proc Eng 5(1):95–100
29. Babalola JO, Olowoyo JO, Durojaiye AO, Olatunde AM, Unuabonah EI, Omorogie MO (2016) Understanding the removal and regeneration potentials of biogenic wastes for toxic metals and organic dyes. J Taiwan Inst Chem Eng 58:490–499
30. Gupta VK, Srivastava SK, Mohan D, Sharma S (1998) Design parameters for fixed bed reactors of activated carbon developed from fertilizer waste material for the removal of some heavy metal ions. Waste Manage 17:517–522
31. Jain AK, Gupta VK, Bhatnagar A, Suhas A (2003) Comparative study of adsorbents prepared from industrial wastes for removal of dyes. Sep Sci Technol 38(2):463–481
32. Gupta VK, Mittal A, Kaur D, Malviya A, Mittal J (2009) Adsorption studies on the removal of colouring agent phenol red

from wastewater using waste materials as adsorbents. J Colloid Interface Sci 337:345–354

33. Gupta VK, Mittal A, Kaur D, Malviya A, Mittal J (2009) Adsorptive removal of hazardous anionic dye 'congo red' from wastewater using waste materials and recovery by desorption. J Colloid Interface Sci 340:16–26

34. Gupta VK, Mittal A, Malviya A, Mittal J (2010) Decoloration treatment of a hazardous triaryl methane dye, light green SF (Yellowish) by waste material adsorbents. J Colloid Interface Sci 342:518–527

35. Khani H, Rofouei MK, Arab P, Gupta VK, Vafaei Z (2010) Multi-walled carbon nanotubes-ionic liquid-carbon paste electrode as a super selectivity sensor: application to potentiometric monitoring of mercury ion (II). J Hazard Mater 183:402–409

36. Gupta VK, Mittal A, Mittal J (2010) Removal and recovery of Chrysoidine Y from aqueous solutions by waste materials. J Colloid Interface Sci 344:497–507

37. Gupta VK, Jain R, Agarwal S, Shrivastava M (2011) Removal of the hazardous dye—Tartrazine by photodegradation on titanium dioxide surface. Mater Sci Eng C 31:1062–1067

38. Gupta VK, Agarwal S, Saleh TA (2011) Synthesis and characterization of alumina-coated carbon nanotubes and their application for lead removal. J Hazard Mater 185:17–23

39. Saleh TA, Gupta VK (2012) Photo-catalyzed degradation of hazardous dye methyl orange by use of a composite catalyst consisting of multiwalled carbon nanotubes and titanium dioxide. J. Colloids Interface Sci. 371:101–106

40. Gupta VK, Nayak A (2012) Cadmium removal and recovery from aqueous solutions by novel adsorbents prepared from orange peel and $Fe_2O_3$ nanoparticles. Chem Eng J 180:81–90

41. Gupta VK, Jain R, Mittal A, Agarwal S, Sikarwar S (2012) Photocatalytic degradation of toxic dye Amaranth on $TiO_2$/UV in aqueous suspensions. Mater Sci Eng C 32:12–17

42. Karthikeyan S, Gupta VK, Boopathy R, Titus A, Sekaran G (2012) A new approach for the degradation of aniline by mesoporous activated carbon as a heterogeneous catalyst: kinetic and spectroscopic studies. J Mol Liq 173:153–163

43. Saleh TA, Gupta VK (2012) Column with CNT/magnesium oxide composite for lead (II) removal from water. Environ Sci Pollut Res 19:1224–1228

44. Gupta VK, Ali I, Saleh TA, Nayak A, Agarwal S (2012) Chemical treatment technologies for waste-water recycling-an overview. RSC Adv 2:6380–6388

45. Gupta VK, Kumar R, Nayak A, Saleh TA, Barakat MA (2013) Adsorptive removal of dyes from aqueous solution onto carbon nanotubes: a review. Adv Colloid Interface Sci 193–194:24–34

46. Saleh TA, Gupta VK (2014) Processing methods, characteristics and adsorption behavior of tire derived carbons: a review. Adv Colloid Interface Sci 211:93–101

47. Gupta VK, Nayak Agarwal A (2015) Bioadsorbents for remediation of heavy metals: current status and their future prospects. Environ Eng Res 20:1–18

48. Mushtaq M, Bhatti HN, Iqbal M, Noreen S (2016) *Eriobotrya japonica* seed biocomposite efficiency for copper adsorption: isotherms, kinetics, thermodynamic and desorption studies. J Environ Manage 176:21–33

49. Tahir MA, Bhatti HN, Iqbal M (2016) Solar red and brittle blue direct dyes adsorption onto *Eucalyptus angophoroides* bark: equilibrium, kinetics and thermodynamic studies. J Environ Chem Eng 4:2431–2439

50. Rashid A, Bhatti HN, Iqbal M, Noreen S (2016) Fungal biomass composite with bentonite efficiency for nickel and zinc adsorption: a mechanistic study. Ecol Eng 91:459–471

51. Hanif MA, Bhatti HN (2015) Remediation of heavy metals using easily cultivable, fast growing, and highly accumulating white rot fungi from hazardous aqueous streams. Desal Water Treat 53:238–248

52. Zhao G, Li J, Ren X, Chen C, Wang X (2011) Few-layered graphene oxide nanosheets as superior sorbents for heavy metal ion pollution management. Environ Sci Technol 45(24):10454–10462

53. Yang S, Chen C, Chen Y, Li J, Wang D, Wang X, Hu W (2015) Competitive adsorption of $Pb^{II}$, $Ni^{II}$, and $Sr^{II}$ ions on graphene oxides: a combined experimental and theoretical study. ChemPlusChem 80:480–484

54. Kennish MJ (1992) Ecology of estuaries: anthropogenic effects. CRC, Boca Raton

55. Ofomaja AE, Naidoo EB (2011) Biosorption of copper from aqueous solution by chemically activated pine cone: a kinetic study. Chem Eng J 175:260–270

56. Stumm W, Morgan JJ (1996) Aquatic chemistry, 3rd edn. Wiley, New York, pp 534–540

57. Sears GW (1956) Determination of specific surface area of colloidal silica by titration with sodium hydroxide. Anal Chem 28:1981–1983

58. Zhan Y, Luo X, Nie S, Huang Y, Tu X, Luo S (2011) Selective separation of Cu(II) from aqueous solution with a novel Cu(II) surface magnetic ion-imprinted polymer. Ind Eng Chem Res 50:6355–6361

59. Arshadi M, Faraji AR, Amiri MJ, Mehravar M, Gil A (2015) Removal of methyl orange on modified ostrich bone waste—a novel organic-inorganic biocomposite. J Col Interface Sci 446:11–23

60. Ho YS (1995) Adsorption of heavy metals from waste streams by peat, PhD Thesis, University of Birmingham, Birmingham, UK

61. Onal Y, Akmil-Basar C, Sarici-Ozdemir C (2007) Investigation kinetics mechanisms of adsorption malachite green onto activated carbon. J Hazard Mater 146:194–203

62. Vimonses V, Lei S, Jin B, Chow CWK, Saint C (2009) Kinetic study and equilibrium isotherm analysis of congo red adsorption by clay materials. Chem Eng J 148:354–364

63. Babalola JO, Koiki BA, Eniayewu Y, Salimonu A, Olowoyo A, Oninla VO, Alabi HA, Ofomaja AE, Omorogie MO (2016) Adsorption efficacy of *Cedrela odorata* seed waste for dyes: non linear fractal kinetics and non linear equilibrium studies. J Environ Chem Eng 4(3):3527–3536

64. Babalola JO, Bamidele TM, Adeniji EA, Odozi NW, Olatunde AM, Omorogie MO (2016) Adsorptive modelling of toxic cations and ionic dyes onto cellulosic extract. Model Earth Syst Environ 2(4):190–204

# 8

# Degradation of organic dye using a new homogeneous Fenton-like system based on hydrogen peroxide and a recyclable Dawson-type heteropolyanion

Abir Tabaï[1] · Ouahiba Bechiri[1] · Mostefa Abbessi[1]

**Abstract** The main objective of this work was to study the degradation of Acid Orange 7 (AO7) dye in aqueous solution by hydrogen peroxide using $HFe_{2.5}P_2W_{18}O_{62}$ $23.H_2O$ as a catalyst. $HFe_{2.5}P_2W_{18}O_{62}$ $23.H_2O$ is a recyclable DAWSON-type heteropolyanion. Effects of various experimental parameters of the oxidation reaction of the dye were investigated. The studied parameters were the initial pH, the initial $H_2O_2$ concentration, the catalyst mass, and the dye concentration. The optimum conditions had been determined, and it was found that efficiency of degradation obtained was about 100%. The optimal parameters were: initial pH 4; $[H_2O_2]_0 = 2$ mM; catalyst mass 0.01 g; concentration of dye 30 mg/L. Infrared spectroscopy analysis of the $HFe_{2.5}P_2W_{18}O_{62}$ $23.H_2O$ catalyst indicates that the catalyst showed good stability for degradation of AO7 even after second cycle.

**Keywords** Toxic dyes · Decolorization · Acid orange 7 · Water treatment · Oxidation · Homogeneous Fenton-like system

## Introduction

The treatment of textile waste waters has always been a serious problem for these industries. These components sometimes are not biodegradable and are toxic enough for aquatic ecosystems [1]. Some classical methods such as adsorption on activated carbon, ozonation, reverse osmosis, ion exchange on synthetic adsorbent resins, flocculation, and decantation are available for removing dye from water. However, these methods have high operating cost [2] or are inefficient due to complex aromatic structure. Other alternatives to degrade recalcitrant organic pollutants are now being studied including advanced oxidation (AOPs) whose characteristic is to generate hydroxyl radicals (˙OH), which are powerful oxidants [3].

These radicals are generated by Fenton's reagent ($Fe^{2+}/H_2O_2$), which has been the subject of many studies in the degradation of organic matter [4]. However, the disadvantage of the method lies in Fenton rejection of a significant amount of sludge $Fe^{2+}$ and $Fe^{3+}$ [5], which requires separation or removal, thus increasing the operational cost. To solve this problem, many studies have focused on the use of a modified Fenton process (Fenton-Like) to treat waste waters. A method which relies on the use iron incorporated in the recyclable compounds [6]. In this work, a new homogeneous Fenton-like system based on hydrogen peroxide and a recyclable Dawson-type heteropolyanion (Fig. 1) [7].

($HFe_{2.5}P_2W_{18}O_{62}$ $23.H_2O/H_2O_2$) system was used for decolorization of Acid Orange 7 (AO7) (also known as Orange II).

Acid Orange 7 is one of the dyes that are produced in large amounts in the world. It is commonly used in pharmaceutical, food, and cosmetics industries. Acid Orange 7 is an azoique dye used in dyeing of silk and wool. For these qualities, Acid Orange 7 is much used in the textile industry. The molecular structure of azo-dye Orange II is shown in Fig. 2. Acid Orange 7 is not amenable to conventional biological treatment [8].

The aim of the study is to examine the influence of different parameters such as the solution pH, the catalyst mass, the concentration of $H_2O_2$, and the initial dye concentration on the degradation of Acid Orange 7 by $H_2O_2$

✉ Ouahiba Bechiri
bechirio@yahoo.fr

[1] Laboratory of Environmental Engineering, Department of Process Engineering, Faculty of Engineering, University of Annaba, P.O. Box 12, 23000 Annaba, Algeria

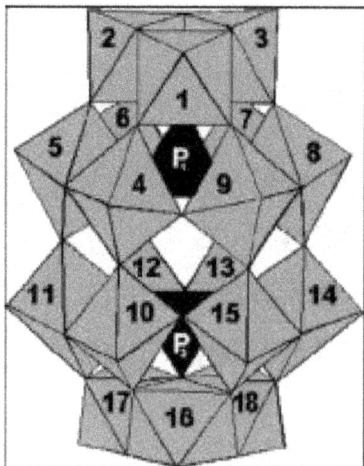

**Fig. 1** Polyhedral representation of the Dawson structure of polyanion

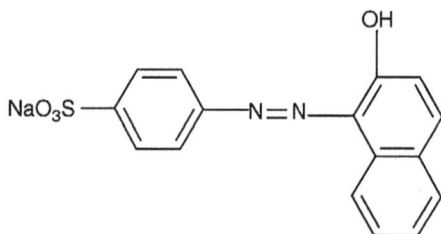

**Fig. 2** Chemical structure of Acid Orange 7 (AO7)

using $HFe_{2.5}P_2W_{18}O_{62}23.H_2O$ as a catalyst. The mechanism of the reaction and catalytic stability were studied.

## Experimental section

### Synthesis of catalyst $HFe_{2.5}P_2W_{18}O_{62}$ 23.H$_2$O [9]

$HFe_{2.5}P_2W_{18}O_{62}$ 23.H$_2$O

5 g (1.088 mol) of $H_6P_2W_{18}O_{62}$ was dissolved in 20 mL of water at room temperature and 0.767 g of solid FeCl$_2$,6-H$_2$O (3.26 mol) was then added. The mixture was then stirred for 10 min. Yellow powder was obtained after five days by slow evaporation.

$H_6P_2W_{18}O_{62}$ was prepared, starting from the substance mother $(K_6P_2W_{18}O_{62}, nH_2O)^{6-}$ by extraction with ether in hydrochloric acid medium according to the methods described in the literature [10].

### Behavior of AO7 oxidation using $HFe_{2.5}P_2W_{18}O_{62}$ 23.H$_2$O/H$_2$O$_2$ system

#### Reagents

The 4-(2-hydroxy-1-naphthylazo) benzene sulfonic acid sodium salt, commonly named orange II or Acid Orange 7

(AO7) (90%) was obtained from Aldrich (H$_2$O$_2$ 35%, W/W) was purchased from Aldrich Chemical Company. All other reagents (NaOH or H$_2$SO$_4$) that used in this study were analytical grade.

#### Procedures and analysis

The experiments were performed in a reactor batch of 500 mL capacity. Various solutions of AO7 were prepared at different concentrations. They were then homogenized by stirring them until the dye is completely dissoluted of the dye. pH was adjusted using 0.1 N H$_2$SO$_4$ or NaOH aqueous solutions. In all experiments 100 mL of dyes (30 mg/L) solution containing the appropriate quantity of catalyst and H$_2$O$_2$ was magnetically stirred at room temperature. The AO7 decreasing concentration of dye was monitored by a 6705 UV visible spectrophotometer JENWAY. The Wave length corresponding to the maximum absorbance is given below: $\lambda_{max} = 486$ nm [11].

The resolution of the wave length and bandwidth was 1 and 0.5 nm. The cell used during the experiments was made of 1 cm thick quartz.

#### Effects of operational parameters on AO7 oxidation

The oxidation of AO7 by H$_2$O$_2$ using ($HFe_{2.5}P_2W_{18}O_{61}$ 23H$_2$O) as a catalyst has been studied according to the following factors: initial pH of the solution, H$_2$O$_2$ concentration, mass of the catalyst ($HFe_{2.5}P_2W_{18}O_{61}$ 23H$_2$O), and dye concentration.

The oxidation efficiency (discoloration) was determined as it is shown below [12]:

$$DE = \left[ (C_i - C_f)/C_i \right] \times 100 \qquad (1)$$

DE: discoloration efficiency; $C_i$: Initial dye concentration; $C_f$: final dye concentration.

*Effect of solution pH*   The pH value of dye solution has significant influence on the catalytic system efficiency because: (1) it can affect the catalyst stability and (2) it influences the catalytic reaction which controls the production rate of hydroxyl radicals [13]. To find the optimum pH for the decolorization of AO7, a series of experiments at initial pH value in the range 3–8 was conducted.

Figure 3 illustrates the effect of pH on the discoloration efficiency of AO7 in water. It was found that the discoloration efficiency of AO7 is strongly pH dependent. The optimal pH is about 4 giving discoloration efficiency equal to 94.25%. This result can be explained by the stability of the catalyst at this pH. Also, H$_2$O$_2$ molecules are unstable in alkaline solution and therefore, the degradation of dye decreases in alkaline solution [14].

The optimal value is chosen pH 4.

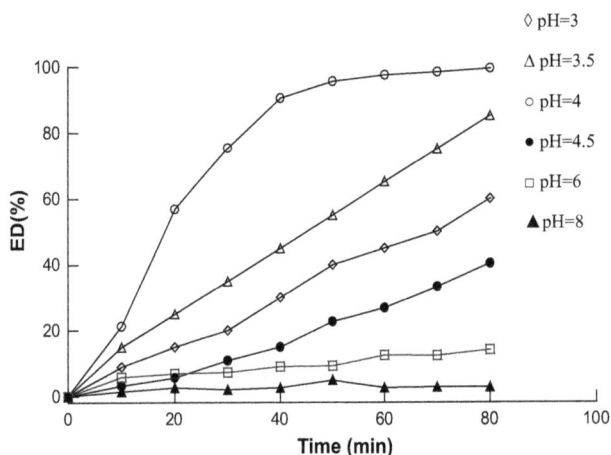

**Fig. 3** Effect of solution pH on AO7 oxidation ($[AO7]_0 = 30$ mg/L, $[H_2O_2]_0 = 0.11$ mM, m ($HFe_{2.5}P_2W_{18}O_{62}$ 23.$H_2O$) = 0.01 g, $T = 25$ °C)

*Effect of the nature of the acid used to adjust the pH* The oxidation of the dye by the Fenton-like process can be influenced by the presence of various ions such as $SO_4^{2-}$, $NO^{3-}$, $Cl^-$, and $PO_4^{3-}$ [15]. To evaluate the influence of these anions, we adjusted the pH of an aqueous solution of AO7 by different acids $H_2SO_4$, $HNO_3$, HCl, and $H_3PO_4$ in similar operating conditions to those previously established.

Figure 4 shows the effect of these acid ions (chloride, sulfate, nitrate, and phosphate) on the dye oxidation. It appears that the presence of phosphate ions inhibits the oxidation of the dye using $HFe_{2.5}P_2W_{18}O_{62}$ 23.$H_2O/H_2O_2$ system.

Sulfate and nitrate ions have virtually no effect on the discoloration of AO7. However, depending on the nature of the acids, the discoloration efficiency is about 94.25, 94.35, 96.09, and 03.06% in the presence of $H_2SO_4$, $HNO_3$, HCl,

and $H_3PO_4$ acids, respectively. These results agree with those found at the degradation of other organic pollutants [16]. We can deduce that $H_2SO_4$, $HNO_3$, and HCl acids lead almost to the same efficiency, but in the presence of $H_3PO_4$ it is too low. The inhibitory effect of phosphate ions may be due to the trapping of $^.OH$ radicals.

*Effect of catalyst mass* The effect of the catalyst mass was investigated, keeping operational parameters identical to those of the above-mentioned experiment. The following catalyst quantities have been used: 0.005; 0.008; 0.01; 0.02 g. The results are illustrated in Fig. 5.

These results show that the discoloration efficiency of AO7 oxidation increases when increasing the catalytic mass up to 0.01 g where the decolorization efficiency is optimal. Beyond this value, there is a decrease of the discoloration efficiency. The excess of the catalyst does not appear to ply a positive role in the AO7 oxidation using $HFe_{2.5}P_2W_{18}O_{62}$ 23.$H_2O/H_2O_2$ system. This result is in good agreement with literature [12, 13, 17]: the decrease of the discoloration efficiency of AO7 when increasing mass of the catalyst can be explained by the presence of side reaction consuming the radicals hydroxyls. The mass of catalyst that gives the best result is 0.01 g.

*Effect of $H_2O_2$ concentration* The effect of $H_2O_2$ concentration on AO7 decolorization was studied in the range 0.04–0.44 mmol. The results are illustrated in Fig. 6.

The results indicate that the degradation of AO7 was increased by increasing the concentration of $H_2O_2$ up to a value of concentration of $H_2O_2$ equal to 2 mM.

Decolorization of certain dyes, mainly azo, with activated hydrogen peroxide in homogeneous systems, using different soluble catalysts (heteropolyanions) was already studied [18].

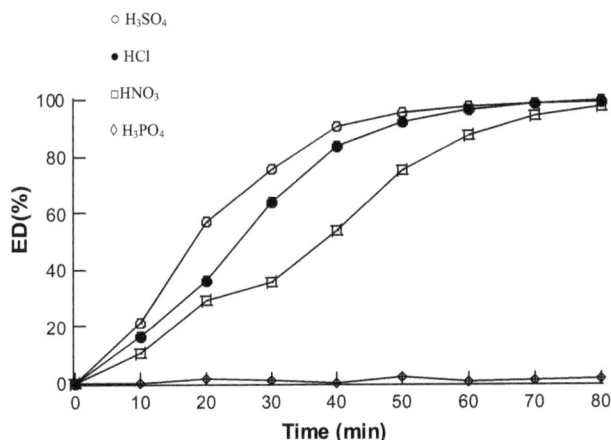

**Fig. 4** Effect of the nature of the acid used to adjust the pH (pH 4, $[AO7]_0 = 30$ mg/L, $[H_2O_2]_0 = 0.11$ mM, m ($HFe_{2.5}P_2W_{18}O_{62}$ 23.$H_2O$) = 0.01 g, $T = 25$ °C)

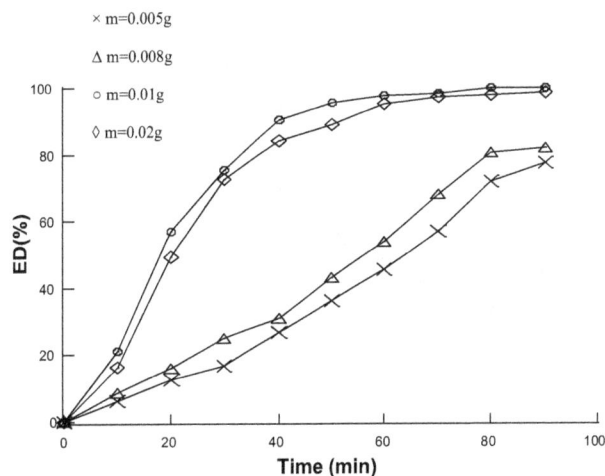

**Fig. 5** Effect of catalyst mass on AO7 oxidation (pH 4, $[AO7]_0 = 30$ mg/L, $[H_2O_2]_0 = 0.11$ mM, $T = 25$ °C)

The activation of hydrogen peroxide by homogeneous catalysts was attributed to the formation of highly active hydroxyl radicals [19]. High concentrated $H_2O_2$ solution undergoes self quenching of OH radicals, with formation of hydroperoxyl radicals $HO_2$. Although $HO_2$ is an effective oxidant itself, its oxidation potential is much lower than that of $\cdot OH$ radicals [20].

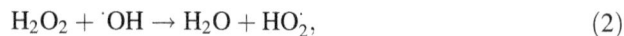

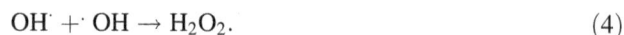

$$H_2O_2 + \cdot OH \rightarrow H_2O + HO_2^{\cdot}, \quad (2)$$

$$HO_2^{\cdot} + \cdot OH \rightarrow H_2O + O_2, \quad (3)$$

$$OH^{\cdot} + \cdot OH \rightarrow H_2O_2. \quad (4)$$

It can be postulated that $H_2O_2$ should be added at an optimum concentration to achieve the best degradation. Hence, 0.2 mM/L of $H_2O_2$ appears as an optimal.

*Radical scavenging experiments and possible reaction mechanism*  In general, $\cdot OH$ radicals are known as a key active species in the catalytic oxidation process. $\cdot OH$ radicals are powerful oxidants for many organic molecules.

Ethanol, methanol, and 2-propanol are known as OH scavengers [21–24]. These reagents were introduced into the reaction system to capture $\cdot OH$ in the oxidation of AO7 process. The results are illustrated in Fig. 7.

These results show that the addition of $\cdot OH$ scavenger reagents causes a sharp decrease in the degradation of AO7, and the discoloration efficiency of AO7 from 100% to 1.6, 2.08, and 2.53% in the presence of methanol, ethanol, and 2-propanol, respectively. According to the above discussion, it is furthermore confirmed that the degradation of AO7 molecules is dependent on the availability of $\cdot OH$ radical species. $\cdot OH$ radicals make a major contribution to degradation of AO7. OH scavenger reagents added consume $\cdot OH$ radical species and less will

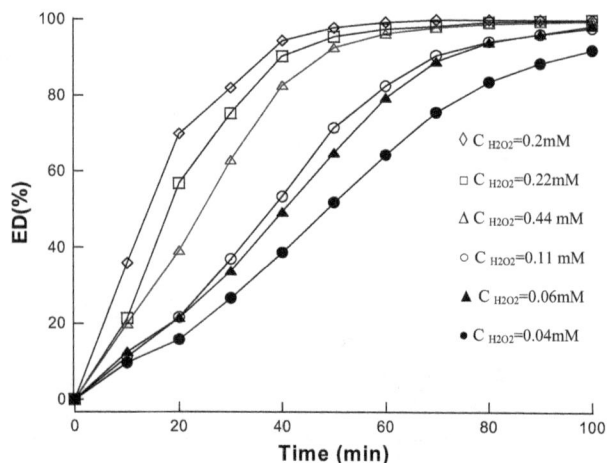

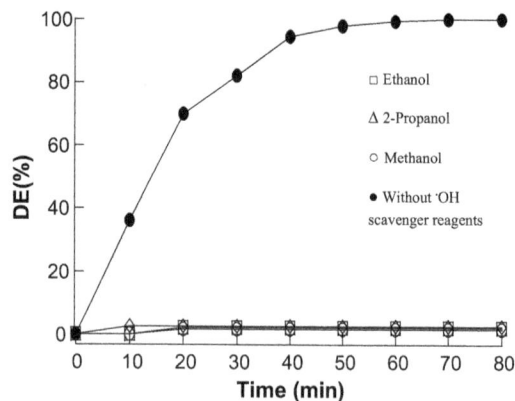

**Fig. 7** Effect of addition of OH scavenger reagents on AO7 oxidation (pH 4, $[AO7]_0 = 30$ mg/L, m $(HFe_{2.5}P_2W_{18}O_{62}$ $23.H_2O) = 0.01$ g, $T = 25\ °C$)

be available for the degradation of the dye molecules, hence the lower discoloration efficiency.

Also, it was shown [25] that the action of $H_2O_2$ on a complex containing $Fe^{3+}$ resulted in the reduction of $Fe^{3+}$ to $Fe^{2+}$ with apparition of $HO_{\cdot2}$.

The action of $H_2O_2$ on the complex of $Fe^{2+}$ leads to the generation of hydroxyl radicals $OH\cdot$. These hydroxyl radicals cause the degradation of the dye.

In agreement with the results found and the mechanism proposed below, we can propose the following mechanism:

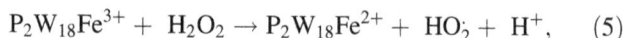

$$P_2W_{18}Fe^{3+} + H_2O_2 \rightarrow P_2W_{18}Fe^{2+} + HO_2^{\cdot} + H^+, \quad (5)$$

$$P_2W_{18}Fe^{2+} + H_2O_2 \rightarrow P_2W_{18}Fe^{3+} + 2\cdot OH. \quad (6)$$

The degradation of AO7 by $H_2O_2$ catalyzed with $HFe_{2.5}P_2W_{18}O_{62}$ $23.H_2O$ heteropolyanion can be well supported by the change in the UV–Vis absorption spectrum over the course of degradation (Fig. 8). The AO7 spectrum was characterized by the chromophore that contained azo linkage [26] in the visible range at 484 and 430 nm corresponding to the hydrazone and azo form. In the ultraviolet region, the absorbance peaks at 230 and 310 nm are due to the benzene and naphthalene rings, respectively. These four characteristic bands were markedly weakened during the degradation reaction, tending to disappear completely after 60 min, without the appearance of new absorption bands in the visible or ultraviolet regions due to destruction of the chromophoric and auxochromic structures by the homogeneous Fenton-like reaction.

*Effect of the AO7 concentration*  It is important from an application point of view to study the dependence of removal efficiency on the initial concentration of the colorant (AO7).

To determine the effect of AO7 concentration on discoloration efficiency, a series of experiments was realized

**Fig. 6** Effect of initial $H_2O_2$ concentration on AO7 oxidation (pH 4, $[AO7]_0 = 30$ mg/L, m $(HFe_{2.5}P_2W_{18}O_{62}$ $23.H_2O) = 0.01$ g, $T = 25\ °C$)

**Fig. 8** UV-Vis spectra of AO7 water solutions during the treatment process with Fe(III)$P_2W_{18}$/$H_2O_2$ system (pH 4, $[H_2O_2]_0 = 0.2$ mM, m ($HFe_{2.5}P_2W_{18}O_{62}$ 23.$H_2O$) = 0.01 g, [AO7] = 30 mg/L, $T = 25$ °C)

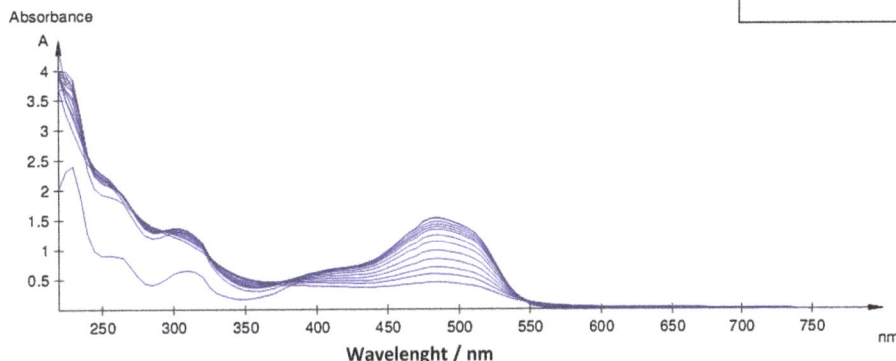

between 5 and 50 mg/L. The variation of the discoloration efficiency of the AO7 dye with $H_2O_2$ and $HFe_{2.5}P_2W_{18}O_{61}$ 23$H_2O$ catalyst as a function of initial dye concentration is represented in Fig. 9. The results show that the discoloration efficiency increases with increasing of dye concentration up to a value equal to 30 mg/L, but above it the discoloration efficiency decreases.

This result can be explained as follows:

At low concentration of dye, the discoloration efficiency is lower (ED 39% for the concentration of AO7 5 mg/L). These results concur with those found in literature [27–29]. This phenomenon can be explained by the effect of $H_2O_2$ concentration is in excess compared to latter and can cause the scavenging effect (the reaction of $H_2O_2$ and $OH^-$ in aqueous solution).

– At higher concentration of dye, the discoloration efficiency decreases (ED 48% for the concentration of AO7 50 mg/L). This phenomenon can be explained by the fact that increasing the initial concentration of dye leads to an increase in the number of molecules of (AO7), while the number of the radical's hydroxyls remains constant ($H_2O_2$ concentration and catalyst kept

constant), thereby causing a decrease in the discoloration efficiency [19, 27].

*Stability and recycling of the catalyst*

After reaction, we have tried to recover the catalyst $P_2W_{18}Fe^{3+}$. The method used is its precipitation, as potassium salt, by adding a mass of KCl in the solution after complete discoloration in a basic medium. The resulting precipitate was characterized by infrared spectroscopy. The spectra are illustrated in Fig. 10.

The spectrum of the recovered heteropolyanion is identical to that of the initially added heteropolyanion. We can therefore conclude that the catalyst remains stable and robust after the reaction.

Figure 11 shows the catalytic performance of recovered catalyst after four cycles through the oxidation of AO7 dye, by hydrogen peroxide.

After second cycle, the catalyst did not exhibit significant loss of catalytic activity and the discoloration efficiency was 100. AO7 degradation decreases only from 100

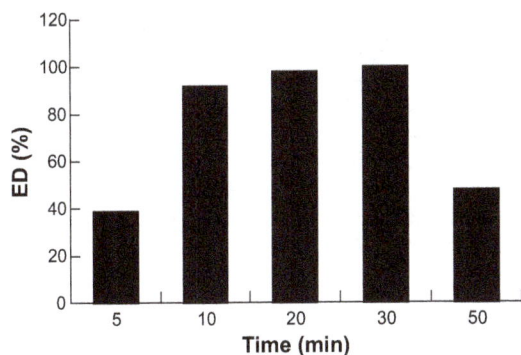

**Fig. 9** Effect of initial dye concentration on AO7 oxidation. (pH 4, $[H_2O_2]_0 = 0.2$ mM, m ($HFe_{2.5}P_2W_{18}O_{62}$ 23.$H_2O$) = 0.01 g, $T = 25$ °C)

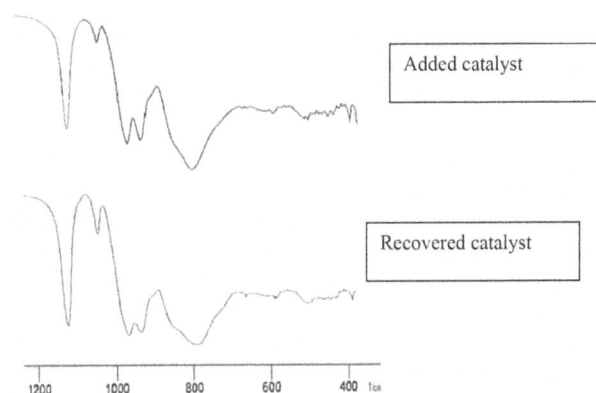

**Fig. 10** IR spectra of $HFe_{2.5}P_2W_{18}O_{62}$ 23.$H_2O$ and recovered $HFe_{2.5}P_2W_{18}O_{62}$ 23.$H_2O$

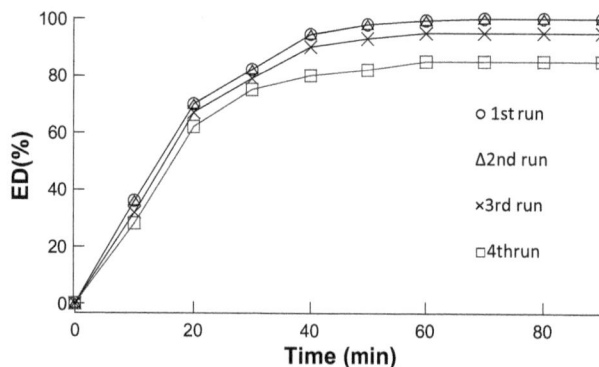

**Fig. 11** Catalytic performance of recovered catalyst after four cycles through the oxidation of AO7 dye by hydrogen peroxide. (pH 4, $[H_2O_2]_0 = 0.2$ mM,  m  $(HFe_{2.5}P_2W_{18}O_{62}$  23.$H_2O) = 0.01$ g, $T = 25$ °C, [AO7] $= 30$ mg/L)

to 95% in three cycles. In four cycles, AO7 degradation decreases from 100 to 89%.

This result is in good agreement with the literature [28]. The loss of activity can be explained by to poisoning of the active catalytic sites due to adsorbed organic species. However, this could be avoided by placing the catalyst in an oven at 180 °C, to enhance the organic solvent.

## Conclusion

Fe(III)-phosphotunstate/$H_2O_2$ system is a new homogeneous Fenton-like system that is capable of oxidizing organic dye compounds. The experimental results show that the initial pH, the initial concentration of $H_2O_2$, Fe(III)$P_2W_{18}$, AO7 concentration, had a great influence on the degradation of AO7 dye. It is approximately 100% dye which has been eliminated the Fe(III)$P_2W_{18}/H_2O_2$ on the following operating conditions: pH 4, catalyst mass 0.01 g, $[H_2O_2]_0 = 2$ m M. The molar ratio $H_2O_2/AO_7 = 0.02$. The possibility of recovery of the catalyst after reaction is verified by adding a mass of KCl in the solution after complete discoloration. The stability of reused catalyst is confirmed after second cycle.

This system constitutes simple and effective method compared to those previously reported for the oxidation of Acid Orange 7 [29, 30].

**Acknowledgements** This work was supported by the Engineering Environmental Laboratory of Badji Mokhtar University (Annaba-Algeria). This work is a tribute to the late Professor Mostefa Abbessi, God rest his soul.

## References

1. Fernández J, Kiwi J, Baeza J, Freer J, Lizama C, Mansilla HD (2004) Orange II photocatalysis on immobilised TiO$_2$, Effect of the pH and H$_2$O$_2$. Appl Catal B Environ 48:205–211

2. Rajeshwar K, Osugi ME, Chanmanee W, Chenthamarakshan CR, Zanoni MVB, Kajitvichyanukul P, Krishnan-Ayer R (2008) Heterogeneous photocatalytic treatment of organic dyes in air and aqueous media. Photochemrev 9:171–192

3. Zhang Y, He C, Deng J, Tu Y, Liu J, Xiong Y (2009) Photo-Fenton-like catalytic activity of nano-lamellar Fe$_2$V$_4$O$_{13}$ in the degradation of organic pollutants. Res Chem Intermed 35:727–737

4. Medien HAA, Khalil SME (2010) Kinetics of the oxidative decolorization of some organic dyes utilizing Fenton-like reaction in water. J King Saud Univ 22:147–153

5. Duesterberg CK, Mylon SE, Waite TD (2008) pH effects on iron-catalyzed oxidation using Fenton's reagent. Environ Sci Technol 42:8522–8527

6. Hsueh CL, Huang YH, Wang CC, Chen CY (2005) Degradation of azo dyes using low iron concentration of Fenton and Fenton-like system. Chemosphere 58:1409–1414

7. Pope MT (1983) Heteropoly and isopoly oxometalates. Springer, New York

8. Kuznetsova NI, Kirillova NV, Kuznetsova LI, Smirnova MY, Likholobov VA (2007) Hydrogen peroxide and oxygen oxidation of aromatic compounds in catalytic systems containing heteropoly compounds. J Hazard Mater 46:569–576

9. Bechiri O, Abbessi M, Belghiche R, Ouahab L (2014) Wells–Dawson polyoxometelates [HP$_2$W$_{18-n}$Mo$_n$O$_{62}$]Fe$_{2.5}$, xH$_2$O; n = 0, 6: synthesis, spectroscopic characterization and catalytic application for dyes oxidation. C R Chim 17:135–140

10. Ciabrini JP, Contant R, Fruchart M (1983) Heteropolyblues: relationship between metal-oxygen-metal bridges and reduction behaviour of octadeca (molybdotungsto) diphosphate anions. Polyhedron 2:1229–1233

11. Djaneye-boundjou G, Amouzou E, Kodom T, Tchakala I, Anodi K, Bawa LM (2012) Photocatalytic degradation of orange II using mesoporous TiO$_2$ (P25) and fenton reactive (Fe/H$_2$O$_2$). IJESMER 1:91–96

12. Bechiri O, Abbessi M, Ouahab L (2012) The oxidation study of methyl orange dye by hydrogen peroxide using Dawson-type heteropolyanions as catalysts. Res Chem Intermed 39:2945–2954

13. Bechiri O, Abbessi M, Samar ME (2013) Decolorization of organic dye (NBB) using Fe(III)P$_2$W$_{12}$Mo$_5$/H$_2$O$_2$ system. Desalin Water Treat 51:31–33

14. Gould DM, Spiro M, Griffith WP (2005) Mechanism of bleaching by peroxides: part 7. The pH dependence of the oxometalate catalysed bleaching of methyl orange. J Mol Cat A Chem 242:176–181

15. Fan HJ, Huang ST, Chung WH, Jan JL, Lin WY, Chen CC (2009) Degradation pathways of crystal violet by Fenton and Fenton-like systems: condition optimization and intermediate separation and identification. J Hazard Mater 71:1032–1044

16. Modirshahla N, Behnajady MA, Ghanbary F (2007) Decolorization and mineralization of C.I. Acid Yellow 23 by Fenton and photo-Fenton processes. Dyes Pigments 73:305–310

17. Mylon SE, Quan S, Waite TD (2010) Process optimization in use of zero valent iron nanoparticles for oxidative transformations. Chemosphere 81:127–131

18. Strukul G (1992) Catalytic oxidations with hydrogen peroxide as oxidant. Kluwer Academic, Dordrecht

19. Lucas MS, Peres JA (2006) Decolorization of the azo dye reactive black 5 by Fenton and photo-Fenton oxidation. Dyes Pigments 71:236–244

20. Sun JH, Shi SH, Lee YF, Sun SP (2009) Fenton oxidative decolorization of the azo dye Direct Blue 15 in aqueous solution. Chem Eng J 155:680–683

21. Tadolini B, Cabrini L (1988) On the mechanism of OH. scavenger action. Biochem J 253:931–933

22. Zheng J, Gao Z, He H, Yang S, Sun C (2016) Efficient degra-

dation of Acid Orange 7 in aqueous solution by iron ore tailing Fenton-like process. Chemosphere 150:40–48

23. Lindsey M, Tarr M (2000) Quantitation of hydroxyl radical during Fenton oxidation following a single addition of iron and peroxide. Chemosphere 41:409–417

24. Overend R, Paraskevopoulos G (1978) Rates of OH radical reactions. 4. Reactions with methanol, ethanol, I-propanol, and 2-propanol at 296 K. J Phys Chem 82:12

25. Ramirez HJ, Costa CA, Madeira LM, Mata G, Vicente MA, Rojas-Cervantes ML, Lopez-Peinado AJ, Martin-Aranda RM (2007) Fenton-like oxidation of Orange II solutions using heterogeneous catalysts based on saponite clay. Appl Catal B 71:44–56

26. Bauer C, Jacques P, Kalt A, Photochem J (2001) Photodegradation of an azo dye induced by visible light incident on the surface of TiO2. Photobiol A Chem 140:87–92

27. Ji F, Li C, Zhang J, Deng L (2011) Efficient decolorization of dye pollutants with LiFe(WO$_4$)$_2$ as a reusable heterogeneous Fenton-like catalyst. Desalination 269:284–290

28. Catrinescu C, Teodosiu C, Macoveanu M, Miehe-Brendle J, Dred RL (2003) Catalytic wet peroxide oxidation of phenol over Fe-exchanged pillared beidellite. Water Res 37:1154–1160

29. Zhang F, Feng C, Li W, Cui J (2014) Indirect electrochemical oxidation of dye wastewater containing acid orange 7 using Ti/RuO$_2$-Pt electrode. Int J Electrochem Sci 9:943–954

30. Wang J, Liu G, Lu H, Jin R, Zhou J (2012) Biodegradation of acid orange 7 and its auto-oxidative decolorization product in membrane-aerated biofilm reactor. Int Biodeterior Biodegrad 67:73–77

# 9

# Biodiesel production from castor oil: ANN modeling and kinetic parameter estimation

Atiya Banerjee[1] · Devyani Varshney[1] · Surendra Kumar[1] · Payal Chaudhary[2] ·
V. K. Gupta[3]

**Abstract** This research work concerns with the transesterification of castor oil with methanol to form biodiesel. As the free fatty acid content in castor oil is more than 1%, an acid catalyst namely, $H_2SO_4$ has been used for esterification. The experimental conditions were determined using central composite design method and the experiments were conducted in a 2 L working volume fully controlled reactor. The input conditions namely, catalyst concentration, methanol to oil molar ratio and temperature were varied, and % fatty acid methyl ester (FAME) content was determined. Based upon the experimental data, an ANN model has been developed which is used to predict %FAME yield for a given set of input conditions. The experimental data and the data predicted by the ANN model have been used to estimate the rate constants of a kinetic model. The ANN model predicts the % FAME yield within ±8% deviation, and the developed kinetic model shows successfully the effect of methanol to oil molar ratio on % FAME yield at 60 °C and 3% (v/v) catalyst loading.

**Keywords** ANN · Castor oil · Biodiesel · Parameter estimation · Differential equation model

✉ Surendra Kumar
skumar.iitroorkee@gmail.com

[1] Department of Chemical Engineering, Indian Institute of Technology Roorkee, Roorkee, Uttarakhand 247667, India

[2] Centre for Transportation Systems, Indian Institute of Technology Roorkee, Roorkee, Uttarakhand 247667, India

[3] Department of Chemistry, Indian Institute of Technology Roorkee, Roorkee, Uttarakhand 247667, India

## Introduction

Global energy consumption has been steadily increasing since the beginning of the millennium and its demand is expected to rise in the coming times. Thus, the general awareness about alternative energy sources has increased manifold, which have a lower impact on the environment as a whole [1]. Sustainable sources of energy are to be explored and used to decrease the overall dependence on available primary sources of energy (fossil fuels) [2]. In this context, biodiesel appears to be a feasible option as its usage renders a reduction of above 90% in emissions of total non-combusted hydrocarbons [3]. An excellent review has been published [4] describing various vegetable oil feedstock sources used for producing biodiesel. Biodiesel is mostly produced from the edible sources such as sunflower [5], soybean [6], and rapeseed [7]. This practice is not favorable towards sustainability because of the fact that it is competitive with food and would increase the cost of both edible oil and biodiesel [8]. Therefore, non-edible oils are preferred for this purpose. Castor oil is one of the most promising non-edible sources for the production of biodiesel [9]. Conventional in situ transesterification of castor oil seeds has been performed by Hincapie et al. [10]. Several catalysts such as KOH, NaOH, $KOCH_3$, HCl, etc. [11] have been tried in its transesterification with ethanol. Reduction in viscosity of castor methyl and ethyl ester blends have been studied [12]. Kinetics of this process has been investigated [13–17] by varying the alcohol to oil molar ratio as well as the temperature. In one such process [18], methanol to oil ratio was varied from 50:1 to 250:1 and temperature from 35 to 65 °C. Optimization of castor oil transesterification reaction to produce biodiesel has been reported by many authors, but optimization of tranesterification reaction by the use of ANN modeling is rare in literature. It may be concluded that these processes have not been investigated in the presence of an acid catalyst from the point of view of kinetic parameter estimation. ANN modeling can be a good approach to optimize acid catalyzed reaction for biodiesel production as it is time saving, and we can optimize biodiesel production on the pilot scale with the experimental data optimized by using ANN. The FFA content in castor oil should be less than 1% for using alkaline catalyst [19]. If the limit of FFA is exceeded, then soap formation occurs which inhibits the suppression of ester [20]. Boucher et al. [21] used significant (50 wt%) amount of $H_2SO_4$ associated with 10.86–93.7 methanol to FFA molar ratio for the oil–fatty acid mixture that contained 2–15% FFAs. Heterogeneous catalyst (sulphonated polystyrene compounds) were employed by Soldi et al. [22] and 85% conversion of fatty acid into methyl ester was reported at a high methanol to oil ratio of 100:1 and reaction time of 18 h. In the present investigation, experiments were conducted in a lab reactor; the experimental conditions were determined by design of experiment method for methanol to oil molar ratio = 6:1–25:1; catalyst amount (vol%) = 1–3; temperature (°C) = 40–60 °C; time duration of experiments = 4 h.

In recent years, artificial neural networks (ANNs) have been widely applied to a wide range of applications such as data prediction [23], fault detection [24], data rectification [25], process control [26], etc. It is capable of handling even incomplete data and is effective in executing fast predictions and even for non-linear generalizations [27, 28]. In 2011, performance of linear and non-linear calibration techniques were compared in predicting biodiesel properties from near IR spectra by Balabin et al. [29]. Evaluation of the intensification of biodiesel production from waste goat tallow was performed using response surface methodology (RSM) and ANN with appreciable modeling efficiency by Chakraborty et al. [30].

This study on castor oil transesterification in presence of acid catalyst aims to use ANN for modeling the experimental data obtained as decided by central composite design (CCD) and then predicting fractional formation profile of FAME at optimized conditions, determined by RSM. Further, the developed ANN model has been used for developing a kinetic model and estimating its rate constants by a suitable parameter estimation method. Broadly, the approach adopted here is to use the available experimental data for ANN model representation without imposing additional requirements on the number of process measurements. The available experimental data and those predicted using ANN model have been used to estimate the kinetic parameters.

## Materials and methods

Materials and methods as discussed in this section have been described in detail in the doctoral thesis of Payal [31] which is concerned with experimental and modeling studies of castor oil transesterification.

### Chemicals

Pharmaceutical grade castor oil was purchased from the local market through the authorized standard chemical supplier of the institute. All chemicals; $H_2SO_4$ (97%), anhydrous methanol, sodium bisulphate ($Na_2SO_4$) were of analytical grade and were purchased from Merck India and used without further purification. Pure standard methyl esters were purchased from Sigma Aldrich, USA.

### Experimental setup

A lab reactor (Applikon, Schiedam, The Netherlands, stirred type, capacity: 3 L, ez control system) was used for conducting the reaction. The reactor was 3 L double jacketed borosilicate glass reactor having 2 L working volume. The temperature was regularly monitored by the display of the system. The reflux system was used to avoid the vaporization of methanol from the reaction system. The temperature of the reaction system was controlled by the circulation of water through outer wall of the reactor vessel. Mechanical stirring was used for the proper mixing of the reaction mixture in the reactor.

### Experimental procedure

Initially 1 L castor oil was transferred into the reactor and heated till it reached the desired temperature. Appropriate quantities of both methanol and catalyst (sulphuric acid) were mixed thoroughly at the preset temperature separately. The above mixture of catalyst and methanol was then transferred to the reactor where the mixing was carried out at a speed equivalent to relative centrifugal force (rcf) = $32.2 \times g$ for the preset reaction temperature and time 4 h. The required mixing intensity (rcf = $32.2 \times g$) was optimized prior to conducting the experiments at constant methanol/oil molar ratio 6:1, catalyst concentration 1% and reaction temperature of 60 °C. The samples were taken up to 4 h regularly at the 20 min time interval and total 12 samples were collected during this time period. The samples were collected in 15 mL centrifuge tubes, filled with 5 mL of distilled water. Shaking and quenching of the samples were done immediately, and the centrifuge tubes with the sample were kept in an ice bath immediately. The samples were washed and centrifuged at rcf $= 894 \times g$ for 20 min to separate ester layer. After centrifuging the sample, two separate layers were formed; the bottom one contained the glycerol and catalyst in water phase while the upper layer is of ester.

## Gas chromatography analysis

Analysis of all samples for methyl ester formation (FAME content) was carried out using the gas chromatograph (GC) (Nucon Gas Chromatograph, 5765, India), equipped with a flame ionization detector (FID) and a capillary column with dimension of 0.55 mm I.D × 10 m length × 0.50 m thickness. Nitrogen was used as carrier gas. The column temperature was kept at 170 °C for 1 min, heated at 10 °C/min up to 240 °C and then it was maintained constant. The temperature of the injector and detector was set at 220 and 240 °C, respectively. Methyl heptadecanoate was used as internal standard for GC. The analysis was done by injecting 1 μL sample into the column. Every sample has been analyzed three times for % FAME yield and the average of three values has been taken.

Quantitative analysis of % ME was done using European standard EN 14103:2003. The % ME yield (or % FAME yield) was calculated using equation

$$\% \text{ FAME yield} = \frac{\sum A - A_{EI}}{A_{EI}} \times \frac{C_{EI} \times V_{EI}}{m} \times 100,$$

where $\sum A$ is the total peak area from methyl esters in $C_{14;0}$–$C_{24;1}$, $A_{EI}$ is the peak area corresponding to methyl heptadecanoate, $C_{EI}$ is the concentration of the methyl heptadecanoate solution (mg/mL), $V_{EI}$ is the volume of methyl heptadecanoate solution (mL) and $m$ is the mass of the sample (mg).

## Experimental design

Central composite design (CCD) is one of the most commonly used techniques for the optimization of experiments. Transesterification of castor oil with ethanol was studied by Cavalcante et al. [17] using a central composite rotatable design. Optimum reaction conditions were determined as oil/ethanol molar ratio of 1:11, catalyst amount of 1.75% KOH, and reaction time of 90 min and 86.32% of biodiesel yield was obtained. To apply this design, the variation

levels for each variable must be specified clearly. Accordingly, the variables are transformed into coded variables bearing the following relationships [32]:

$$X_j = (x_j - x_j^0)/\Delta x_j, \tag{1}$$

$$x_j^0 = \left(x_j^{\max} + x_j^{\min}\right)\Big/2, \tag{2}$$

$$\Delta x_j = \left(x_j^{\max} - x_j^{\min}\right)\Big/2, \tag{3}$$

where $X_j$: coded value of variable, $x_j^0$: basic level, $x_j$: actual value, $\Delta x_j$: level of variation.

Central composite design is a factorial design with center points, augmented with a group of axial points (also called star points). As the distance of the factorial point from the center of the design space is defined as ±1 unit for each factor, the distance of a star or axial point from the center of the design is ±α with (α) > 1. In this study, the CCD was used to optimize three parameters (methanol/oil molar ratio, catalyst amount, and temperature) for enhancing the % FAME yield. In CCD, the total number of experimental combination was $2^k + 2k + n$, where 'k' is the number of independent variables and 'n' is the number of repetitions of experiments at the central axis point to reduce the pure error [16, 33]. Totally, 20 experiments were required for this work. Here, the run numbers 18, 19, and 20 were not considered because the experiment was not feasible to be performed under these conditions as either catalyst amount or temperature was low [34] and 6 were repeated experiments. The dependent variables for this study were % FAME yield, $X_c$ (%) and the independent variables selected were: methanol/oil molar ratio ($x_1$), catalyst amount ($x_2$), and temperature ($x_3$). The range and levels of individual variable factor have been given in Table 1. The experimental data of the FAME yield for various catalyst amounts, molar ratios, and temperatures have been given in Table 2a, b. These data (columns A–J in Table 2a, b) have also been used to train the neural network for the kinetic modeling of the process. Properties of the castor oil FAME produced were also measured and are given in Table 3 [31, 35]. Except viscosity and water content, all other properties of the FAME were found to be satisfactory and within standard range of values [36]. Therefore, this biodiesel may be used for blending with other fuels.

| Table 1 Independent variables and levels used for CCD | S. no. | Variables | Coded symbols | Range and levels | | |
|---|---|---|---|---|---|---|
| | | | | −1 | 0 | +1 |
| | 1 | Methanol/oil molar ratio | $x_1$ | 6 | 15.5 | 25 |
| | 2 | Catalyst amount (vol%) | $x_2$ | 1 | 2 | 3 |
| | 3 | Temperature (°C) | $x_3$ | 40 | 50 | 60 |

**Tables 2** Experimental results on % FAME formation with time

| S. no. | Time (min) | % formation of FAME, $X_c$ | | | | | |
|---|---|---|---|---|---|---|---|
| | | A 1/60/25:1 | B 1/60/6:1 | C 3/60/25:1 | D 3/60/6:1 | E 1/40/25:1 | F 1/40/6:1 |
| (a) | | | | | | | |
| 1 | 0 | 0 | 0 | 0 | 0 | 0 | 0 |
| 2 | 20 | 7.50 | 13.81 | 31.50 | 16.88 | 5.32 | 5.95 |
| 3 | 40 | 9.51 | 21.46 | 40.79 | 20.58 | 6.95 | 7.80 |
| 4 | 60 | 12.96 | 24.93 | 47.06 | 24.63 | 8.45 | 9.35 |
| 5 | 80 | 16.82 | 28.81 | 54.59 | 26.75 | 12.23 | 10.54 |
| 6 | 100 | 21.59 | 31.92 | 59.79 | 29.04 | 13.94 | 12.09 |
| 7 | 120 | 32.04 | 35.96 | 63.06 | 31.59 | 16.94 | 15.15 |
| 8 | 140 | 36.90 | 39.79 | 67.05 | 37.95 | 18.45 | 18.99 |
| 9 | 160 | 40.32 | 41.71 | 69.09 | 44.79 | 20.18 | 21.95 |
| 10 | 180 | 46.65 | 42.05 | 71.09 | 48.09 | 22.65 | 24.65 |
| 11 | 200 | 53.05 | 44.19 | 73.53 | 50.05 | 26.94 | 26.56 |
| 12 | 220 | 62.18 | 46.63 | 75.59 | 52.76 | 28.42 | 28.19 |
| 13 | 240 | 72.55 | 48.29 | 76.85 | 55.27 | 31.45 | 29.95 |

| S. no. | Time (min) | % formation of FAME, $X_c$ | | | | | |
|---|---|---|---|---|---|---|---|
| | | G 3/40/25:1 | H 3/40/6:1 | I 2/50/15.5:1 | J 3.68/50/15.5:1 | K 3/45/12:1 | L 3/60/18:1 |
| (b) | | | | | | | |
| 1 | 0 | 0 | 0 | 0 | 0 | 0 | 0 |
| 2 | 20 | 7.23 | 16.28 | 9.93 | 33.45 | 18.41 | 20.50 |
| 3 | 40 | 9.59 | 19.15 | 14.56 | 38.56 | 23.95 | 34.16 |
| 4 | 60 | 12.81 | 23.92 | 21.72 | 42.69 | 26.49 | 40.15 |
| 5 | 80 | 18.12 | 24.56 | 26.92 | 46.44 | 28.85 | 45.96 |
| 6 | 100 | 22.42 | 26.65 | 30.59 | 51.56 | 30.91 | 50.55 |
| 7 | 120 | 23.59 | 27.05 | 34.49 | 55.69 | 33.18 | 54.72 |
| 8 | 140 | 28.65 | 28.53 | 38.95 | 58.91 | 35.39 | 57.86 |
| 9 | 160 | 31.95 | 29.75 | 43.65 | 60.08 | 37.98 | 60.98 |
| 10 | 180 | 34.03 | 30.07 | 45.96 | 62.15 | 40.18 | 63.85 |
| 11 | 200 | 36.95 | 33.05 | 47.56 | 63.95 | 42.59 | 65.97 |
| 12 | 220 | 44.15 | 35.84 | 49.62 | 65.67 | 45.19 | 66.89 |
| 13 | 240 | 47.36 | 37.71 | 52.25 | 67.89 | 46.15 | 67.56 |

1/60/25:1 means catalyst amount in % v/v, temperature in °C and methanol to oil molar ratio respectively

**Table 3** Properties of FAME produced from castor oil

| S. no. | Property | Determined in present study | Standard value/range Canoira et al. [36] |
|---|---|---|---|
| 1 | Kinematic viscosity (mm$^2$/s at 40 °C) | 18.55 | 3.5–5.00 |
| 2 | Density (kg/m$^3$) | 920.5 | 860–900 |
| 3 | Acid Value (mg KOH/g) | 0.25 | <0.50 |
| 4 | Flash point (°C) | 170 | >120 |
| 5 | Water content (%) | 0.006 | <0.0005 or <500 mg/kg |
| 6 | Calorific value (MJ/kg) | 39.5 | 37.27–38.22 |
| 7 | Oxidation stability at 110 °C (h) | 8 | >6.0 |
| 8 | Sulfur content (mg/kg) | 0.3 | <10 |

## Modeling of experimental observations through ANN

ANNs are useful for the study of complex phenomena for which we have appropriate data but a poor understanding of the mathematical relationship between them [37]. There are several modeling strategies in ANNs, which have various applications in designing and analyzing existing processes. Choosing the right network architecture is one of the most important tasks prior to the modeling process. A feed-forward neural network architecture was selected for the model development owing to its powerful non-linear mapping ability between inputs and outputs [38], which consisted of an input layer, an intermediate hidden layer and an output layer. In a typical feed-forward network every node in each layer is connected to all the nodes in the following layer. The neurons in the hidden layer often consist of sigmoidal neurons. A sigmoid function is a bounded differentiable real function that is defined for all real input values and has a positive derivative at each point. Tangent sigmoid function was preferred as the activation function in the hidden layer for all neural network computations.

$$\tan \text{sig}(x) = 2/(1 + \exp(-2x)) - 1. \tag{4}$$

A total of 156 experimental datasets were used to train (130 datasets; columns A–J in Table 2a, b) and to test (26 datasets; columns K and L in Table 2b) the performance of the developed ANN. Each dataset includes four input variables namely, methanol to oil molar ratio (mr), catalyst amount (c), temperature (T) and time (t). Scaling of the input and output variable is usually recommended [39]. Therefore, each input variable was scaled in the range {0, 1} by dividing them with their maximum values, respectively, in Table 4. These scaled variables were applied to the input neurons to carry out the ANN modeling. The available data was distributed into three different sets for the training method, i.e., 70% for the training set, 15% for the validation set and remaining 15% for the test set.

The Levenberg–Marquardt algorithm works on the principle of error-back propagation, and was used to train the neural network [40]. Mean square error (MSE) was chosen as the performance function for observing

deviations between experimental data and calculated data from the network's output layer.

For the present study, MSE was given by:

$$\text{MSE} = \sum_{k=1}^{N} \left( \frac{X_k^{\text{exp}}}{100} - \frac{X_k^{\text{cal}}}{100} \right)^2 \bigg/ N, \tag{5}$$

where $N$: number of datasets, $X_k^{\text{exp}}$: experimental value of the FAME percentage yield, $X_k^{\text{cal}}$: calculated value of the FAME percentage yield.

### Parameter estimation

Parameter estimation has been carried out for the data obtained experimentally and others generated by using ANN model. Several physical and kinetic parameters appear in the process model developed for transesterification of castor oil. The physical parameters, like the reactor volume and species densities were fixed based on the process knowledge. Each kinetic parameter has a pre-exponential factor and activation energy associated with it. The activation energies (E) have been fixed as the temperature (T) and catalyst concentration was taken to be the same for all the molar ratios.

$$k = k_0 e^{\left(-\frac{E}{\text{RT}}\right)}, \tag{6}$$

where $k$: reaction constant at given temperature ($T$), $k_0$: reaction constant at reference temperature.

Various parameter estimation techniques such as maximum likelihood estimation [41], prediction error minimization (PEM) [42], trust-region SQP algorithm [43] etc. have been used previously by researchers to estimate parameters of a model. Further, Hosten and Emig [44] have developed sequential experimental design procedures for precise parameter estimation in ordinary differential equations. In this work, parameter estimation technique based on a prediction error minimization (PEM) framework has been used. A central point in the PEM approach is to design a predictor, which is a function that returns a predicted value of the output of the system for given parameter values and a sequence of measurements. Comparison of the predicted value with an actual measurement gives the prediction error, which can be seen as a function of the system parameters. The prediction errors are squared and summed together. This function minimization with respect to the parameters is a way of finding good parameter estimates. Similar works have been performed by Mjalli and Ibrehem [42] and Mallikarjunan et al. [45] where they have used optimal hybrid modeling approach for polymerization reactors using PEM technique.

In this work, the parameters values were estimated using MATLAB 7.6 (R2008a). The MATLAB code consisted of optimization tool 'fminsearch'. The differential equation was solved symbolically using 'dsolve' command.

**Table 4** Maximum values of input and output variables

| S. no. | Variables | Maximum value |
|--------|-----------|---------------|
| 1 | Molar ratio (–) | 25:1 |
| 2 | Catalyst amount (% v/v) | 3.68 |
| 3 | Temperature (°C) | 60 |
| 4 | Time (min) | 240 |
| 5 | FAME yield (%) | 100 |

$$X_{c_{i,j}} = g(t_{i,j}, k_1, k_2, C_{A0_i}, M_i). \tag{7}$$

The function which has been minimized is as follows:

$$\sum_{i=1}^{N} \sum_{j=1}^{M} \times \left[ X_{c_{i,j}} - g(t_{i,j}, k_1, k_2, C_{A0_i}, M_i) \right]^2, \tag{8}$$

where $X_{c_{i,j}}$: values obtained experimentally and predicted through ANN, $M$: number of data points and $N$: number of molar ratios.

The function 'fminsearch' finds the minimum of a scalar function of several variables, starting at an initial estimate. This is generally referred to as unconstrained nonlinear optimization. 'fminsearch' uses the simplex search method. This is a direct search method that does not require numerical or analytic gradients. If $n$ is the length of $x$, a simplex in $n$-dimensional space is characterized by the $n + 1$ distinct vectors that are its vertices. In two-dimensional space, a simplex is a triangle; in three-dimensional spaces, it is a pyramid. At each step of the search, a new point in or near the current simplex is generated. The function value at the new point is compared with the function's values at the vertices of the simplex and, usually, one of the vertices corresponding to worst value of function is replaced by the new point, giving a new simplex. This step is repeated until the size of the simplex is less than the specified tolerance [46].

## Results and discussion

### Training of the ANN network

The feed-forward network was developed by training it with various number of combinations of sigmoidal neurons in one and two hidden layers. The MSE was calculated for all uni-layered architecture and a decreasing behavior was observed for the training MSE when hidden layer size was increased (Fig. 1a).

When hidden neurons are increased beyond a certain level, over fitting occurs and network adapts to the noisy training data [47]. Over fitting easily leads to the disturbance of the network and predictions often lie outside the range of considered variables in a multilayered network even if the input data are totally noise free [48].

Different weight initializations have been used to train to prevent the network from converging to a local minimum which gives erroneous results. The optimal value of the number of neurons employed in the single hidden layer was set at 12 beyond which training and testing MSE start to diverge uncontrollably on further increase in the number of hidden neurons. Since, we have four different scaled input variables and one variable (fractional formation of FAME) as our desired output, our chosen topology for the FF-ANN bears a 4–12–1 relationship (Fig. 2).

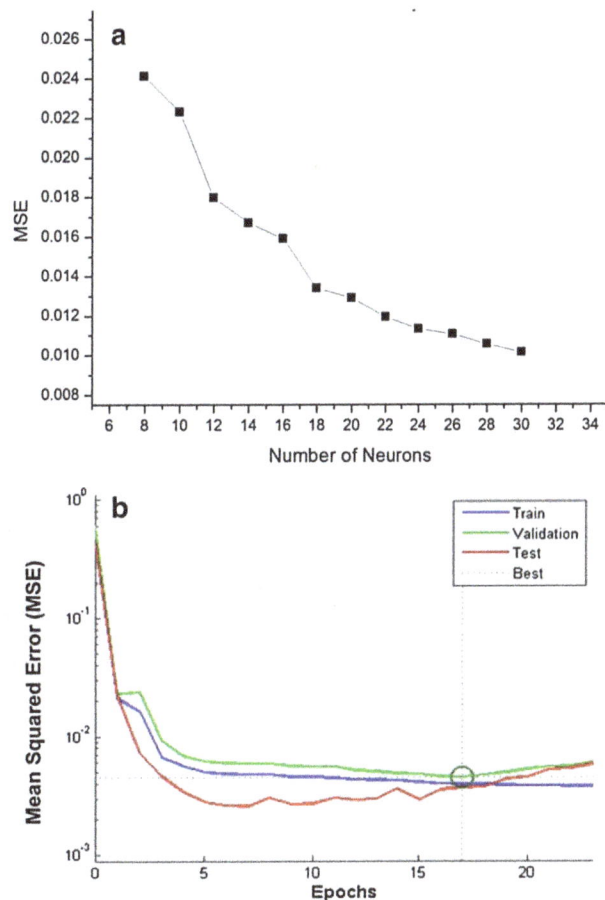

**Fig. 1 a** Training MSE behaviour for uni-layered topology. **b** Performance plot for the chosen architecture

Training of the chosen network was stopped when validation gradient started to overshoot. The best validation performance that gave the overall global minima of the system, was reported as 0.0045476 for the network at 17th epoch during the training (one epoch describes the time it takes to train the network once using the back propagation algorithm), Fig. 1b. The desired outputs were obtained from the output layer of the network.

An overall regression coefficient of 0.953 was obtained for the network which showed that the network had been satisfactorily trained (Fig. 3). A few outliers were observed which are common in any neural network development due to various factors such as experimental error, observational error etc. This trained network was further used for the validation and testing procedures (Sects. 3.2, 3.3).

### ANN model validation

The optimized ANN model was validated using two different sets of experimental data other than those used in developing ANN model having 13 data each. For this purpose, MATLAB 8.1(2013a) neural network toolbox [49] was

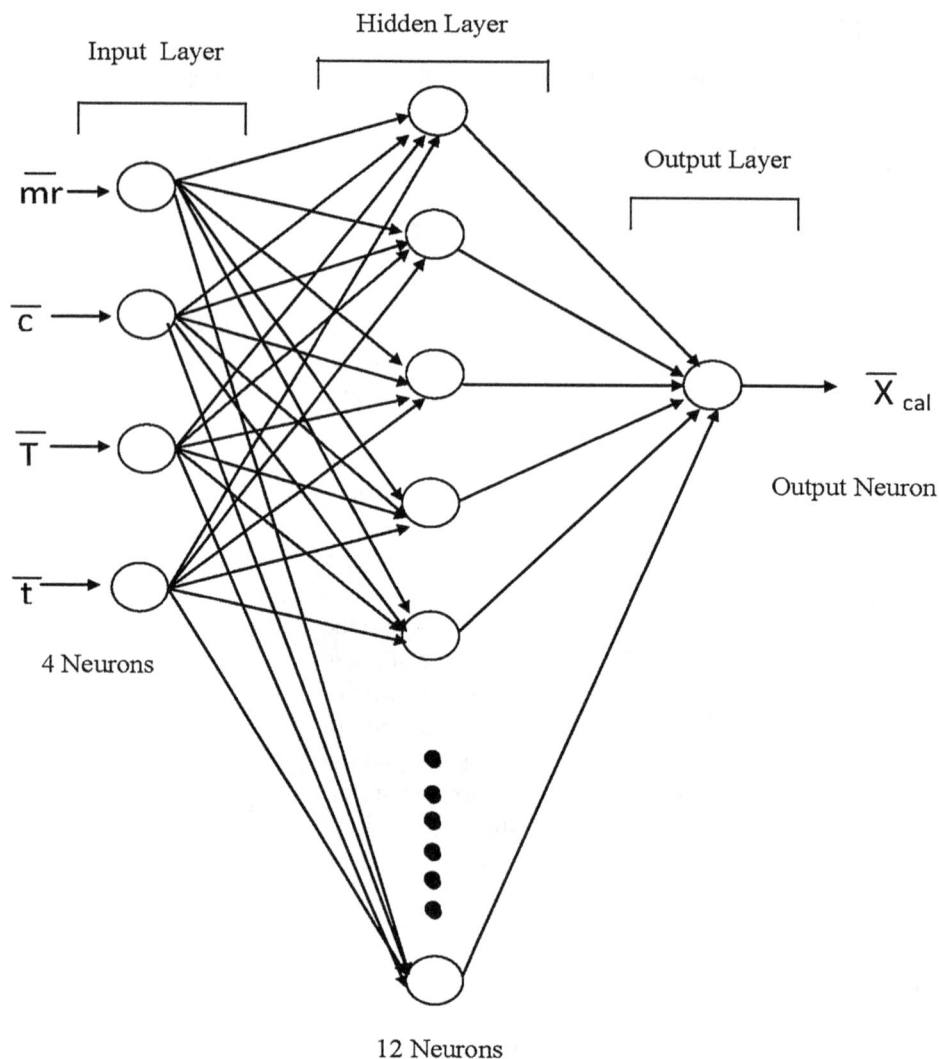

**Fig. 2** The 4–12–1 network topology used for modeling

employed. These additional data were obtained by conducting experiments on the same lab reactor at 45 and 60 °C, respectively (columns K and L in Table 2b). Comparison between ANN predictions and experimental data for these two separate validation sets show that all the model predictions are within $\pm 8\%$ deviation. The three-layered FF-ANN with 12 hidden neurons (4–12–1) was capable of predicting the output of the process and yielded a very good approximation to the castor oil transesterification process.

From the above investigation, it may be concluded that the network accounts well for the variation in input variables and the developed ANN model can be successfully applied in predicting the FAME yield at conditions in the range of considered variables.

**Predictions using ANN model**

The ANN model which was trained and tested against experimental data (see Sects. 3.1, 3.2) was used to predict

FAME yields at different input datasets. It was considered appropriate to develop a kinetic model at optimized conditions, determined earlier for the lab reactor used [31]. Therefore, predictions were made using the developed ANN model at the conditions (catalyst = 3% v/v, temperature = 60 °C, time = 0–4 h) to study the effect of molar ratio on FAME yield. Predictions for different molar ratios 9:1, 12:1, and 15:1 are shown in Fig. 4. The results obtained for these conditions from the ANN model have been used in the following Sect. (3.4) to develop the kinetic model.

**Kinetic modeling**

*General reaction model*

In the preceding section, profiles for fractional formation of FAME with time have been obtained using developed ANN model for methanol to oil molar ratios namely, 9:1,

**Fig. 3** Regression plots for the network for **a** training, **b** validation, **c** test, **d** overall

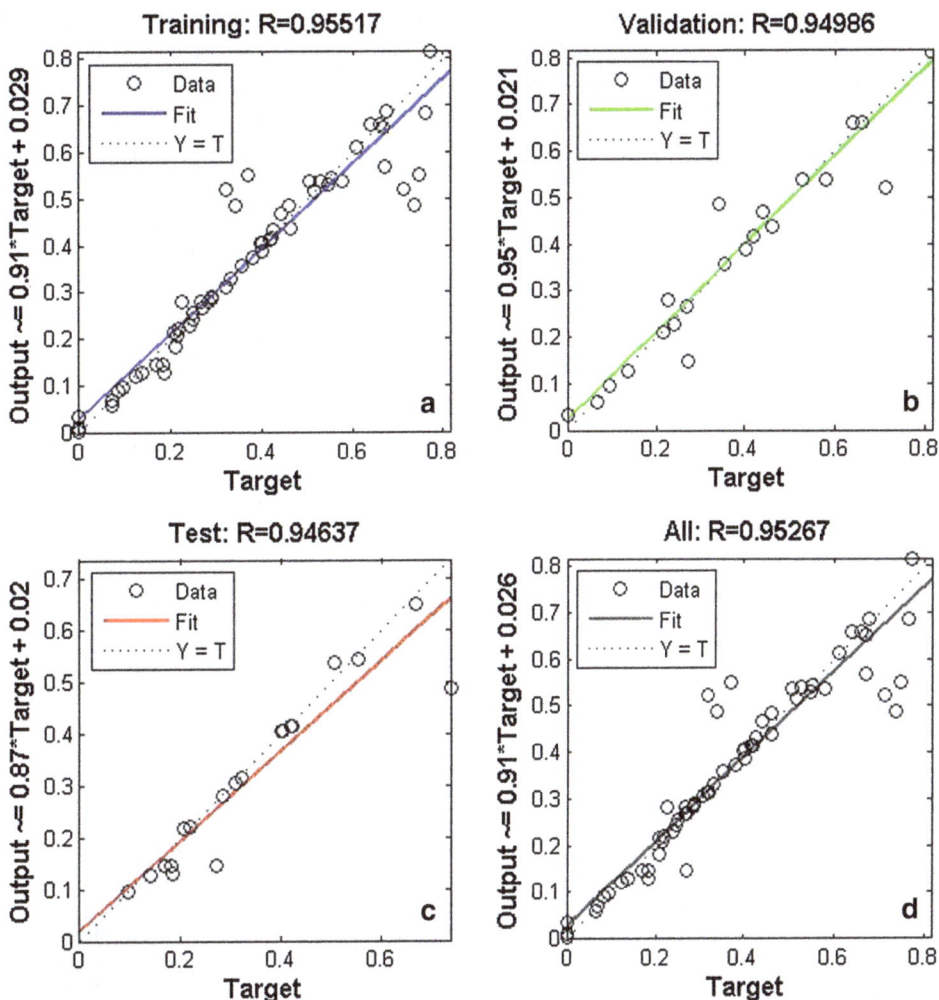

**Fig. 3** Regression plots for the network for **a** training, **b** validation, **c** test, **d** overall

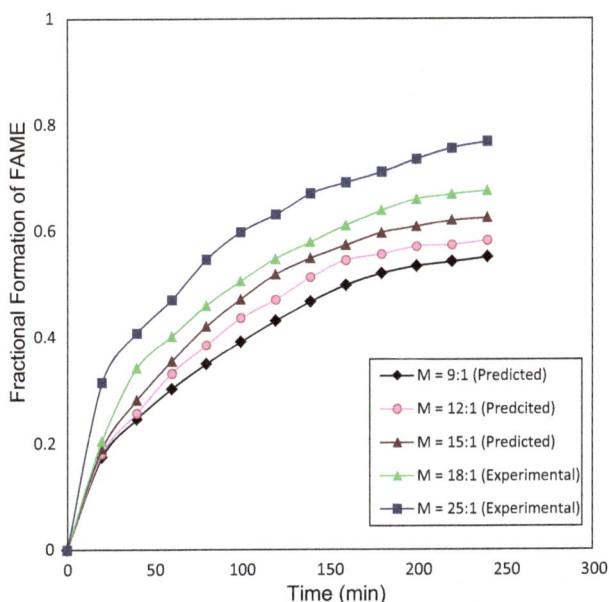

**Fig. 4** ANN model predictions for intermediate molar ratios

12:1, and 15:1, and experimental profiles for molar ratio 18:1 and 25:1 are already available at optimum catalyst amount and temperature. This section concerns the development of suitable kinetic model on the basis of these five profiles so that effect of change in molar ratio on fractional formation of FAME can be studied. The $C_{A0}$ values corresponding to different molar ratios used for fitting are given in the Table 5.

Following kinetic model was used to fit the data obtained by ANN. The kinetic model is in general reversible reaction model, in which both forward and backward reactions are second order, first order with respect to each of the reactants and products.

$$\mathrm{d}X_c/\mathrm{d}t = k_1 C_{A0}(3 - X_c)(M - X_c) - k_2 C_{A0} X_c^2, \tag{9}$$

with $X_c = 0$ at $t = 0$. Where $C_{A0}$ initial concentration of castor oil (mol/L), $X_c$: fractional formation of FAME, $M$: molar ratio of methanol to castor oil, $k_1$, $k_2$: rate constants (L min/mol).

**Table 5** $C_{A0}$ values corresponding to different molar ratios

| S. no. | Molar ratio (M) | $C_{A0}$ (gmol/L) | Catalyst (vol%/vol) |
|---|---|---|---|
| 1 | 9:1 | 0.733668 | 3% |
| 2 | 12:1 | 0.67393 | 3% |
| 3 | 15:1 | 0.622769 | 3% |
| 4 | 18:1 | 0.57927 | 3% |
| 5 | 25:1 | 0.49771 | 3% |

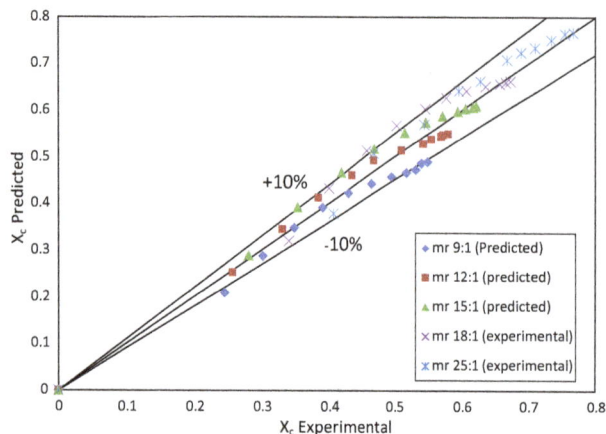

**Fig. 5** Comparison of $X_c$ predicted by kinetic model with $X_c$ experimental

*Estimation of parameters*

The values of $k_1$ and $k_2$ were estimated using the procedure discussed in Sect. 2. The values estimated are $k_1 = 0.000262$ and $k_2 = 0.023178$ (L min/mol), respectively. The correlation coefficients and normalized % standard deviations for different molar ratios i.e. 9:1, 12:1, 15:1, 18:1 and 25:1 were obtained as (0.933453, 8.783124%), (0.989036, 3.637967%), (0.994593, 2.760923%), (0.983285, 4.831601%), and (0.974017, 5.589103%), respectively. Higher values of correlation coefficient (>0.93) and lower values of normalized % standard deviation (<9%) testify the goodness of fit of the kinetic model. To compare the accuracy of model, all the predicted values of fractional formation of FAME, $X_c$ Predicted, by kinetic model, and its experimental values and generated values by ANN model, $X_c$ experimental, have been plotted in Fig. 5 for all five molar ratios. From this figure, it is obvious that most of the predictions lie within ±10% deviation.

## Conclusion

In this research work, experiments have been conducted at 12 sets of operating conditions. A FF-ANN model (4–12–1) has been developed on the basis of these data. The model is

capable of predicting fractional formation of FAME ($X_c$) at other operating conditions. In this work, $X_c$ have been computed using the developed ANN model at 9:1, 12:1 and 15:1 methanol to oil molar ratio at 60 °C and 3% catalyst loading. Further, a kinetic model has also been developed using the experimental and computed data at 60 °C and 3% catalyst loading. The kinetic model considers the transesterification reaction to be a second order reversible, first order with respect to each of the reactants and products, and is applicable to methanol to oil molar ratios varying between 9:1 and 25:1. Estimated values of kinetic constants $k_1$ and $k_2$ are 0.000262 and 0.023178 L min/mol, respectively.

**Acknowledgements** Authors are grateful to the Ministry of Human Resource Development (MHRD), Government of India, New Delhi for providing the financial assistance during the course of this research work.

## References

1. Oliveira FCC, Brandao CRR, Ramalho HF, da Costa LAF, Suarez PAZ, Rubim JC (2007) Adulteration of diesel/biodiesel blends by vegetable oil as determined by Fourier transform (FT) near infrared spectrometry and FT-Raman spectroscopy. Anal Chim Acta 587:194–199
2. Baptista P, Felizardo P, Menezes JC, Correia MJN (2008) Multivariate near infrared spectroscopy models for predicting the iodine value, CFPP, kinematic viscosity at 40 C and density at 15 C of biodiesel. Talanta 77:144–151
3. Knothe G, Sharp CA, Ryan TW (2006) Exhaust emissions of biodiesel, petrodiesel, neat methyl esters, and alkanes in a new technology engine. Energy Fuel 20:403–408
4. Leung DYC, Wu X, Leung MKH (2010) A review on biodiesel production using catalyzed transesterification. Appl Energy 87:1083–1095
5. Porte AF, CdS Schneider Rd, Kaercher JA, Klamt RA, Schmatz WL, da Silva WLT (2010) Sunflower biodiesel production and application in family farms in Brazil. Fuel 89:3718–3724
6. Gomes MCS, Arroyo PA, Pereira NC (2011) Biodiesel production from degummed soybean oil and glycerol removal using ceramic membrane. J Membr Sci 378:453–4561
7. Xin J, Imahara H, Saka S (2008) Oxidation stability of biodiesel fuel as prepared by supercritical methanol. Fuel 87:1807–1813
8. Kansedo J, Lee KT, Bhatia S (2009) *Cerbera odollam* (sea mango) oil as a promising non-edible feedstock for biodiesel production. Fuel 88:1148–1150
9. Berman P, Nizri S, Wiesman Z (2011) Castor oil biodiesel and its blends as alternative fuel. Biomass Bioenergy 35:2861–2866
10. Hincapie H, Mondragon F, Lopez D (2011) Conventional and in situ transesterification of castor seed oil for biodiesel production. Fuel 90:1618–1623
11. Meneghetti SMP, Meneghetti MR, Wolf CR, Silva EC, Lima GES, Coimbra MDA, Soletti JI, Carvalho SHV (2006) Ethanolysis of castor and cottonseed oil: a systematic study using classical catalysts. JAOCS 83:819–822
12. Thomas TP, Birney DM, Auld DL (2012) Viscosity reduction of castor oil esters by the addition of diesel, safflower oil esters and additives. Ind Crops Prod 36:267–270
13. Pradhan S, Madankar CS, Mohanty P, Naik SN (2012) Optimization of reactive extraction of castor seed to produce biodiesel using response surface methodology. Fuel 97:848–855

14. Ramezani K, Rowshanzamir S, Eikani MH (2010) Castor oil transesterification reaction: a kinetic study and optimization of parameters. Energy 34:4142–4148
15. Lopez JM, Garcia Cota TDNJ, Monterrosas EEGR, Martineza N, de la Cruz Gonzalez VM, Flores JLA, Ortega YR (2011) Kinetic study by 1H nuclear magnetic resonance spectroscopy for biodiesel production from castor oil. Chem Eng J 178:391–397
16. Jeong GT, Park DH (2009) Optimization of biodiesel production from castor oil using response surface methodology. Appl Biochem Biotechnol 156:431–441
17. Cavalcante KSB, Penha MNC, Mendonça KKM, Louzeiro HC, Vasconcelos ACS, Maciel AP, de Souza AG, Silva FC (2010) Optimization of transesterification of castor oil with ethanol using a central composite rotatable design (CCRD). Fuel 89:1172–1176
18. Madankar CS, Pradhan S, Naik SN (2013) Parametric study of reactive extraction of castor seed (Ricinus communis L.) for methyl ester production and its potential use as bio lubricant. Ind Crops Prod 43:283–290
19. Tiwari AK, Kumar A, Raheman H (2007) Biodiesel production from jatropha oil (Jatropha curcas) with high free fatty acids: an optimized process. Biomass Bioenergy 31:569–575
20. Canakci M, Gerpen JV (2001) Biodiesel production from oils and fats with high free fatty acids. Trans ASAE 44:1429–1436
21. Boucher MB, Unker SA, Hawley KR, Wilhite BA, Stuart JD, Parnas RS (2008) Variables affecting homogeneous acid catalyst recoverability and reuse after esterification of concentrated omega-9 polyunsaturated fatty acids in vegetable oil triglycerides. Green Chem 10:1331–1336
22. Soldi RA, Oliveira ARS, Ramos LP, Cesar-Oliveira MAF (2009) Soybean oil and beef tallow alcoholysis by acid heterogeneous catalysis. Appl Catal A Gen 361:42–48
23. Barton R (1995) Analysis and rectification of data from dynamic chemical processes via artificial neural networks. Ph. D Dissertation, The University of Texas at Austin, Austin, Texas
24. Suewatanakal W (1993) A comparison of fault detection and classification using ANN with traditional methods. Ph. D. Dissertation, The University of Texas at Austin, Austin, Texas
25. Karjala TW (1995) Dynamic data rectification via recurrent neural networks. Ph. D Dissertation, The University of Texas at Austin, Austin, Texas
26. MacMurray JC(1993) Modelling and control of a packed distillation column using artificial neural networks. M.S. Thesis, The University of Texas at Austin, Austin, Texas
27. Rajendra M, Jena PC, Raheman H (2009) Prediction of optimized pretreatment process parameters for biodiesel production using ANN and GA. Fuel 88:868–875
28. Daponte P, Grimaldi D (1998) Artificial neural networks in measurements. Measurement 23:93–115
29. Balabin RM, Lomakina EI, Safieva RZ (2011) Neural network (ANN) approach to biodiesel analysis: analysis of biodiesel density, kinematic viscosity, methanol and water contents using near infrared (NIR) spectroscopy. Fuel 90:2007–2015
30. Chakraborty R, Sahu H (2013) Intensification of biodiesel production from waste goat tallow using infrared radiation: process evaluation through response surface methodology and artificial neural network. Appl Energy 114:827–836
31. Payal PhD (2014) Thesis on castor oil transesterification: experimental and modelling studies. Indian Institute of Technology Roorkee, Roorkee
32. Lazíc ZR (2004) Design of experiments in chemical engineering. Wiley, Oxford
33. Yuan X, Liu J, Zeng G, Shi J, Tong J, Hunag G (2008) Optimization of conversion of waste rapeseed oil with high FFA to biodiesel using response surface methodology. Renew Energy 33:1678–1684
34. Kilic M, Uzun BB, Putun E, Putun AE (2013) Optimization of biodiesel production from castor oil using factorial design. Fuel Process Technol 111:105–110
35. Chaudhary P, Kumar B, Kumar S, Gupta VK (2015) Transesterification of castor oil with methanol—kinetic modeling. Chem Prod Process Model 10(2):71–80
36. Canoira L, Galean JG, Alcantara R, Lapuerta M, Contreras RG (2010) Fatty acid methyl esters (FAMEs) from castor oil: production process assessment and synergistic effects in its properties. Renew Energy 35:208–217
37. Gorr WL, Nagin D, Szczypula J (1994) Comparative study of artificial neural network and statistical models for predicting student grade point averages. Int J Forecast 10:17–34
38. Haykin S, Networks Neural (1999) A comprehensive foundation, 2nd edn. Prentice-Hall, Upper Saddle River
39. Hojjat H, Etemad SG, Bagheri R, Thibault J (2011) Thermal conductivity of non-Newtonian nanofluids: experimental data and modeling using neural network. Int J Heat Mass Transf 54:1017–1023
40. Hagan MT, Menhaj M (1994) Training feed forward networks with the Marquardt algorithm. IEEE Trans Neural Netw 5:989–993
41. Bindish R, Rawlings JB (2003) Parameter estimation for industrial polymerization processes. AIChE J 49:2071–2078
42. Mjalli FS, Ibrehem AS (2011) Optimal hybrid modelling approach for polymerization reactors using parameter estimation techniques. Chem Eng Res Des 89:1078–1087
43. Arora N, Biegler LT (2004) Parameter estimation for a polymerization reactor model with a composite-step trust-region NLP algorithm. Ind Eng Chem Res 43:3616–3631
44. Hosten LH, Emig G (1975) Sequential experimental design procedures for precise parameter estimation in ordinary differential equations. Chem Eng Sci 30:1357–1364
45. Mallikarjunan V, Pushpavanam S, Immanuel CD (2010) Parameter estimation strategies in batch emulsion polymerization. Chem Eng Sci 65:4967–4982
46. Lagarias JC, Reeds JA, Wright MH, Wright PE (1998) Convergence properties of the Nelder–Mead simplex method in low dimensions. SIAM J Optim 9(1):112–147
47. Cavas L, Karabay Z, Alyuruk H, Dogan H, Demir GK (2011) Thomas and artificial network models for the fixed-bed adsorption of methylene blue by a beach waste Posidonia oceanic (L.) dead leaves. Chem Eng J 171:557–562
48. Himmelblau DM (2008) Accounts of experiences in the application of artificial neural networks in chemical engineering. Ind Eng Chem Res 47:5782–5796
49. Beale M, Hagan M, Demuth H (2013) MATLAB Neural network toolbox user's guide (R2013A)

# 10

# Eco-friendly treatment of textile dye from aqueous solution using encapsulated biosorbent matrix beads: kinetics and breakthrough analysis

Radhakrishnan Kannan[1] · Sethuraman Lakshmi[1] · Natarajan Aparna[1] · Sivaraman Prabhakar[2] · Wilson Richard Thilagaraj[1]

**Abstract** In this study, *Moringa oleifera*-encapsulated alginate beads (MEA) were synthesized using Ca-alginate for the removal of anionic blue-FFS (AB-FFS) dye from an aqueous solution. The synthesized beads were characterized by scanning electron microscopy (SEM), Fourier transform infrared spectroscopy (FT-IR) and Brunauer–Emmett–Teller (BET) methods. The influences of initial pH, initial dye concentration and temperature were established in batch mode. The equilibrium biosorption obeyed both the Langmuir and Freundlich isotherm models with $Q_{max}$ of MEA of about 76.92 mg/g at 343 K. Based on thermodynamic data, it was found that the biosorption process was endothermic and spontaneous. Fixed bed column studies were conducted using different parameters such as flow rate and bed height of the biosorbent. The data obtained were fitted with well-established models, namely Thomas and Yoon–Nelson. Desorption of beads could be achieved by a minimum of five successive cycles without significant loss of initial dye concentration in the batch mode. Experimental results indicate that MEA appears to be a promising biosorbent material used for treating the textile wastewater.

**Keywords** Breakthrough curve · Column · Encapsulation · Isotherms · Kinetics

✉ Wilson Richard Thilagaraj
thilagaraj.richard@gmail.com

[1] Department of Biotechnology, School of Bioengineering, SRM University, Kattankulathur 603203, Tamil Nadu, India

[2] Department of Chemical Engineering, School of Bioengineering, SRM University, Kattankulathur 603203, Tamil Nadu, India

## Introduction

Many dyestuffs were used in textile dyeing and printing processes, which are non-biodegradable. Among the dyes, anionic dyes are carcinogenic and have the highest risk of diseases due to its toxic in nature. The presence of ethylene glycol in AB-FFS dye is responsible for producing a variety of illnesses such as proteinuria, uremia and anuria which are toxic to the liver and kidney [1, 2]. In recent days, treating the anionic dyes becomes a big challenge and several methods failed to produce satisfactory results.

Based on previous reports, adsorption methods do offer the most efficient technique used for the removal of dyes and pigments from industrial effluents. Adsorbents such as activated carbon [3], pith carbon [4], bagasse fly ash [5], rice husk [6] and cashew nut shell [7] have been used for dye decolorization under the batch mode. All these adsorbents had limited success over the industrial effluent treatments. Moreover, equilibrium data from batch mode provides fundamental information, but not applicable for real-time process. Therefore, there is a need to investigate the fixed bed column study to obtain engineering data for scaling up from laboratory-scale to pilot-scale level. Recently, a variety of natural adsorbents such as pistachio hull waste [8], dried *Moringa oleifera* seeds [9], orange peel cellulose [10] and *Artocarpus heterophyllus* [11] have been reported for the removal of heavy metals and dyes.

Earlier studies have confirmed that *Moringa oleifera* seed as biosorbent, which has polypeptides of molecular weight ranging from 6 to 16 kDa and its isoelectric pH value 10, contains high zeta potential, and cationic and anionic properties. Considering these properties, *Moringa oleifera* could be used as a good biosorbent when it is encapsulated. Our previous studies have reported that encapsulated *Moringa oleifera* beads are effectively used for the removal of heavy metals [9]. Hence, this present study is focused on the performance of encapsulated *Moringa oleifera* seed powder by cross-linking with polymer Ca-alginate, to obtain stable beads and used for its biosorption efficiency for AB-FFS solution. The biosorbent was investigated in batch as well as fixed bed column packed with MEA with respect to flow rate and bed height. Eventually, the isotherm, kinetics and breakthrough curve analysis for the removal of AB-FFS dyes were investigated.

## Materials and methods

### Preparation of dye solutions

The Anionic Blue-FFS dye was manufactured by Dynamic Industries Ltd, India. It is an anionic dye which is referred to as Anionic Blue-15 with a chemical formula: $C_{42}H_{46}N_3NaO_6S_2$ and molecular weight: 775.95. This product is commercially available as CI No: 42665. The chemical structure of the Anionic Blue-FFS dye has attributed to Lecotan Blue-AC, Triacid Blue-B, Raviramine Blue-BS and Dinacid Brilliant Blue-B, which is shown in Fig. 1. Synthetic AB-FFS dye solutions were prepared by adding the desired amount of dye in double-distilled water to obtain a stock solution of 1000 mg/L. The working solution was diluted to the required concentration of the experiments.

### Preparation of biosorbent

Dried *Moringa oleifera* seeds were de-shelled and fresh white kernels were washed several times with de-ionized water and air dried. Then, kernels were pulverized and converted into a fine powder with an approximate size range from 300 to 500 μm. MEA were prepared using 2 % of alginate solution was mixed with kernel powder in the ratio of 1:5 to form a suspension solution which was converted

into Ca-alginate beads by a phase inversion technique using 3 % of calcium chloride solution with the help of the peristaltic pump. The flow rate of the pump was adjusted to 250 spherical beads per hour. After encapsulation, MEA in the calcium chloride solution was incubated at room temperature for 2 h for stability, which led to the complete replacement of sodium ions by calcium ions. Eventually, the beads were dried and washed with 99 % of ethanol for complete sterilization for storage long periods [9]. In addition, the physical appearances and properties of the MEA have been characterized such as BET surface area, porosity, bulk density and stability as shown in Table 1.

### Instrumental analysis

The surface morphology of MEA was identified by scanning electron microscopy (SEM) using an Agilent Technology scanning probe microscope. The surface functional groups of MEA were determined by Fourier transform infrared reflection (FTIR) measurements using an ALPHA FT-IR spectrometer. The zeta potential curve of the MEA was analyzed using a (Zetasizer 2000) Malvern Zeta meter.

### Batch biosorption studies

Batch biosorption experiments were carried out in 250 ml Erlenmeyer flasks containing 100 ml of AB-FFS initial dye concentrations (25, 50, 75 and 100 mg/L) with the desired amount of MEA (1 g/100 mL). The effect of pH was investigated on the dye in the range of 2–10 pH and the effect of temperature was investigated in the range of 323–343 K. The pH of the solution was adjusted using 0.1 M of HCl and 0.1 M of NaOH solutions. The flasks were agitated in the orbital shaker (Orbitek, Scigenics Biotech) at 100 rpm. Samples were collected at regular intervals. The residual concentration of dyes in the supernatant of each sample was measured using UV–VIS spectrometer at 560 nm using JASCO, UV–670. One set of experiment was conducted using Ca-alginate beads, which is free from *M.oleifera*. The equilibrium dye adsorbate $q_e$ (mg/g) was calculated using the following formula (1):

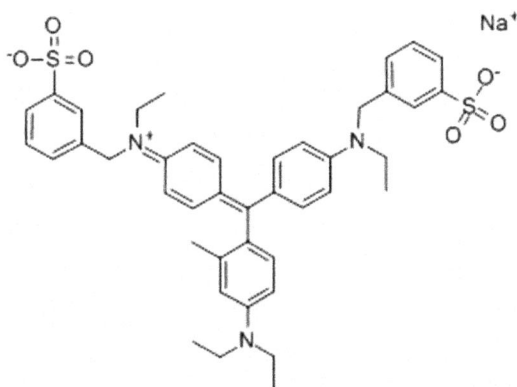

**Fig. 1** Structure of anionic blue-FFS dye

**Table 1** The physical appearance and properties of MEA beads

| Parameters | Values |
| --- | --- |
| Porosity (%) | 40–45 |
| Bulk density (g/cm$^3$) | 1.5 |
| Stability | Stable and rigid |
| Mean diameter (mm) | 1.75–2.25 |
| BET surface area (m$^2$/g) | 14.5 |
| $q_{max}$ (mg/g) | 0.0167 |
| Pore volume (cm$^3$/g) | 0.05 |

$$q_e = (C_o - C_e/m)V, \tag{1}$$

where $C_o$ and $C_e$ are the initial and final dye concentration in mg/L; $V$ is the volume of the dye solution (L); $m$ is the dry weight of the biosorbent (g).

## Fixed bed column studies (FBC)

The fixed bed column reactor was made up of Perspex tubes of 4.5 cm internal diameter and 55 cm in height and the beads (MEA) packed with different heights such as 5, 10 and 15 cm. In this experiment, a known concentration of AB-FFS dye was pumped at different flow rates to a known height of the adsorbent. The particle size of the adsorbent used in the experiment was $2.0 \pm 0.1$ mm. The samples were collected from the column at regular intervals and the absorbance of the color was measured using a UV–VIS spectrometer at 560 nm. The continuous process flow diagram is shown in Fig. 2.

## Computational statistics and mathematical modeling

Batch experiments were analyzed in triplicate ($N = 3$) and the data reveals the mean values which represent the error bars. Regression, the correlation coefficient, standard deviation was calculated using SSPC PC + TM statistical package, 1983. Multiple mean comparisons using least significant difference were computed using the significance level ($p < 0.05$). To investigate the data, two inherent

models like Langmuir and Freundlich were used to describe the mechanism of biosorption and to find the maximum biosorption capacity of the adsorbents at different temperatures. The results from the effect of temperature were used to calculate the thermodynamic parameters such as free energy, enthalpy and entropy. Eventually, to access the kinetic parameters, the data with different concentrations results were analyzed using pseudo-first-order, pseudo-second-order and intraparticle diffusion models. To consider, the best fit model to describe the performance of the breakthrough curves, regression coefficients ($R^2$) can be fit between experimental and theoretical values of Thomas and Yoon–Nelson equations.

## Desorption studies

For batch desorption experiments, deionized water and 0.1 M of analytical-grade $NH_4Cl$, EDTA, $CH_3COOH$, HCl, $HNO_3$ and NaOH were used. In this study, a series of 250 ml of Erlenmeyer flasks containing 100 mL of desorption solution was mixed with 10 g of dye-loaded MEA at room temperature. The mixed solution was agitated in an orbital shaker (Orbitek, Scigenics biotech) at 200 rpm for 1 h. The MEA was removed from the solution and centrifuged at 1000 rpm for 5 min and the desorbed concentrations were analyzed using a UV-spectrometer. After each cycling experiments, beads were washed with distilled water thrice and then treated again in 100 mg/L concentration of AB-FFS dye.

**Fig. 2** Continuous process flow diagram

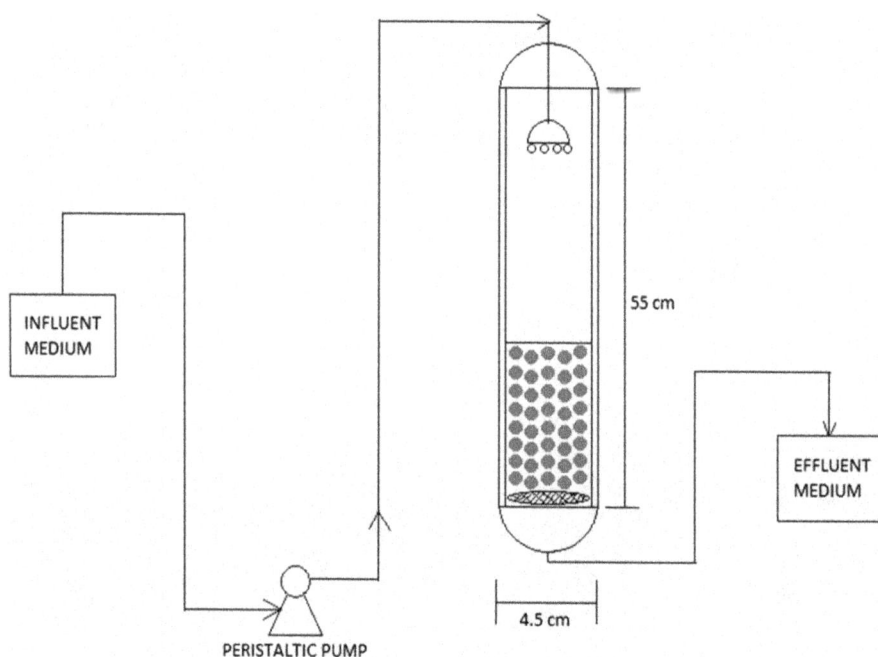

# Results and discussion

## Characterization of biosorbent

The morphology of MEA was investigated before and after biosorption using SEM analysis. From Fig. 3a, the micrograph of dried MEA reveals, the presence of the porous matrix structure which has a rough and a heterogeneous surface area for biosorption. The same bead was examined after biosorption as shown in Fig. 3b. Almost all the porous matrix structures completely disappeared and it clearly depicts that the functional groups present in the MEA provide active sites for AB-FFS to bind and there is much difference in the adsorbent structure (i.e., absence of the porous matrix).The FTIR spectra of MEA before and after biosorption were analyzed as shown in Fig. 3c. A characteristic band at 2902 and 2862 $cm^{-1}$ is attributed to C–H of $CH_2$ groups of proteins and at 1800–1600 $cm^{-1}$ is attributed to the C=O bond stretching that defines the presence of a carbonyl group of the fatty acid [12]. The peak that appears at 1627 $cm^{-1}$ may be attributed to the N-H (amine) groups present in the protein portions of the MEA. On the other hand, an FTIR spectrum of MEA after biosorption shows the disappearance of the two peaks, which is due to the mechanism of cross-linking of MEA which could be either by de-protonation or ionic interaction [9]. In addition, the point of zero charge ($pH_{zpc}$) was examined for MEA beads, and the results showed that the $pH_{zpc}$ is 4.9–5.1 as shown in Fig. 3d. It can be used to explain the influence of pH on biosorption. When the pH value is lower than $pH_{zpc}$, the surface charge of the MEA is positive and hence the biosorption capacity decreases. Similarly, the value of pH is greater than $pH_{zpc}$, and the surface charge of the MEA has a higher neagative charge, which results in higher attraction of the dyes.

## Batch biosorption studies

### Effect of initial pH and pHzpc

The pH varies the surface charge properties of the reactant so that the adsorbate or adsorbent can bind to the active site to undergo reactions. The effect of pH on the biosorption capacity of the AB-FFS dye with concentration of 100 mg/L was studied in the range of 2–10. The influence of pH on the biosorption can be explained on the basis of point of zero charge ($pH_{zpc}$), which is the point at which the net charge of the biosorbent is zero. From Fig. 4a, it can be seen that the degree of AB-FFS biosorption onto MEA decreased from 95.01 to 17.53 % when the pH was increased from 2 to 10. It can be explained by the pHzpc of the adsorbent and the nature of the AB-FFS dye (anionic). The $pH_{zpc}$ of the MEA

is around 4.9–5.1, which explains that the surface of the adsorbent is positively charged at under pH 5.1 [9]. At lower pH, large amount of $H^+$ ions is present on the surface of the biosorbents, which favors strong electrostatic attraction and hence maximun rate of biosorption. As the increasing pH (>5) value, the biosorption decreased gradually due to the $OH^-$ ions forms the negative charge on the surface of the adsorbents which do not bind to the anionic adsorbate, thus reduces the rate of biosorption due to the electrostatic repulsion [13]. Thus, a maximum AB-FFS biosorption was attained at an optimum pH of 2. Further, the rest of the experiments were conducted at pH 2.

### Effect of initial dye concentration

The study of initial dye concentration influences the performance of MEA and provides information about driving force to overcome all mass transfer resistance between the solid and aqueous phase. The biosorption of AB-FFS concentration was investigated at 25–100 mg/L up to 500 min for all the concentrations using 1.0 g of MEA at room temperature under pH 2. From Fig. 4b, the equilibrium biosorption was observed at 300 min and was found to be a maximum of 98.5 ± 0.5 % for 25 mg/L and minimum of 90.85 ± 1.25 % for 100 mg/L, respectively. At lower AB-FFS concentrations, the dye molecules present in the aqueous solution interacted with the active sites available on the solid surface, facilitating higher biosorption. Consequently, the increase of initial AB-FFS concentration resulted in a reduction of its biosorption percentage due to the limited number of binding sites and more dye molecules were left unabsorbed in the solution resulting in lower rate of biosorption [8, 14].

### Effect of temperature and thermodynamic studies

The effect of temperature is an important biosorption parameter to investigate the stability of MEA. From Fig. 4c, it was found that biosorption rate was increased with the increase in temperature. As the temperature increased from 323 to 343 K, it was observed that biosorption porosity increased with a decrease in viscosity of the solution and thereby increased the diffusion rate [8]. Endothermic chemical interaction and disaggregation lead to an increase in temperature. From this experiment, it was found that MEA biosorption was quite good and stable even at higher temperature [15, 16]. Thermodynamic parameters such as Gibbs free energy change ($\Delta G^o$), standard enthalpy change ($\Delta H^o$) and standard entropy change ($\Delta S^o$) are calculated to understand more about the effect of temperature on the biosorption. Biosorption experiments were calculated at different temperatures (323, 333 and 343 K) using the following equations:

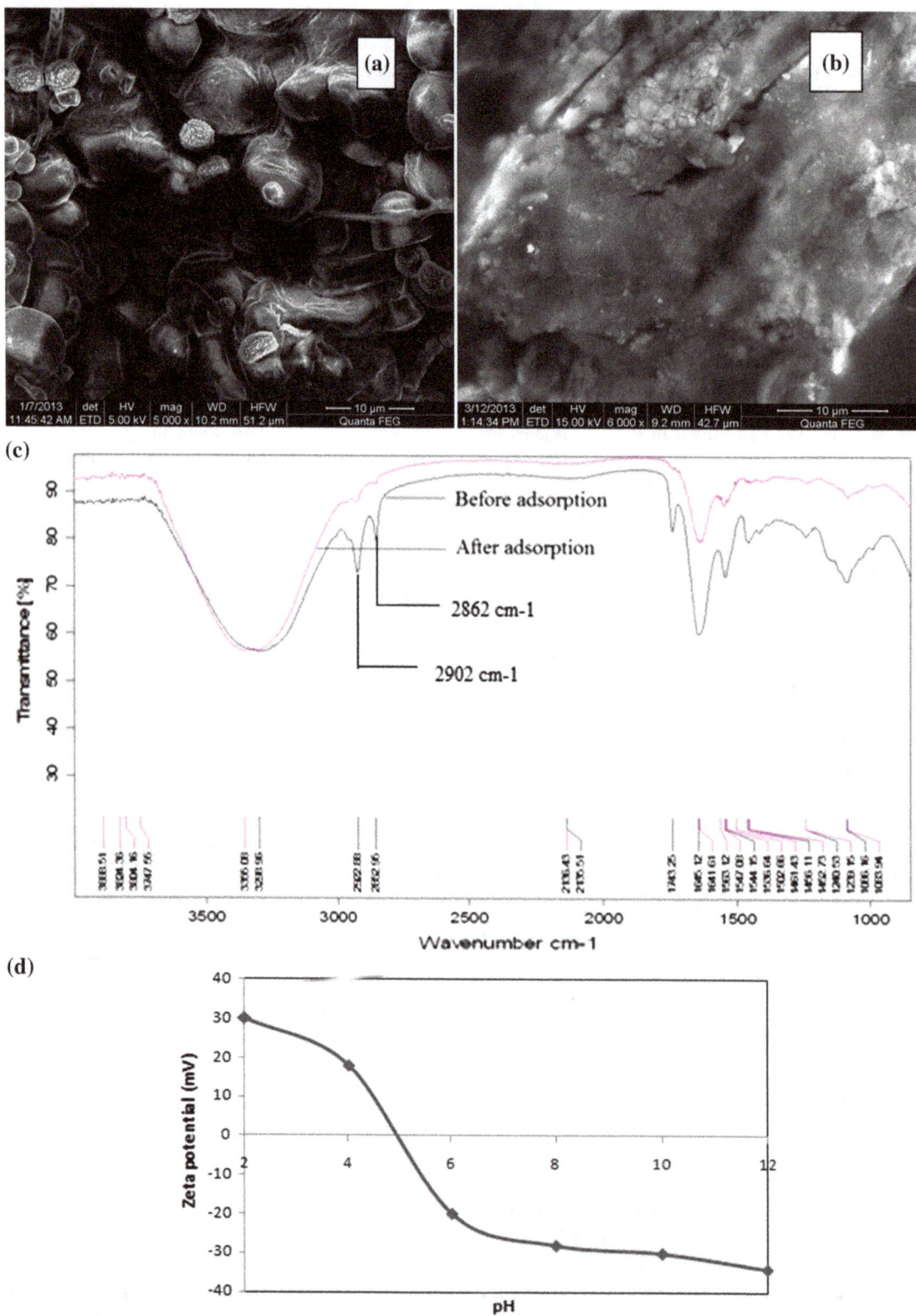

Fig. 3 SEM images before biosorption (a) and after biosorption (b). c FTIR spectra before and after biosorption. d Zeta potential–pH profile

**(a)**

**(b)**

**(c)**

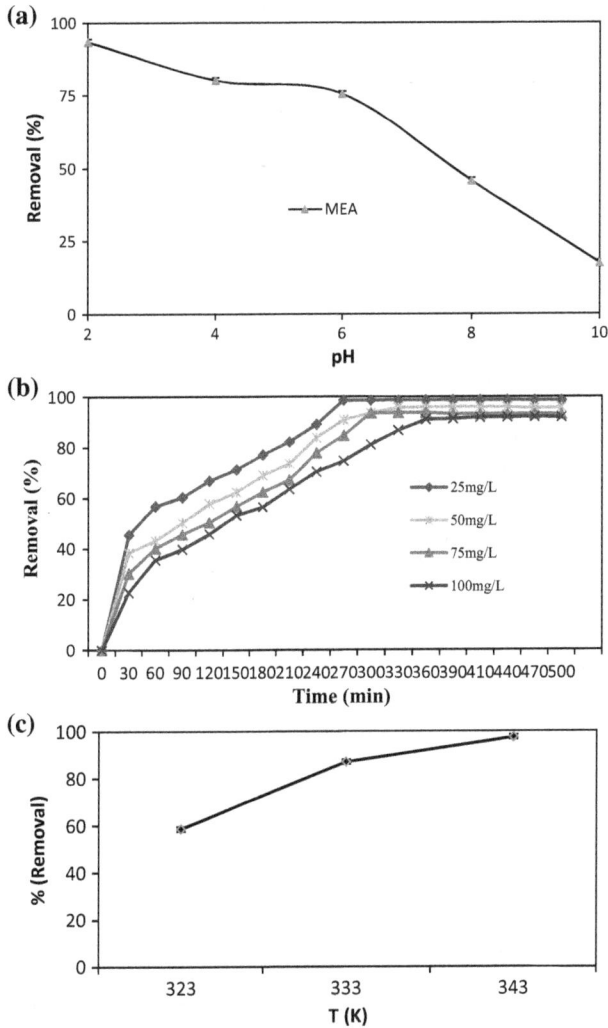

**Fig. 4** Batch biosorption of AB-FFS onto MEA: **a** influence of pH (2–10), **b** influence of initial dye concentrations (25–100 mg/L) and **c** influence of temperature (323, 333 and 343 K). Experimental conditions: synthetic AB-FFS solution volume: 100 mL; contact time: 8 h; biosorbent mass: 1 g; agitation speed: 100 rpm; *Error bars* represent SD

$$K_c = C_a/C_e, \qquad (2)$$

$$\Delta G^\circ = -RT \ln K_c, \qquad (3)$$

$$\Delta G^\circ = \Delta H^\circ - T\Delta S^\circ, \qquad (4)$$

where $K_c$ is the distribution coefficient for the adsorption; $C_a$ is the amount of dye (mg) adsorbed on the adsorbent per liter of the solution at equilibrium and $C_e$ is the equilibrium concentration (mg/L) of the dye in the solution. $R$ is the universal gas constant (8.314 J/mol K) and $T$ (°K) the absolute temperature. The standard enthalpy ($\Delta H^\circ$) and entropy ($\Delta S^\circ$) of adsorption were determined from the Vant Hoff Eq. (5):

$$\ln K_c = [(\Delta S^\circ)/R - \Delta H^\circ/RT]. \qquad (5)$$

$\Delta H^\circ$ and $\Delta S^\circ$ were obtained from the slope and intercept of the Vant Hoff plot of $\ln K_c$ versus $1/T$. The values of $\Delta G^\circ$, $\Delta H^\circ$ and $\Delta S^\circ$ under different temperature are listed in Table 2. The negative values of $\Delta G^\circ$ reveal that the biosorbent was spontaneous. The values of Gibbs free energy decreases with an increase in temperature which signifies that MEA is favorable for biosorption. The positive value of $\Delta H^\circ$ and $\Delta S^\circ$ indicates that the biosorption is an endothermic reaction and there is randomness at the solid–solution interface [17].

**Biosorption isotherm modeling**

To perform the characterization of the biosorption behavior of the dye, the equilibrium adsorption isotherms on MEA at pH 2 with different temperatures were obtained for the AB-FFS dye. Therefore, these equilibrium data are an essential source for practical design and fundamental understanding behavior of the carrier matrix for MEA. There are few nonlinear regression equations well described by various models. In this study, Langmuir [18] and Freundlich [19] model was chosen and it provides a scrupulously accurate method to reveal the linearity fitting and to explain how the AB-FFS dye interacts with MEA. The equations are listed as follows:

The general form of the Langmuir model is given by

$$C_e/q_e = 1/q_{max}K_L + C_e/q_{max}, \qquad (6)$$

where $q_{max}$ (mg/g) is the maximum biosorption capacity at equilibrium; $K_L$ (L/mg) is the Langmuir binding constant; $q_e$ gives the value of the sorbed capacity at equilibrium time (mg/g); $C_e$ is the equilibrium concentration of the adsorbed ions (mg/L). The values of $q_{max}$ and $K_L$ were estimated from the slope and intercept of the linear plots of $C_e/q_e$ against $C_e$ using values from the batch experiments.

The general form of the Freundlich model is given as

$$\ln q_e = \ln K_{Fr} + 1/n \ln C_e, \qquad (7)$$

where the $K_{Fr}$ (mg$^{1-1/n}$ L$^{1/n}$ g$^{-1}$) value gives the relative biosorption capacity of the adsorbent and $1/n$ is the dimensionless Freundlich adsorption intensity value. The values of $K_{Fr}$ and $n$ can be determined by plotting $\ln q_e$ versus $\ln C_e$ resulting in a straight line with a slope of $n$ and an intercept of $\ln K_{Fr}$. A higher value of $K_{Fr}$ estimates provide higher affinity toward the ions and the value of $1/n$ lies between $0.1 < 1/n < 1$ representing favorable biosorption [20].

The detailed parameters of Langmuir and Freundlich isotherm equations are listed in Table 3. From the Langmuir isotherm, the maximum monolayer adsorption capacity of MEA increased from 18.18 to 76.92 (mg/g) with the increase of the solution temperature from 323 to

**Table 2** Thermodynamic parameters of AB-FFS onto MEA at different temperatures

| Dye concentration (mg/L) | $\Delta G^{\circ}$(kJ/mol) | | | $\Delta H^{\circ}$ (kJ/mol) | $\Delta S^{\circ}$ (J/mol K) |
|---|---|---|---|---|---|
| | 323 K | 333 K | 343 K | | |
| 25 | −4331 | −8416 | −11,093 | 89.14 | 284 |
| 50 | −1889 | −7378 | −9268 | | |
| 75 | −7277 | −6285 | −9096 | | |
| 100 | −3007 | −5924 | −8697 | | |

**Table 3** Isotherm parameters for the biosorption of AB-FFS dye onto MEA

| MEA (K) | Langmuir | | | | Freundlich | | |
|---|---|---|---|---|---|---|---|
| | $q_m$ (mg/g) | $K_L$ (L/mg) | $R^2$ | $R_L$ | $K_{Fr}$ mg$^{(1-/n)}$L$^{1/n}$/g | $n$ | $R^2$ |
| 323 | 18.18 | 0.028 | 0.935 | 0.41 | 0.431 | 1.315 | 0.998 |
| 333 | 23.8 | 0.039 | 0.957 | 0.46 | 2.488 | 0.654 | 0.998 |
| 334 | 76.92 | 0.091 | 0.995 | 0.52 | 2.557 | 0.781 | 0.989 |

343 K, respectively, which confirms that the process is an endothermic reaction. The correlation coefficients ($R^2$) are listed in Table 3, strongly indicating that the biosorption of AB-FFS onto MEA follows both the Langmuir and Freundlich isotherm model.

MEA biosorption was further analyzed in terms of separation factor; $R_L$ is a dimensional parameter which is defined as $R_L = (1/1 + K_L C_o)$, derived from the Langmuir equation and $C_i$ is the initial dye concentration. $R_L$ indicates the biosorption process to be either favorable ($0 < R_L < 1$), unfavorable ($R_L > 1$), linear ($R_L = 1$) or irreversible ($R_L = 0$). From Table 3, $R_L$ values gradually increase from 0 to 1, which indicates the process is more favorable, and higher temperature may enhance biosorption process.

## Biosorption kinetics modeling

To examine the controlling mechanism of biosorption as well as to understand the behavior of the AB-FFS dye onto the beads, the three kinetics model was investigated: pseudo-first order kinetics, pseudo-second order kinetics followed by Weber and Morris intraparticle diffusion model. In this study, the kinetics of biosorption of two different concentrations of AB-FFS dye (50 and 100 mg/L) onto MEA was carried out with the pseudo-first-order kinetics by Lagergren's [21] and pseudo-second-order model by Ho and McKay's [22] followed by Weber and Morris's intraparticle diffusion model [23].

For the biosorption of AB-FFS, the first-order kinetics can be represented by Eq. 8:

$$\log (q_e - q_t) = \log q_t - K_1 t/2.303, \tag{8}$$

where $q_e$ and $q_t$ are the amounts of AB-FFS dye adsorbed (mg/g) at equilibrium and at time $t$ (h), respectively, and $K_1$ is the adsorption rate constant (min$^{-1}$). The Lagergren

first-order kinetic constant ($K_1$) and the theoretical value of ($q_e$) can be obtained from the linear plots of log ($q_e - q_t$) against time for 50 and 100 mg/L of initial dye concentrations as shown in Fig. 5a. From the Fig. 5a, the plot was found to be linear and the correlation coefficient ($R^2$) was greater than 0.9, which indicates that pseudo-first order were appropriate for the use of AB-FFS dye onto MEA. The rate constant ($K_1$) and $q_e$ (exp) values along with the correlation coefficient ($R^2$) are listed in Table 4. The data was further validated using second-order kinetics, which can be expressed by Eq. 9:

$$t/q_t = 1/K_2 q_e^2 + t/q_e. \tag{9}$$

For the pseudo-second-order kinetic model, the plots of $t/q_t$ against time at 50 and 100 mg/L initial dye concentrations are shown in Fig. 5b. The value of qe (mg/g) and second-order kinetic constants k$^2$ (g/mg min) along with correlation coefficients of MEA are listed in Table 4. In addition from Table 4, it was noticed that the theoretical predicted $q$ values were not concordant with the experimental values and $R^2$ values were not in accordance, suggesting that the use of MEA on AB-FFS dye did not fit well with the second-order kinetics. Further, Weber and Morris found a kinetic process of liquid–solid adsorption of intraparticle diffusion and mass action. To identify the diffusion mechanism, the relationship between $q_t$ and $t^{1/2}$ could be written in the form of Eq. (10) given by:

$$q_t = K_p t^{\frac{1}{2}} + C, \tag{10}$$

where $q_t$ (mg/g) is the amount of dye adsorbed at various times ($t$), $K_P$ is the intraparticle diffusion rate constant (mg/g min$^{0.5}$) and $C$ is the intercept of the line which is directly proportional to the boundary layer of thickness. This multilinear Eq. 10 involves three steps. The first step is the

**(a)**

**(b)**

**(c)**

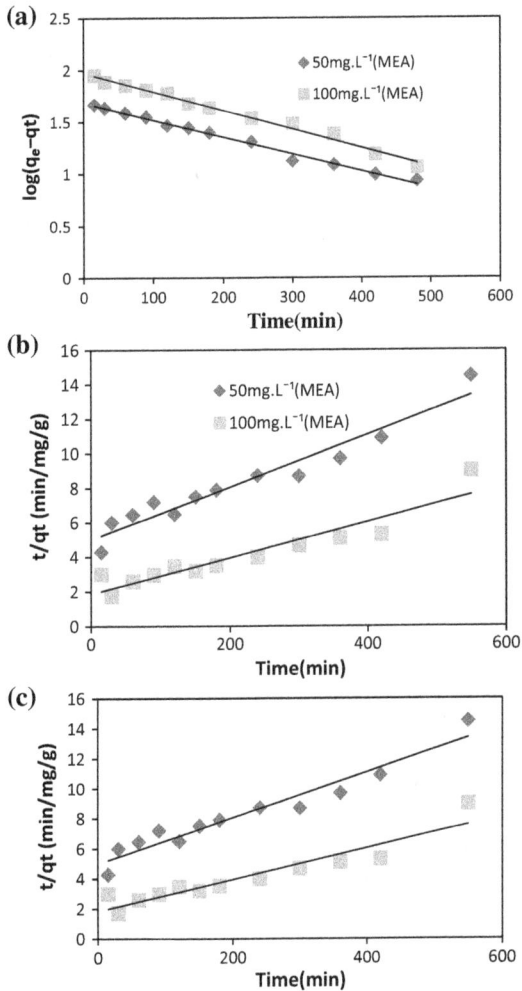

**Fig. 5** Pseudo-first-order model (**a**), Pseudo-second-order model (**b**) and intraparticle diffusion model (**c**) for batch adsorption of AB-FFS dye onto MEA. Experimental parameters—*dosage* 1 g, *volume* 100 ml, *dye concentration* 50 and 100 mg/L, *temperature* 273 K, *pH* 2, *reaction time* 500 min

immediate process that allows quick adsorption, the second step is the intraparticle diffusion stage, whereas the final stage is the equilibrium step in which the solute moves slowly from macropores to micropores causing a slow adsorption rate [8, 22]. A plot of $q_t$ against $t^{1/2}$ should be linear, from which $K_P$ and $C$ can be calculated from the slope and intercept of the plot. According to Weber and Morris, if the intraparticle diffusion is the rate-limiting

step, it is necessarily the plot of $q$ versus $t^{1/2}$ which passes through the origin.

The intraparticle diffusion of AB-FFS onto MEA at an initial dye concentration of 50 and 100 mg/L is illustrated in Fig. 5c. It was observed that the straight line of the intraparticle region of MEA of 100 mg/L did not pass through the origin, while the rest pass through the origin. Moreover, it is clear from Table 4 that MEA diffusion of 100 mg/L has a larger intercept $C$ value, which relates high boundary layer resistance and the surface adsorption in the rate-limiting step. According to McKay, the boundary layer thickness retards the intraparticle diffusion. Table 4 clearly explains that the intraparticle parameter $K_P$ values increase along with increasing dye concentrations. Thus, the increase of AB-FFS concentrations results in an increase in the diffusion rate of dye into pores of the beads. Especially, MEA increases regularly for 50 mg/L (5.09 mg/g min$^{0.5}$) and 100 mg/L (9.25 mg/g min$^{0.5}$) with low $K_P$ value.

## Fixed bed column study

The performance of the MEA biosorbent was tested in a fixed bed column, to check the feasibility of the MEA in real-time effluents. The AB-FFS solution (100 mg/L) was fed through the column of different bed heights (5 and 15 cm) with two different flow rates (1, 2 and 5 mL/min) and the samples were collected at regular intervals.

### Effect of bed height

The effect of bed height is an important parameter for calculating the design of the FBC performance. At a constant initial AB-FFS concentration (100 mg/L) and flow rate (2 mL/min), the breakthrough curve for biosorption of the AB-FFS onto MEA of two different bed heights (BH) such as 5 cm (12.25 g) and 15 cm (36.75 g), respectively, is shown in Fig. 6. From Fig. 6a, both the breakthrough times, $t_b$ and exhaustion time $t_e$, were found to increase with increasing bed height, whereas the shape of the breakthrough was slightly different with the variation of bed height. From Fig. 6a, an earlier breakthrough and exhaustion time was attained in BH—5 cm, ($t_b = 33$th min and $t_e = 340$th min) and BH—15 cm ($t_b = 150$th min and $t_e = 420$th min), respectively. An increase in dye uptake

**Table 4** Kinetic parameters for the biosorption of AB-FFS dye onto MEA

| MEA | Pseudo-first-order kinetics | | | | Pseudo-second-order kinetics | | | Intraparticle diffusion | | |
|---|---|---|---|---|---|---|---|---|---|---|
| Parameters | $q_e$ (exp) (mg/g) | $q_e$ (cal) (mg/g) | $k_1$ (min$^{-1}$) | $R^2$ | $q_e$ (cal) (mg/g) | $k_2$ (g/mg min) | $R^2$ | $k_p$ (mg/g min$^{0.5}$) | $C$ | $R^2$ |
| 50 mg/L | 47.8 | 47.863 | 0.1428 | 0.990 | 66.66 | 0.2001 | 0.932 | 5.09 | 2.93 | 0.993 |
| 100 mg/L | 94.55 | 93.118 | 0.1756 | 0.987 | 100 | 0.5391 | 0.876 | 9.25 | 7.74 | 0.982 |

**(a)**

**(b)**

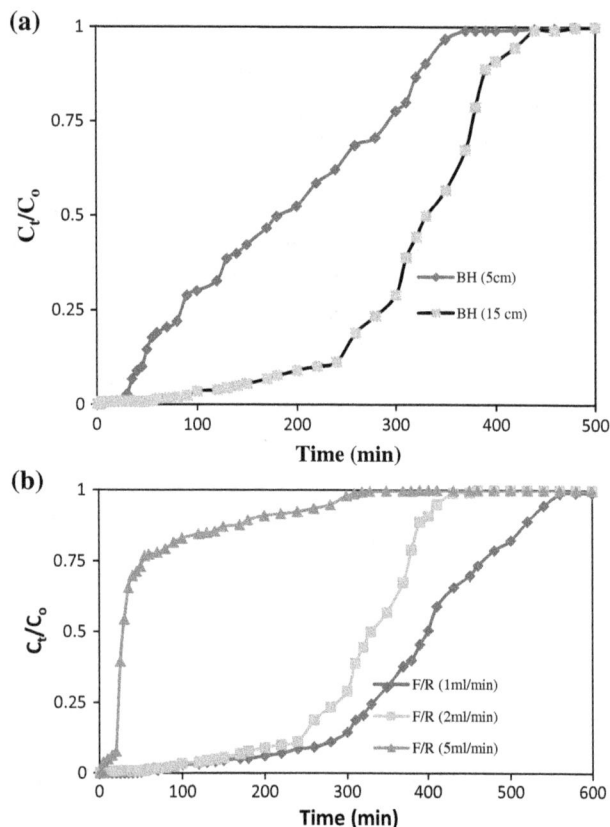

**Fig. 6** **a** Breakthrough curves for biosorption of AB-FFS onto MEA at different bed heights (BH-5 and 15 cm with constant flow rate $-2$ mL min$^{-1}$ and initial concentration $-100$ mg/L). **b**. Breakthrough curves for biosorption of AB-FFS onto MEA at different flow rates (1, 2 and 5 mL/min with constant BH-15 cm and initial concentration $-100$ mg/L)

was observed at elevated bed height due to the increase in the amount of the MEA. As the bed height increases, simultaneously, the mass transfer zone (MTZ) also increases in the column, which moves downward from the entrance of the bed to the exit.

## Effect of flow rate

The effect of varying flow rate is an important parameter which determines the contact time of the dye with the biosorbent in the FBC. The breakthrough curves $C_t/C_o$ against time (min) for three different flow rates (1, 2 and 5 mL/min) with constant bed height of 15 cm were investigated and shown in Fig. 6b. From the figure, as the flow rate increases, the breakthrough curve becomes steeper and the biosorbent achieves early saturation at 5 mL/min. Similarly, at lower flow rate, a longer contact time with the shallow biosorption zone results in a higher uptake of AB-FFS. From Fig. 6b, earlier breakthrough and exhaustion time were attained flow rates (F/R) 1 mL/min ($t_b = 170$th min and $t_e = 540$th min), 2 mL/min

($t_b = 140$th min and $t_e = 410$th min) and 5 mL/min ($t_b = 10$th min and $t_e = 280$th min), respectively.

## Modeling of breakthrough curves

Fixed bed column data obtained were further analyzed for their breakthrough behaviors using two different mathematical equation models such as those of Thomas [24] and Yoon–Nelson [25]. The biosorption performance was assessed at an initial concentration ratio, $C_t/C_o > 0.05$, consequently, 5 % breakthrough until $C_t/C_o > 0.95$ that is, 90 % breakthrough for dye decolorization by considering water quality and operating limits of MTZ of the column [26].

## Applications of the Thomas model

The experimental data were fitted to the Thomas model to determine the maximum dye biosorption capacity of the column ($q_o$) and the Thomas rate constant ($k_{Th}$), as shown in Table 5. Further, this model is based on the assumption that the process follows Langmuir isotherms of equilibrium with no axial dispersion and the rate driving force which obeys the second-order reversible reaction kinetics [27]. The linearized form of Thomas model Eq. (11) can be expressed as follows:

$$\ln\left(\frac{C_o}{C_t} - 1\right) = \frac{k_{Th}\, q_o\, m}{v} - \frac{k_{Th}\, C_o}{vt}, \qquad (11)$$

where $k_{Th}$ (mL/mg/min) is the Thomas rate constant, $q_o$ (mg/g) is the equilibrium adsorbate uptake per gram of the biosorbent, $C_o$ and $C_t$ (mg/L) are the inlet and outlet concentrations (g), $m$ is the mass of the adsorbent in the column, and v (mL/min) stands for flow rate. The value $C_t/C_o$ is the ratio of the outlet to inlet effluent concentrations. By plotting the linear plots of ln $[(C_o/C_t) - 1]$ against time ($t$), the rate constant value ($k_{Th}$) was determined and the maximum capacity of biosorption ($q_o$) was obtained from the slope and intercepts using values from the column experiments. The regression coefficient ($R^2$) and the relative constants such as $q_o$ (mg/g) and $k_{Th}$ values were calculated from the experimental data as shown in Table 5. The regression coefficients ($R^2$) were between 0.8 and 0.9, which showed that the experimental data fitted the Thomas model well. From Table 5, in general, it was observed that by increasing the flow rate, the biosorption capacity ($q_o$) decreased, but the values of the rate constant ($k_{Th}$) increased. Further, by extending the bed height, the values of $q_o$ decreased and the $k_{Th}$ value increased significantly. In addition, from Table 5, it was found that 15 cm of bed height (BH) with 1 mL/min F/R gives maximum biosorption values ($q_o$, 1543 mg/g).

**Table 5** Thomas and Yoon–Nelson model parameters for the removal of AB-FFS using MEA at different conditions using linear regression analysis

| Bed height (cm) | Flow rate (mL/min) | Thomson model | | | | | Yoon–Nelson model | | | | |
|---|---|---|---|---|---|---|---|---|---|---|---|
| | | $K_{TH}$ (mL/min mg) | $q_{exp}$ (mg/g) | $q_{the}$ (mg/g) | $\varepsilon$ % | $R^2$ | $K_{YN}$ (/min) | $\tau_{exp}$ (min) | $\tau_{the}$ (min) | $\varepsilon$ % | $R^2$ |
| 5 | 2 | 0.00046 | 447 | 458 | 2.35 | 0.867 | 0.022 | 180.01 | 182.09 | 2.35 | 0.867 |
| 15 | 1 | 0.00015 | 1505 | 1543 | 4.01 | 0.785 | 0.015 | 203.1 | 205.7 | 4.01 | 0.785 |
| 15 | 2 | 0.00047 | 801 | 806 | 5.35 | 0.902 | 0.023 | 190 | 191.9 | 5.35 | 0.903 |
| 15 | 5 | 0.001 | 367 | 380 | 5.01 | 0.805 | 0.019 | 94.5 | 96 | 5.01 | 0.806 |

## Application of the Yoon–Nelson model

A theoretical model developed by Yoon–Nelson was useful to investigate the breakthrough behavior of AB-FFS on MEA. In addition, this model was derived based on the assumption that the rate of decrease in the probability of biosorption for each adsorbate molecule is proportional to the probability of adsorbate biosorption and the probability of an adsorbate breakthrough on the biosorbent [25, 28]. The linearized model of a single component system is expressed as in Eq. 12:

$$\ln\left(\frac{Ct}{C_0 - C_t}\right) = k_{YN}\ t - \tau\, k_{YN}, \qquad (12)$$

where, $k_{YN}$ (1/min) is the Yoon–Nelson rate constant and $\tau$ (min) is the time required for 50 % adsorbate breakthrough. The values of $k_{YN}$ and $\tau$ were estimated from the slope and intercepts of the linear graph between $\ln[(C_t)/(C_0 - C_t)]$ versus time $t$ at different flow rates with different bed heights and the values of $k_{YN}$ and $\tau$ are illustrated in Table 5. From Table 5, the values of $k_{YN}$ were found to increase with higher F/R and lower BH; however, it decreased with lower F/R and higher BH. Nevertheless, $\tau$ (the time required for 50 % breakthrough) was higher at lower F/R and higher BH. In addition, from Table 3, it was found that the $\tau$ (min) values of 205 min and 223 min were obtained at 15 BH with F/R 1 mL/min and 15 BH with F/R 2 mL/min. The values of the regression coefficient ($R^2$) are listed in Table 5 and it was found that almost all the values of $R^2$ were between the range of 0.8 and 0.9. Furthermore, there is good concurrence between the predicted and experimental data which provide the best fit to the Yoon–Nelson model.

## Desorption studies

To improve the cost-effectiveness in industries, reusability of the biosorbent is an important factor in practical applications for dye removal from wastewaters. The reusability of MEA beads can be determined by its biosorption performance in consecutive biosorption/resorption cycles under batch mode and the desorption efficiencies are compared in Fig. 7. From the figure, the use of deionized water was found to be negligible (<5 %). Further, the use of NH₄Cl was resorbed by only <20 %, whereas CH₃COOH and EDTA give 50–60 % of the desorption efficiency. The use of acids such as HNO₃ and HCl gives more than 90 % de-resorption efficiency which can be attributed to the fact that the AB-FFS dyes react faster with acids than bases. On the other hand, during the desorption process, NaOH reacts with calcium-alginate beads and it was completely dissolved due the replacement of calcium and sodium ions. Overall, HNO₃ and HCl was found to have higher resorption efficiencies and similar work has been reported previously [29, 30]. The overall MEA adsorbent was found to be cost-effective and feasible for treatment of industrial wastewater treatment.

## Conclusion

This research work introduced the encapsulated *Moringa oleifera* beads as a potential biosorbent for dye removal in the batch as well as in continuous column studies which gave more than 90 % of removal efficiency. The maximum

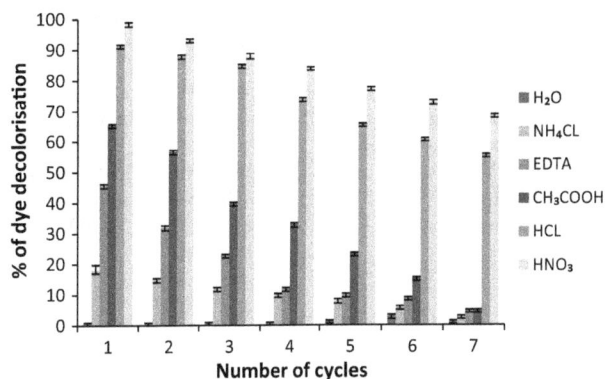

**Fig. 7** Desorption efficiency of AB-FFS dye onto MEA using different desorbing agents (concentration: 0.1 M)

biosorption capacity of AB-FFS dye calculated from Langmuir isotherm, which can be well fitted onto MEA of 18.18, 23.80 and 76.92 mg/g, respectively, was studied from 323 to 343 K. The thermodynamic parameter results reveal that the biosorbent was spontaneous and it was an endothermic reaction process. Among the kinetic model studies, the biosorption kinetics favor both pseudo-first-order and intraparticle diffusion model. In fixed bed column studies, data obtained were fitted with well-established models, namely Thomas and Yoon–Nelson. Overall, this research showed that MEA is a potential and promising alternative biosorbent for removal of color from textile effluents. Moreover, dried *Moringa oleifera* is a biological waste material that is easily viable and biodegradable.

# References

1. Lin SH, Lou CC (1996) Treatment of textile wastewater by foam flotation. Environ Technol 17:841–849
2. Kaneko S (1982) Adsorption of several dyes from aqueous solutions on silica-containing-oxide gels. Sep Sci Technol 17:1499–1510
3. Juang RS, Swei SL (1996) Effect of dye nature on its adsorption from aqueous solutions onto activated carbon. Sep Sci Technol 31:2143–2156
4. Namasivayam C, Kavitha D (2003) Adsorptive removal of 2-clorophenol by low-cost pith carbon. J Hazard Mater 98:257–270
5. Gupta VK, Jain D, Mohan S, Sharma M (2000) Removal of basic dyes (Rhodamine-B and Methylene blue) from aqueous solutions using bagasse fly ash. Sep Sci Technol 35:2097–2113
6. Saha P (2010) Study on the removal of methylene blue dye using chemically treated rice husk. Asian J Water Environ Poll 7:2
7. Ponnusamy SK, Subramaniam R (2013) Process optimization studies of Congo red dye adsorption onto cashew nut shell using response surface methodology. Int J Ind Chem 4:17
8. Moussavi G, Khosravi R (2011) The removal of cationic dyes from aqueous solutions by adsorption onto pistachio hull waste. Chem Eng J Res Des 89:2182–2189
9. Radhakrishnan K, Lakshmi S, Radha P, Aparna N, Vishali S, Thilagaraj WR (2016) Biosorption of heavy metals from actual electroplating wastewater using encapsulated Moringa oleifera beads in fixed bed column. Desalin Water Treat 57:3572–3587
10. Lai Y, Thirumavalavan M, Lee J (2010) Effective adsorption of heavy metals ions (Cu$^{2+}$, Pb$^{2+}$, Zn$^{2+}$)from aqueous solution by immobilization of adsorbents on Ca-alginate beads. Toxicol Environ Chem 92:697–705
11. Radhakrishnan K, Aparna N, Thilagaraj WR (2015) Effective and ecofriendly nano biosorbent for treatment of textile wastewater. Res J Chem Environ 19:14–23
12. Araujo CST, Melo EI, Alves VN, Coelho NMM (2010) *Moringa oleifera* Lam. seeds as a natural solid adsorbent for removal of Ag in Aqueous solutions. J Brazilian chem soc 21:1727–1732
13. Nassar NN (2010) Kinetics, mechanistic, equilibrium, and thermodynamic studies on the adsorption of acid red dye from waste water by γFe$_3$O$_4$ nanoadsorbents. Sep Sci Technol 48:1092–1110
14. Xi Y, Shen Y, Yang F, Yang G, Liu C, Zhang Z, Zhu D (2013) Removal of azo dye from aqueous solutionby a new biosorbent prepared with *Aspergillus nidulans* cultured in tobacco wastewater. J Taiwan Ins Chem Eng 44:815–820
15. Bouhamed F, Elouear Z, Bouzid J (2012) Adsorption removal of copper (II) from aqueous solutions on activated carbon prepared from Tunisian date stones: equilibrium, kinetics and thermodynamics. J Taiwan Inst Chem Eng 43:741–749
16. Qiu B, Cheng X, Sun D (2012) Characteristics of cationic Red X-GRL biosorption by anaerobic activated sludge. Bioresour Technol 113:102–105
17. Inbaraj BS, Sulochana N (2006) Use of jackfruit peel carbon (JPC) for adsorption of rhodamine-B, a basic dye from aqueous solution. Indian J Chem Technol 13:17–23
18. Langmuir I (1916) The constitution and fundamental properties of solids and liquids. J Am Chem Soc 38:2221–2295
19. Freundlich HMF (1906) Over the adsorption in solution. Z Phys Chem 57:385–471
20. Murugan M, Subramanian E (2006) Studies on defluoridation of water by Tamarind seed, and unconventional biosorbent. J Water Health 4:453–461
21. Lagergren S (1898) About the theory of so-called adsorption of soluble substances. Kungliga Svenska Vetenskapsakademiens. Hand linger 24:1–39
22. Ho YS, McKay G (1999) Pseudo-second order model for sorption processes. Process Biochem 34:451–465
23. Weber WJ, Morris JC (1963) Kinetics of adsorption on carbon from solution. J Sanit Eng Division 31–59
24. Thomas HC (1948) Chromatography: a problem in kinetics. Ann N Y Acad Sci 49:161–182
25. Yoon YH, Nelson JH (1984) Application of gas adsorption kinetics. I. a theoretical model for respirator cartridge service life. Am Ind Hyg Assoc J 45:509–516
26. Malkoc E, Nuhoglu Y, Abali Y (2006) Cr (VI) adsorption by waste acorn of *Quercus ithaburensis* in fixed beds: prediction of breakthrough curves. Chem Eng J 119:61–68
27. Bhaumik M, Setshedi Maity KA, Onyango MS (2013) Chromium (VI) removal from water using fixed bed column of polypyrrole/Fe$_3$O$_4$ nanocomposite. Sep Purifi Technol 69:11–19
28. Chowdhury ZZ, Zain SM, Rashid AK, Rafique RF, Khalid K (2013) Breakthrough curve analysis for column dynamics sorption of Mn(II) ions from wastewater by using Mangostana garcinia peel-based granular-activated carbon. J chem 1–8. doi:10.1155/2013/959761
29. Kalavathy MH, Miranda LR (2010) Moringa oleifera—a solid phase extractant for the removal of copper, nickel and zinc from aqueous solutions. Chem Eng J 158:188–199
30. Liu L, Wan Y, Xie Y, Zhai R, Zhang B, Liu J (2012) The removal of dye from aqueous solution using alginate-halloysite nanotube beads. Chem Eng J 187:210–216

11

# Process standardization and kinetics of ethanol driven biodiesel production by transesterification of ricebran oil

Inkollu Sreedhar[1] · Yandapalli Kirti Kishan[1]

**Abstract** In this study, trans-esterification of rice bran oil employing a heterogeneous catalyst like CaO has been conducted in a batch reactor with ethanol. The optimal set of various critical reaction parameters are found to be temperature of 70 °C, reactant molar ratio 9:1, agitation speed of 600 rpm, catalyst mount of 3 wt % to maximize the biodiesel yield. For the heterogeneous system possibly influenced by reaction and pore diffusion, the controlling regime has been found based on the experimental data and well established theoretical estimations. The kinetic model along with the relevant parameters viz., reaction order, rate constants and Arrhenius parameters have been estimated.

**Keywords** Trans-esterification · Biodiesel · Kinetics · Controlling regime · Process standardization

## List of symbols

### Variables

| | |
|---|---|
| $a_m$ | Specific surface area of catalyst (m$^2$/gm) |
| $A$ | Frequency factor (min$^{-1}$) |
| $c_A$ | Concentration of triglyceride in liquid phase (mol/l) |
| $c_{A_0}$ | Initial concentration of triglyceride in liquid phase (mol/l) |
| $c_{A,s}$ | Concentration of triglyceride on the interfacial solid liquid area (mol/l) |
| $c_B$ | Concentration of ethanol in liquid phase (mol/l) |
| $c_R$ | Concentration of FAEE in liquid phase (mol/l) |
| $C$ | Integration constant |
| $D$ | Molecular diffusion coefficient (m$^2$/s) |
| $D_{\text{eff}}$ | Effective diffusion coefficient (m$^2$/s) |
| $E_a$ | Activation energy (KJ/mol) |
| $k$ | Pseudo-first order reaction rate constant (min$^{-1}$) |
| $k_{\text{ad}}$ | Ethanol adsorption rate constant (min$^{-1}$) |
| $k_{\text{app}}$ | Apparent process rate constant (min$^{-1}$) |
| $k_{s,A}$ | Triglyceride mass transfer coefficient towards catalyst surface (m/min) |
| $k_{\text{mt,A}}$ | Volumetric triglyceride mass transfer coefficient (min$^{-1}$) |
| $m_{\text{cat}}$ | Mass of heterogeneous catalyst (g) |
| $M$ | Molecular weight of solvent |
| $Q$ | Instantaneous concentration of adsorbed ethanol (mol/gm) |
| $Q_{\text{max}}$ | Maximum concentration of adsorbed ethanol (mol/gm) |
| $(-r_A)$ | Rate of triglyceride consumption (mol/(l min)) |
| $(-r_B)$ | Rate of ethanol consumption (mol/(l min)) |
| $R$ | Gas constant (J K$^{-1}$ mol$^{-1}$) |
| $R_p$ | Catalyst particle radius (m) |
| $t$ | Time (min, s) |
| $T$ | Temperature (K) |
| Th | Thiele modulus |
| $V$ | Volume of reaction mixture (cm$^3$) |
| $V_m$ | Molal volume of solute at normal boiling point (cc/g mol) |
| $x$ | Association parameter of solvent |
| $X_A$ | Degree of triglyceride conversion |

### Greek symbols

| | |
|---|---|
| $\varepsilon_p$ | Catalyst particle porosity |
| $\tau_p$ | Catalyst particle tortuosity |

✉ Inkollu Sreedhar
isreedhar2001@yahoo.co.in

[1] Department of Chemical Engineering, BITS Pilani Hyderabad Campus, Hyderabad, India

$\eta$    Viscosity of solution (cp)
$\theta$   Fraction of the catalyst available active specific surface

**Table 1** Properties of rice bran oil

| Property | Refined rice bran oil |
| --- | --- |
| Moisture (%) | 0.1–1.15 |
| Density | 0.913–0.920 |
| refractive index | 1.4672 |
| Saponification value | 187 |
| Free fatty acids (%) | 0.15–0.2 |
| Kinematic viscosity @ 40 °C (cSt) | 43.52 |
| Cetane number | 50.1 |
| Cloud point (°C) | 13 |
| Pour point (°C) | 1 |
| Flash point (°C) | 316 |
| Calorific value (MJ/kg) | 41.1 |

# Introduction

Rapid industrialization and increasing population at a global level has led to a tremendous upswing in the energy utilization [1, 2]. Hence the focus has been shifted in employing renewable energy resources due to the anticipated exhaustibility of the fossil reserves besides the growing concern for environmental quality. Biodiesel has become the researchers' spotlight for some time now due to its significant attributes like green and non-toxic nature leading to significant reduction in greenhouse gases, renewability and biodegradability [3, 4].

Conventionally, biodiesel is produced by transesterification of triglyceride feedstock (vegetable oils/animal fats/waste cooking oils) with methanol or other short chain alcohols in the presence of a catalyst. Many studies have been reported on this reaction employing various homogeneous and heterogeneous catalysts with different types of oils viz., palm, sunflower, rapeseed, soyabean oils. Heterogeneous catalysts like zeolites, metal oxides, heteropolyacids, hydrotalcites are preferred to the homogeneous catalysts like KOH or NaOH due to their reusability, eco-friendly nature and easy separation of products [5–8]. On the other hand, the homogeneous catalysis route though relatively rapid and result in higher yields, is not recommended due to commercial reasons as it demands high quality crude oils and an expensive and toxic downstream processing due to undesired side reactions like saponification and hydrolysis occurring. Hence there has been a growing impetus on the heterogeneous synthesis, identifying novel and effective catalysts and cost effective feedstock as the main hurdle in the commercialization of biodiesel production by transesterification reaction has been the high cost of raw materials which contributes to 70 % of total cost [9]. Various research studies have been reported on this topic like process standardization with reference to various critical parameters like nature of vegetable oil and alcohol, reactant mole ratio, nature and amount of catalyst, reaction time, agitation rate and temperature and employing different reaction conditions like microwave irradiation, ultrasonic assisted and supercritical conditions to maximize biodiesel yield and reaction rate [10–19].

In this work, biodiesel synthesis has been carried out in a batch reactor using ricebran oil and ethanol in the presence of heterogeneous catalyst, CaO. Various properties of Rice Bran oil have been listed in Table 1. The influence of various process parameters has been investigated to maximize the biodiesel yield and kinetic studies have been conducted at optimal conditions to estimate the rate constants. Ricebran oil has been selected for our study as it is derived from agricultural waste which is abundantly available in most of the rice producing countries and has a significant potential as an alternative cost-effective feedstock when compared to those derived from cereal or seed sources like sunflower, soyabean, canola etc. [20–23].

# Materials and methods

The reactants for the transesterification reaction viz., rice bran oil (100 % pure. Priya brand), ethanol (99 % pure) and extra pure CaO catalyst (size <150 nm) are procured from Hi-Media Labs, Hyderabad and are used as is. The experimental set-up (Fig. 1) employed for the reaction consists of three necked borosil glass reactor of 1 l capacity immersed in a constant temperature oil bath with a propeller inserted at the centre to agitate the reaction mixture. The reactants of predetermined quantities are fed to the reactor and the reaction is conducted to produce the desired product of biodiesel. The reaction scheme for the transesterification reaction is shown in Fig. 2. Biodiesel yield is estimated using a Gas Chromatograph (Agilent make, 7820A model with HP88 column) fitted with HP88 column. The conditions employed for GC analysis are FID temperature: 260 °C; Oven temperature initially at 125 °C raised to 145 °C at a rate of 8 °C/min and then maintained at that temperature for 26 min after which it is increased to 220 °C at a rate of 2 °C/min and then kept steady for 1 min. 0.2 µl of the product sample from biodiesel rich layer after quenching it in an ice bath to arrest any further reaction is injected along with solvent hexane into GC to get various peaks from which biodiesel yield is estimated. Experiments have been conducted to study the influence of

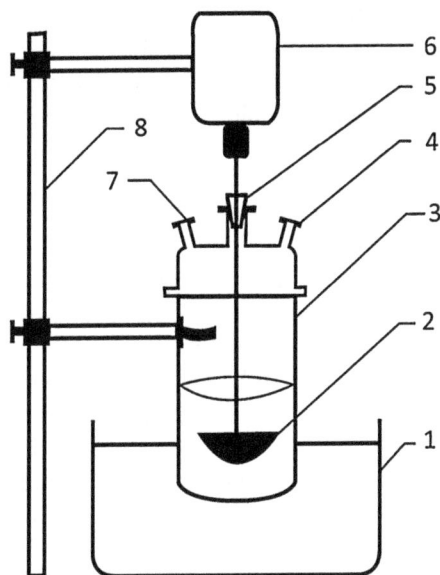

**Fig. 1** The apparatus for transesterification reaction: *1* digitally controlled water bath, *2* impeller blade, *3* three necked glass flask, *4* inlet for reactants, *5* rubber cork, *6* stirrer, *7* inlet for temperature measurement, *8* retort stand

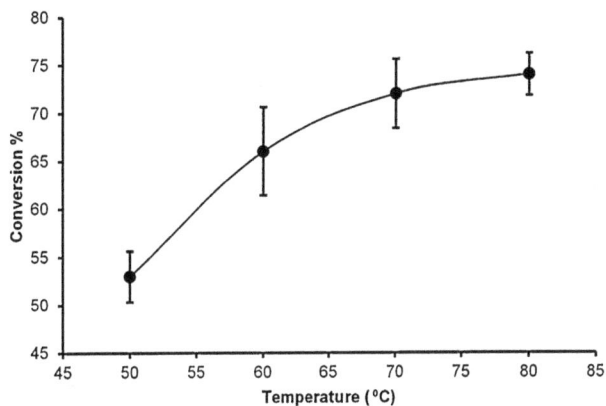

**Fig. 3** Influence of temperature on conversion at 600 rpm, 3 wt % catalyst, ethanol/oil molar ratio 9:1, reaction time 2 h

molar ratio constant. The results are given in Fig. 3 which a positive influence of temperature on the biodiesel yield giving more than 70 % at 70 °C. Our results are found to be in agreement with those reported by Wu et al. [24, 25].

### Effect of molar ratio of reactants

In order to examine the effect of molar ratio of ethanol to rice bran oil on the yield, reactions are carried out using three different mole ratios 6:1, 9:1, 12:1 keeping all other parameters constant and the results obtained are shown in Fig. 4. In trans-esterification reaction, 1 mol of vegetable oil reacts with 3 mol of ethanol to give 3 mol of biodiesel and 1 mol of glycerol and this was a reversible reaction. So, excess of any of the reactant (more than stoichiometric ratio, 3:1) should shift the equilibrium to the right pushing the forward reaction achieving enhanced yield of 72 % till 9:1 mole ratio. Yield was found to decrease to 64 % at 12:1 mol ratio due to the relative decrease in the catalyst amount vis-a-vis the reaction mixture and hence the possible triggering of reverse reaction by the enhanced formation of glycerol which reacts with biodiesel formed reducing its yield to 64 % [16].

various critical parameters viz., reaction temperature (50, 60, 70 and 80 °C); reactant mole ratio (ethanol to oil ratios of 6:1, 9:1 and 12:1); agitation speed (400, 600 and 800 rpm); catalyst weight percent (1.5, 3, 4.5 and 6 %) and reaction time (30, 60, 90 and 120 min) on biodiesel yield.

## Results and discussion

### Effect of reaction temperature

Temperature has been reported to be one of the critical parameters in the transesterification reaction for biodiesel synthesis. Experiments have been carried out at three different temperatures of 50, 60 and 70 °C keeping all other parameters viz., agitation speed, catalyst amount, reactant

**Fig. 2** Transesterification reaction

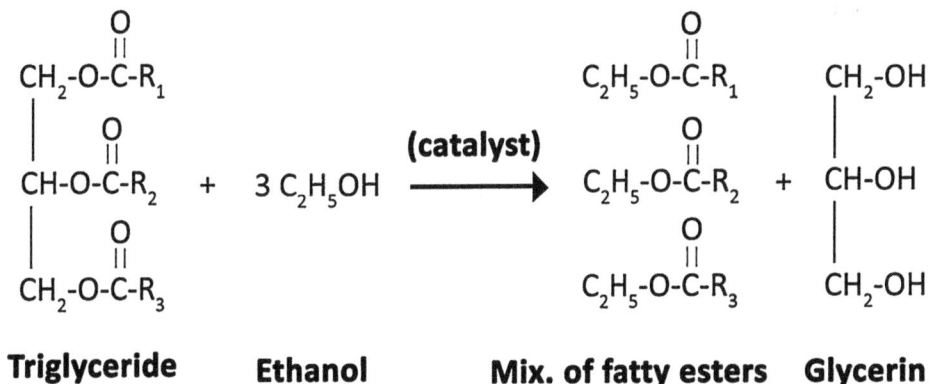

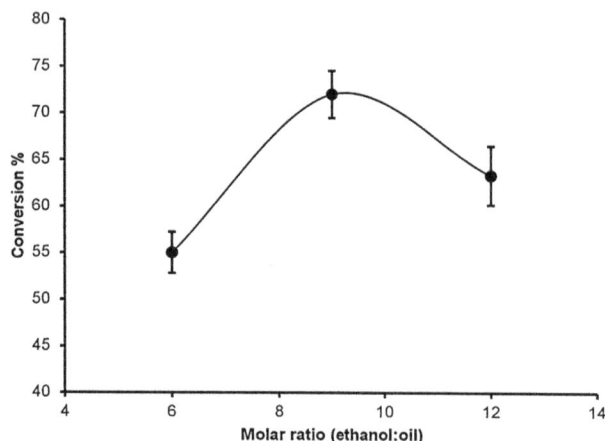

**Fig. 4** Influence of reactant molar ratio on conversion at 600 rpm, 3 wt % catalyst, temp. 70 °C, reaction time 2 h

Hence 9:1 is found to be the optimal mole ratio of ethanol to oil to achieve maximum yield. Similar results have been reported by Wang et al. [26].

**Effect of agitation speed**

Experiments have been conducted at various stirrer speeds of 400; 600 and 800 rpm to understand its influence on the biodiesel yield maintaining all other parameters constant. Figure 5 shows the results of this study according to which it was clear that 600 rpm was optimum giving maximum yield of biodiesel, 71 %. The results employing two different catalyst amounts 3 and 6 % were shown in the Figure. It was found that the 600 rpm was optimum for both the cases and that 3 % catalyst gave relatively higher yields when compared to 6 %. The existence of optimum rpm could be explained by the swirling effect at higher speeds in unbaffled vessels that leads to concentration of

solid instead of the desired suspension [27]. This effect of swirling would be more pronounced with higher catalyst amount which also would increase the solution viscosity reducing the reactant mobility and hence the yield as is evident from the Fig. 5.

**Effect of catalyst amount**

To understand the effect of catalyst amount on the biodiesel yield, experiments were conducted using various amounts of 1.5 %. 3, 4.5 and 6 % on weight basis. The results are given in Fig. 6 below according to which 3 % was the optimum amount of catalyst that gave highest biodiesel yield of 72 %. From the plot, it is clear that the biodiesel yield increased with amount of catalyst up to certain extent i.e., 3 % beyond which the yield is observed to decrease. This can be attributed to various factors like increase in solution viscosity that makes stirring ineffective, increasing importance of mass transfer than the catalyst amount under alkaline conditions and more pronounced swirling effect at higher speeds. Similar results have been reported by, Lin et al. [28] and Luengnaruemitchai et al. [29].

**Effect of reaction time**

The transesterification reaction under optimal conditions as above was conducted for 2 h with samples taken every 15 min i.e., at intervals of 15, 30, 45, 60, 75, 90, 105 and 120 min. The results obtained are shown in the Fig. 7 according to which the biodiesel yield increased with time up to 90 min achieving 70 % yield after which there is no significant change in the yield as the reaction could have reached near equilibrium conversion. It is also evident from the plot that it required about 30 min to achieve most conversion after which there is only a slight increase.

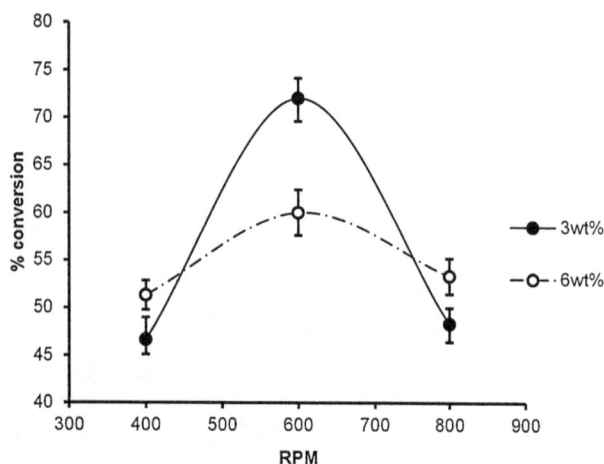

**Fig. 5** Influence of agitation speed on conversion at 70 °C, reaction time 2 h, 3 wt % and 6 wt % catalyst, 9:1 molar ratio of ethanol to oil

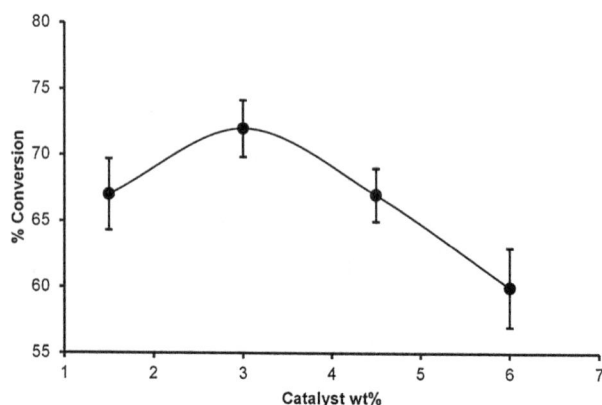

**Fig. 6** Influence of catalyst wt % on conversion at 600 rpm, ethanol/oil molar ratio 9:1, temp. 70 °C, reaction time 2 h

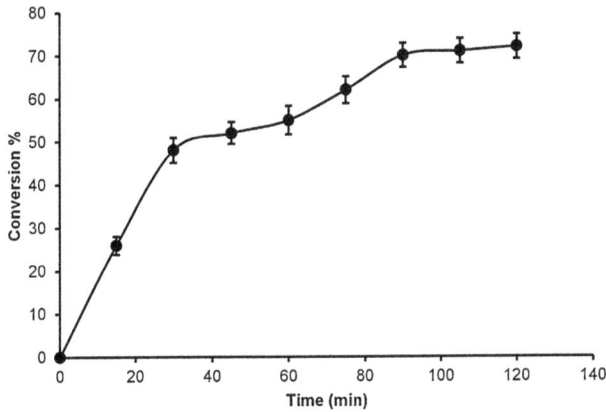

**Fig. 7** Influence of reaction time on conversion at 600 rpm, ethanol/oil molar ratio 9:1, temp 70 °C, 3 wt % catalyst

Similar trends have been reported on the influence of time on percent conversion [30–32].

### Recyclability studies

The biodiesel synthesis at the optimal conditions of 70 °C, 90 min; 3 wt % catalyst; 9:1 mol ratio of reactants (ethanol to oil) and the agitation speed of 600 rpm has been conducted up to three cycles by recycling the CaO catalyst after regenerating through filtration and calcination. The biodiesel yield of 72 % has been successfully achieved in all the three cycles within a range of 5 %. This reinforces the selection of calcium oxide as a stable catalyst in the transesterification reaction without undergoing major physico-chemical changes. Lee et al. [33] too observed that CaO and other mixed oxides were stable up to multiple cycles in the esterification and transesterification reactions.

### Reaction kinetics

*Theoretical approach*

The overall transesterification of rice bran oil can be represented in following stoichiometric equation:

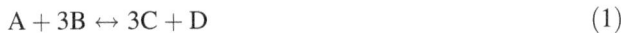

$$A + 3B \leftrightarrow 3C + D \tag{1}$$

where A is triglyceride, B is ethanol, C is Fatty acid ethyl ester and D is glycerol.

The transesterification reaction by heterogeneous catalyst occurs in a three phase system with two immiscible liquid phases (Rice bran oil and Ethanol) and solid (calcium oxide) as catalyst. Along with the transesterification, some side reactions also occur such as saponification of glycerides, ethyl esters and neutralization of fatty acids by catalyst. The following assumptions were used to model the process of transesterification.

Assumptions:

1. The transesterification reaction occurs between glycerides and ethoxide ions adsorbed on catalyst surface.
2. The mass transfer rate of ethanol towards catalyst surface and its adsorption rate onto surface of catalyst do not limit the overall rate of reaction.
3. The mass transfer rate of glycerides towards catalyst surface influences the overall reaction rate in the initial reaction period.
4. The mass transfer rate of glycerides towards the catalyst surface determines its adsorption rate onto active sites of catalyst surface during the initial reaction period.
5. Ethanol adsorption onto catalyst surface follows pseudo-first order kinetics.
6. When the external mass transfer limitation gets negligible in the later reaction period, the rate of reaction between triglycerides and ethoxide ion influences the overall reaction rate.
7. The reaction of triglyceride follows pseudo-first order reaction.
8. The rate of desorption of transesterification products from catalyst surface and their mass transfer rates into reaction mixture do not influence overall reaction rate.
9. The internal diffusion rate will not influence the transesterification reaction.
10. The reaction mixture is assumed to be homogeneous throughout i.e., its composition and catalyst distribution are uniform since the reaction is performed in a batch reactor.
11. The neutralization of free fatty acids is assumed to be negligible since the refined rice bran oil used in the reaction had around 0.15–0.2 % of free fatty acids.

The process of transesterification catalyzed by heterogeneous catalyst contains several steps, hence rate limiting step should be determined by comparing their rates. The adsorption rate of ethanol onto the catalyst surface area is equal to the rate of its concentration increase on the catalyst surface.

$$\frac{m_{cat}}{V} \frac{dQ}{dt} = k_{ad}(Q_{max} - Q) \frac{m_{cat}}{V} \tag{2}$$

where $Q$ and $Q_{max}$ are the instantaneous and maximum adsorbed ethanol concentrations on the surface of catalyst, $k_{ad}$ is the rate constant for ethanol adsorption, $m_{cat}$ is the mass of catalyst, $V$ is the volume of reaction mixture, and $t$ is the time.

For mass balance of ethanol on surface of catalyst, rate of ethanol depletion in the bulk phase is determined by the mass transfer of ethanol from bulk phase to active sites of catalyst surface and the rate of adsorption on the catalyst surface.

$$(-r_B) = \left(-\frac{dc_B}{dt}\right) - \frac{m_{cat}}{V}\frac{dQ}{dt} \qquad (3)$$

where $c_B$ is ethanol concentration adsorbed in the liquid phase and $(-r_B)$ is the rate of ethanol depletion reaction rate.

Since ethanol depletion rate is equal to rate of formation of Fatty acid ethyl esters (FAEE):

$$(-r_B) = \frac{dc_R}{dt} \qquad (4)$$

where $c_R$ is the FAEE concentration in the liquid phase and the Eq. (3) can be transformed into

$$\frac{dQ}{dt} = \left[\left(-\frac{dc_B}{dt}\right) - \frac{dc_R}{dt}\right]\frac{V}{m_{cat}} \qquad (5)$$

From the assumptions (3), (4) and (7) the mass transfer rate of triglyceride is equal to its reaction rate:

$$(-r_A) = -\frac{dc_A}{dt} = k_{s,A}.\theta.a_m.\left(c_A - c_{A,s}\right).\frac{m_{cat}}{V} = k.c_{A,s} \qquad (6)$$

where $k_{s,A}$ is the mass transfer coefficient of triglyceride, $\theta$ is the fraction of the available active surface area of catalyst, $a_m$ is the active surface area of catalyst, $c_A$ and $c_{A,s}$ are triglyceride concentrations in liquid phase and adsorbed on catalyst surface per liquid phase volume respectively and $k$ is the reaction rate constant of pseudo-first order reaction.

Introducing a new term, volumetric triglyceride mass transfer constant $k_{mt,A}$ as follows:

$$k_{mt,A} = k_{s,A}.\theta.a_m.\frac{m_{cat}}{V} \qquad (7)$$

Now the Eq. (6) becomes

$$(-r_A) = k_{mt,A}.\left(c_A - c_{A,s}\right) = k.c_{A,s} \qquad (8)$$

By rearranging the equation, easily measurable $c_A$ is used to express not measurable $c_{A,s}$:

$$c_{A,s} = \frac{k_{mt,A}}{k_{mt,A} + k}.c_A \qquad (9)$$

This can be rearranged to:

$$-\frac{dc_A}{dt} = \frac{k.k_{mt,A}}{k_{mt,A} + k}.c_A \qquad (10)$$

which could be represented as

$$-\frac{dc_A}{dt} = k_{app}.c_A \qquad (11)$$

where $k_{app}$ is the apparent rate constant which is a function of both mass transfer and chemical reaction rate. From Eq. (10) there are possibilities of two extreme situations

during transesterification reaction. Based on assumption (4), if $k_{mt,A} \ll k$ in the initial reaction period:

$$k_{app} = k_{mt,A} \qquad (12)$$

Which means in the initial reaction period, the mass transfer rate limits the overall reaction rate which and also depends on amount of catalyst according to Eq. (7) i.e.,

$$-\frac{dc_A}{dt} = k_{mt,A}.c_A \qquad (13)$$

In later stages of reaction, the adsorbed ethanol concentration decreases on catalyst surface and at the same time fraction of active catalyst surface available for triglyceride adsorption and volumetric triglyceride mass transfer coefficient increase. When $k_{mt,A} \gg k$:

$$k_{app} = k \qquad (14)$$

Which means the overall rate depends on the chemical reaction rate between adsorbed ethanol and triglyceride molecules i.e.,

$$-\frac{dc_A}{dt} = k.c_A \qquad (15)$$

The triglyceride concentration can be expressed in terms of conversion as follows:

$$c_A = c_{A_0}(1 - X_A) \qquad (16)$$

Now, Eq. (11) can be written as

$$\frac{dX_A}{dt} = k_{app}(1 - X_A) \qquad (17)$$

After integration, following equation is obtained:

$$-\ln(1 - X_A) = k_{app}t + C \qquad (18)$$

where $C$ is the integration constant. Hence, both reaction rate and mass transfer follow first order kinetics with a different rate constant ($k_{app} = k$ and $k_{app} = k_{mt,A}$ respectively).

**Internal mass transfer limitation**

If the internal mass transfer resistance exists, then it will contribute to the reaction rate. This can be verified theoretically by calculating the Thiele modulus. If the value of Thiele modulus is less than 0.4, it implies that the resistance offered by catalyst pores is negligible.

Thiele modulus for a spherical particle is given by

$$Th = \frac{R_p}{3}\sqrt{\frac{k}{D_{eff}}} \qquad (19)$$

where $R_p$ is radius of particle, $k$ is the rate constant of pseudo-first order reaction, $D_{eff}$ is effective diffusion coefficient. In our work the rate constant was found to be $0.0103\ \text{min}^{-1}$ and the largest average particle size is

15.3 μm. The effective diffusion coefficient can be calculated by the following equation:

$$D_{\text{eff}} = \frac{D\varepsilon_p}{\tau_p} \tag{20}$$

where $D$ is the molecular diffusion coefficient, $\varepsilon_p$ porosity of catalyst particle, $\tau_p$ is the catalyst particle tortuosity. The following equation can be used to calculate the molecular diffusion coefficient of triglycerides through ethanol.

$$D = 7.4 \times 10^{-8} \left( \frac{(xM)^{0.5}T}{\eta V_m^{0.6}} \right) \tag{21}$$

where $x$ is the association parameter of solvent (1.5 for ethanol), $M$ is molecular weight of solvent, $T$ is temperature in K, $\eta$ is viscosity of solution in centipoise, $V_m$ is molal volume of solute at normal boiling point in cc/g.mol.

The temperature of best case scenario is 70 °C (343 K), $\eta$ of the solution is 4.388 centipoise, molecular weight of solvent is 46 g/mol, molal volume of solute is 950.711 cc/g.mol. At these conditions value of D was found out to be $7.849 \times 10^{-7}$ m$^2$/s and $D_{\text{eff}} = 1.046^{-7}$ m$^2$/s, from which we can calculate the value of Thiele modulus which was calculated to be $8.001 \times 10^{-4}$. Hence the Thiele modulus is far below its lower limit of 0.4 and hence the internal mass transfer can be neglected.

*Experimental approach*

From equation (18), the plot of $-\ln(1 - X_A)$ vs $t$ are drawn for different reaction scenarios and were compared with first order reaction line, and the corresponding correlation coefficients were found to be 0.9346, 0.9439 and 0.932 for 70, 60 and 50 °C reaction respectively while keeping other reaction parameters same at ethanol/oil molar ratio of 9:1, 600 rpm and 3 wt % catalyst. While comparing with second order reaction, the correlation coefficient was found to be 0.634 (Fig. 8) much lesser than first order assumption. Hence it follows pseudo-first order reaction.

First order rate constants are estimated at three different temperatures of 50, 60 and 70 °C by plotting '$-\ln(1 - X_A)$ vs $t$'. These plots are shown in Fig. 9 and the corresponding $k$ values are 0.0059, 0.0082 and 0.0103 min$^{-1}$ which were employed to calculate the Arrhenius parameters.

$$k = Ae^{\frac{-E_a}{RT}} \tag{22}$$

Activation energy for the reaction, $E_a$ and the frequency factor A are estimated using Arrhenius Eq. (19) and making a plot of ln k vs $1/T$ (Fig. 10).

The Arrhenius parameters estimated are

$$E_a = 25.723 \text{ KJ/mol} \quad A = 86.091 \text{ min}^{-1}$$

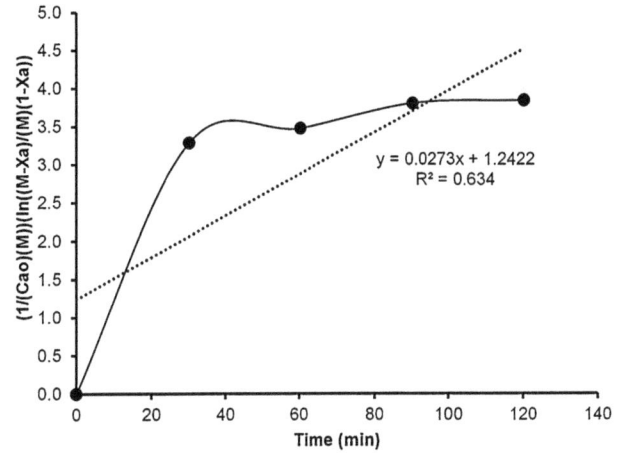

**Fig. 8** Second order reaction assumption at 600 rpm, ethanol/oil molar ratio 9:1, temperature 70 °C, 3 wt % catalyst

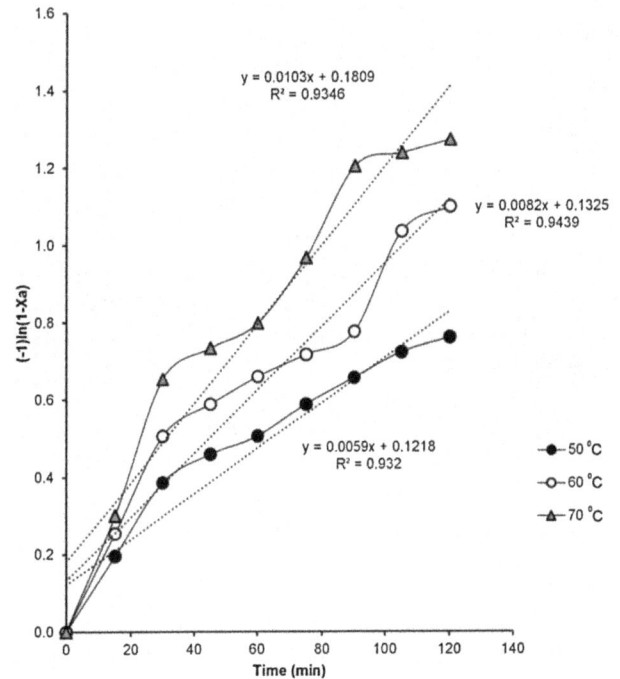

**Fig. 9** First order reaction assumption at 600 rpm, ethanol/oil molar ratio 9:1, 3 wt % catalyst and temperatures 50, 60, 70 °C

## Conclusions

In this work, biodiesel production through transesterification reaction using ethanol and not so edible and cost-effective oil of rice bran oil has been standardized with reference to various critical parameters viz., temperature, reactant mole ratio, catalyst amount, agitation speed, reaction time to achieve maximum yield. Our studies

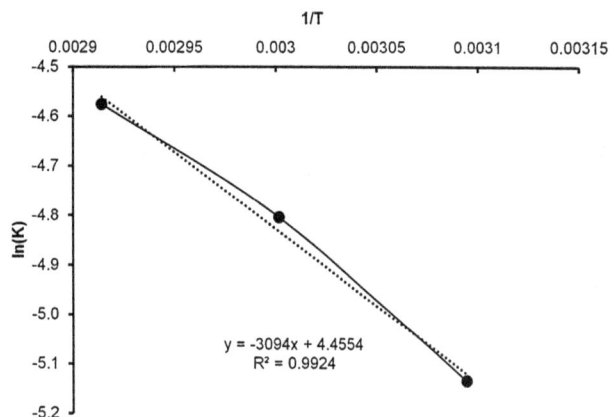

**Fig. 10** Arrhenius plot for three temperatures, 50, 60 and 70 °C

indicate that ethanol too can be used in place of methanol as alcohol in the reaction to achieve reasonable yields. The kinetic studies conducted show that the transesterification reaction is very fast with major reaction completed during the initial 30 min after which there is only a marginal increase up to 90 min. For the multiphase reaction, the controlling regime is estimated from experimental and theoretical validations and then a suitable kinetic model has been formulated that fits well with the experimental data. All the kinetic parameters have been estimated. In future, this reaction has to be further studied in identifying other critical parameters like pressure, reactor configuration, vapour phase reaction, mixing intensity with various impellers etc.

**Acknowledgments**   The authors wish to thank the management of BITS Pilani Hyderabad Campus for providing necessary facilities to carry out this research work.

# References

1. Hoffmann U (2011) Some reflections on climate change, green growth illusions and development space. Paper 205, United Nations Conference on Trade and Development (UNCTAD), December 2011. http://www.unctad.org/en/PublicationsLibrary/osgdp2011d5_en.pdf. Accessed 10 Feb 2015
2. Walker DA (2015) Biofuels—for better or worse? Ann App Biol 156(3):319–327
3. Chouhan APS, Sarma AK (2011) Modern heterogeneous catalysts for biodiesel production: a comprehensive review. Renew Sust Energ Rev 15(9):4378–4399
4. Atadashi I, Aroua M, Abdul AA, Sulaiman N (2012) The effects of water on biodiesel production and refining technologies: a review. Renew Sust Energ Rev 16(5):3456–3470
5. Salamatinia B, Mootabadi H, Bhatia S, Abdullah AZ (2010) Optimization of ultrasonic assisted heterogeneous biodiesel production from palm oil: a response surface methodology approach. Fuel Process Technol 91:441–448
6. SathyaSelvabala V, Selvaraj DK, Kalimuthu J, Periyaraman PM, Subramanian S (2011) Two-step biodiesel production from

7. Kim MJ, Park SM, Chang DR, Seo G (2010) Transesterification of triacetin, tributyrin, and soybean oil with methanol over hydrotalcites with different water contents. Fuel Process Technol 91:618–624
8. Ngamcharussrivichai C, Totarat P, Bunyakiat K (2008) Ca and Zn mixed oxide as a heterogeneous base catalyst for transesterification of palm kernel oil. Appl Catal A 341:77–85
9. Zhang Y, Dubé MA, McLean DD, Kates M (2003) Biodiesel production from waste cooking oil: 2. Economic assessment and sensitivity analysis. Bioresour Technol 90:229–240
10. Saydut A, Duz MZ, Kaya C, Kafadar AB, Hamamci C (2008) Transesterified sesame (Sesamum indicum L.) seed oil as a biodiesel fuel. Bioresour Technol 99:6656–6660
11. Rashid U, Anwar F (2008) Production of biodiesel through optimized alkaline-catalyzed transesterification of rapeseed oil. Fuel 87:265–273
12. Kawashima A, Matsubara K, Honda K (2008) Development of heterogeneous base catalysts for biodiesel production. Bioresour Technol 99:3439–3443
13. Lang X, Dalai AK, Bakhshi NN, Reaney MJ, Hertz PB (2001) Preparation and characterization of bio-diesels from various bio-oils. Bioresour Technol 80:53–62
14. Mendow G, Veizaqa NS, Querini CA (2011) Ethyl ester production by homogeneous alkaline transesterification: influence of catalyst. Bioresour Technol 102:6385–6391
15. Chen J, Wu W (2003) Regeneration of immobilized Candida antarctica lipase for transesterification. J Biosci Bioeng 95:466–469
16. Vyas AP, Verma JL, Subrahmanyam N (2010) A review on FAME production processes. Fuel 89:1–9
17. Song ES, Lim JW, Lee HS, Lee YW (2008) Transesterification of RBD palm oil using supercritical methanol. J Supercrit Fluids 44:356–363
18. Ozturk G, Kafadar AB, Duz MZ, Saydut A, Hamamci C (2010) Microwave assisted transesterification of maize (Zea mays L.) oil as a biodiesel fuel. Energ Explor Exploit 28(1):47–57
19. Coluccy JA, Borrero E, Alape F (2005) Biodiesel from an alkaline transesterification reaction of soybean oil using ultrasonic mixing. Am Oil Chem Soc 82:525–530
20. Shiu PJ, Gunawan S, Hsieh WH, Kasim NS, Ju YH (2010) Biodiesel production from rice bran by a two-step in situ process. Bioresour Technol 10:984–989
21. Ju YH, Vali SR (2005) Rice bran oil as a potential resource for biodiesel: a review. J Sci Ind Res 64:866–882
22. Lin L, Ying D, Chaitep S, Vittayapadung S (2009) Biodiesel production from crude rice bran oil and properties as fuel. Appl Energ 86:681–688
23. Zullaikah S, Lai CC, Vali SR, Ju YH (2005) A two-step acid-catalyzed process for the production of biodiesel from rice bran oil. Bioresour Technol 96:1889–1896
24. Wu H, Zhang J, Liu Y, Zheng J (2014) Biodiesel production from Jatropha oil using mesoporous molecular sieves supporting $K_2SiO_3$ as catalysts for transesterification. Fuel Process Technol 119:114–120
25. Wu H, Zhang J, Wei Q, Zheng J, Zhang J (2013) Transesterification of soybean oil to biodiesel using zeolite supported CaO as strong base catalysts. Fuel Process Technol 109:13–18
26. Wang Y, Ou S, Liu P, Zhang Z (2007) Preparation of biodiesel from waste cooking oil via two-step catalyzed process. Energ Convers Manag 48(1):184–188
27. McCabe WL, Smith JC, Harriot P (2004) Unit operations in chemical engineering, 7th edn. McGraw Hill Chemical Engineering Series, New York

28. Lin L, Ying D, Chaitep S, Vittayapadung S (2009) Biodiesel production from crude rice bran oil and properties as fuel. Appl Energy 86:681–688

29. Noiroj K, Intarapong P, Luengnaruemitchai A, Jai-In S (2009) A comparative study of $KOH/Al_2O_3$ and KOH/NaY catalysts for biodiesel production via transesterification from palm oil. Renew Energy 34:1145–1150

30. Fernandez MB, Tonetto GM, Crapiste G, Damiani DE (2007) Kinetic of the hydrogenation of sunflower oil over alumina supported palladium catalyst. Int J Chem React Eng 5(A10):1–22

31. Veljković VB, Stamenković OS, Todorović ZB, Lazić ML, Skala DU (2009) Kinetics of sunflower oil methanolysis catalyzed by calcium oxide. Fuel 88:1554–1562

32. Shahbazi MR, Khoshandam B, Nasiri M, Ghazvini M (2012) Biodiesel production via alkali-catalyzed transesterification of Malaysian RBD palm oil—characterization, kinetics model. J Taiwan Inst Chem Eng 43:504–510

33. Lee AF, Bennett JA, Manayila JC, Wilson K (2014) Heterogeneous catalysis for sustainable biodiesel production via esterification and transesterification. Chem Soc Rev 43:7887–7916

# 12

# Synthesis, characterization and electrochemical properties of poly (phenoxy-imine)s containing carbazole unit

İsmet Kaya[1] · Sebra Çöpür[1] · Hatice Karaer[1,2]

**Abstract** Several new Schiff base polymers were synthesized via oxidative polymerization method in an aqueous alkaline medium in the presence of NaOCl as an oxidant and were confirmed by FT-IR, $^1$H-NMR, $^{13}$C-NMR and UV–Vis spectroscopic techniques. Furthermore, cyclic voltammetry measurements were carried out and the HOMO–LUMO energy levels and electrochemical band gaps ($E_g'$) were calculated. Additionally, the optical band gaps ($E_g$) were determined using their UV–Vis spectra of the materials. The morphologic properties of the polymers were investigated by scanning electron microscopy. In addition, the number average molecular weight ($M_n$), weight average molecular weight ($M_w$) and polydispersity index values of the polymers were determined by gel permeation chromatography technique. Electrical conductivity measurements of the doped (with iodine) and undoped polymer related to doping time were carried out by four-point probe technique using a Keithley 2400 electrometer. Their thermal behaviors were determined by TG–DTA and DSC measurements. The synthesized compounds were soluble in common solvents such as DMF, THF and DMSO. Photoluminescence properties of the polymers were determined in different concentrations of DMF solvent.

✉ İsmet Kaya
  kayaismet@hotmail.com

[1] Polymer Synthesis and Analysis Laboratory, Department of Chemistry, Çanakkale Onsekiz Mart University, 17020 Çanakkale, Turkey

[2] Department of Chemistry, Faculty of Sciences, Dicle University, 21280 Diyarbakır, Turkey

**Keywords** Carbazole · Fluorescence · Thermal analysis · Poly(phenoxy-imine) · Band gaps

## Introduction

Poly(imine)s, known as Schiff base polymers or poly (azomethine)s or also named polyazines (when hydrazine is used as diamine compound) [1] which are of great interest to researchers because of to their potential applications and advantageous properties. Recently, polyazomethines have attracted much attention of both industries and academia and they have been widely investigated for their electrochemical properties, thermal stability, fluorescence, intrinsic conductivity [2].

Polyimines conjugated polymers have claimed the attention of researchers because of their potentially advantageous electronic applications, such as their electrical properties and environmentally stability, with acceptable mechanical strength [3]. Polyazomethines are conducting polymers [4] that usually show an optical absorption band in the visible region owing to their extended delocalization of the $\pi$ electrons along the polymer backbone. Upon doping with suitable dopants, charge carriers, namely bipolaron and polaron, are formed in the conjugated backbone. This class of polymers was primarily found to be electroactive as well as semiconductive materials [5, 6], and their conductivity could be increased by doping with a dopant like iodine [2]. Furthermore, Schiff base polymers have been become increasingly interesting in the field of optical materials since they possess great potential for device applications like light-emitting diodes, photovoltaic cells and thin film transistors [7].

Poly(azomethine)s including conjugated bonding and active hydroxyl group have been studied for more than

60 years, and used in several fields [8]. The oxidative polymerization method is simply the reaction of compounds including –OH groups and active functional groups (–CHO, –NH$_2$, –COOH) in their structure with the oxidants like air oxygen NaOCl, H$_2$O$_2$ an in the aqueous alkaline medium [9].

Carbazole-containing polymers are of great interest owing to their several potential for applications in organic electronics, such as organic solar cells, organic field effect transistors (OFET) and organic light-emitting devices (OLED), etc. [10].

In this study, new Schiff bases were synthesized by condensation reaction of 4-diethylaminosalicylaldehyde, 3,4-dihydroxybenzaldehyde and 2,4-dihydroxybenzaldehyde compounds with 3-amino-9-ethyl-carbazole. Then, these products were polymerized via oxidative polycondensation method in an aqueous alkaline medium in the presence of NaOCl as an oxidant. The structures of all compounds were confirmed by FT-IR, UV–Vis, $^1$H-NMR

and $^{13}$C-NMR measurements. Thermal stabilities of all compounds were determined by TG–DTA and DSC measurements. Also, the conductivity and photoluminescence (PL) properties of polymers were determined from four-point probe technique and spectrofluorophotometer measurements, respectively.

## Experimental

### Materials

3-Amino-9-ethyl-carbazole, ethyl alcohol, ethyl acetate, chloroform, *N,N*-dimethylacetamide, *N,N*-dimethylformamide, dimethyl sulfoxide, acetonitrile and sodium hypochlorite (NaOCl, 37%) were supplied from Merck Chemical Co. (Germany). 4-Diethylamino salicylaldehyde (Alfa Aesar), 3,4-dihydroxybenzaldehyde (Fluka) and 2,4-dihydroxybenzaldehyde were supplied from Acros.

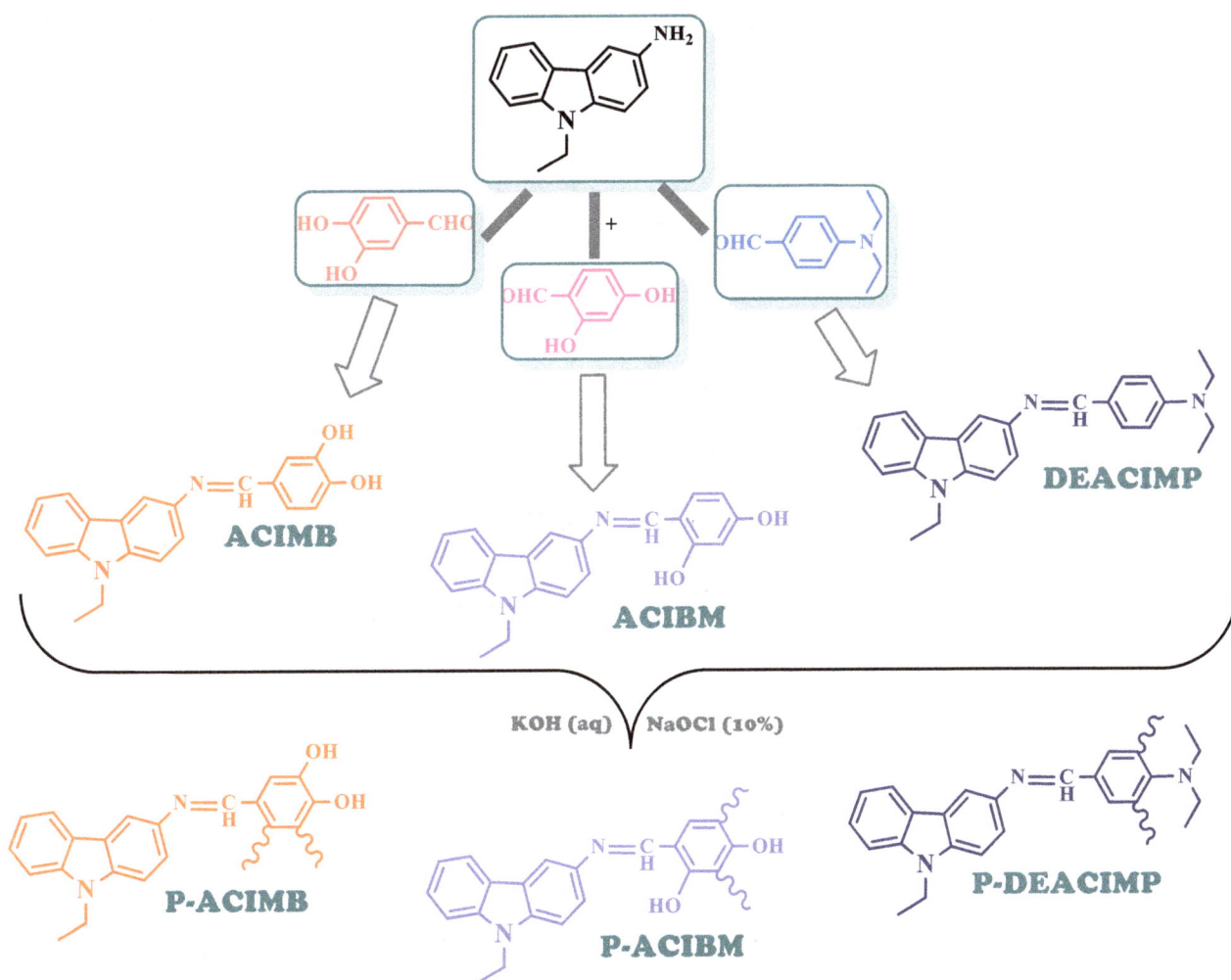

**Scheme 1** Syntheses of Schiff bases and their polymers

## Synthesis of the monomers (DEACIMP, ACIMB and ACIBM)

The synthesis of 5-(diethylamino)-2-[(3-amino-9-ethyl-carbazole) imino methyl] phenol (DEACIMP) was synthesized according to the literature [11] (Scheme 1) as follows: 4-diethylamino salicylaldehyde (1.77 g) and 3-amino-9-ethyl-carbazole (1.95 g) were dissolved in 20 mL absolute ethanol in two separate beakers, which were then mixed. This mixture was refluxed for 5 h in a two-necked flask and cooled to room temperature. The precipitate formed was filtered, washed with ethanol and then dried under reduced pressure. The same procedure was used to obtain 4-[(3-amino-9-ethyl-carbazole) imino methyl] benzene-1,2-diol (ACIMB) and 4-[(3-amino-9-ethyl-carbazolyl) benzene imine methyl]-1,3-diol (ACIBM) the 3,4-dihydroxybenzaldehyde (1.38 g), 3-amino-9-ethyl-carbazole (1.95 g) and 2,4-dihydroxy-benzaldehyde (1.38 g), 3-amino-9-ethyl-carbazole (1.95 g) were used for synthesis the ACIMB and ACIBM, respectively. The yields of DEACIMP, ACIMB and ACIBM compounds were found to be 80, 85, 89, respectively.

## General synthesis procedure of P-DEACIMP, P-ACIMB and P-ACIBM polymers

P-DEACIMP, P-ACIMB and P-ACIBM were synthesized through the oxidative polycondensation of DEACIMP, ACIMB and ACIBM with aqueous solution of NaOCl (10%). P-DEACIMP was synthesized through oxidative polycondensation of 5-(diethylamino)-2-[(3-amino-9-ethyl-carbazole) imino methyl] phenol using aqueous solution of NaOCl (10%). The 5-(diethyl-amino)-2-[(3-amino-9-ethyl-carbazole) imino methyl] phenol (0.385 g) was dissolved in an aqueous solution of 25 mL KOH (1.0 M) placed into a 100-mL three-necked round-bottom flask, which was fitted with a condenser, thermometer. Furthermore, a funnel containing NaOCl, which was added dropwise over about 30 min, was equipped. The reaction mixture was stirred at 100 °C for 24 h, cooled to room temperature and then 25 mL HCl (1.0 M) was added to solution. The P-DEACIMP was washed with water for the separation from mineral salt. The polymers were dried at 60 °C in an oven for 24 h [8, 12, 13].

The same procedure was used to obtain the P-ACIMB and P-ACIBM but the 4-[(3-amino-9-ethyl-carbazole) iminemethylbenzene]-1,2-diol (ACIMB) (0.33 g) and 4-[(3-amino-9-ethyl-carbazole) iminomethylbenzene]-1,3-diol (ACIBM) (0.33 g) were used for synthesis the P-ACIMB and P-ACIBM, respectively (Scheme 1). The yields of P-DEACIMP, P-ACIMB and P-ACIBM compounds were found to be 70, 72, 76, respectively.

## Characterization techniques

A PerkinElmer spectrum one FT-IR system was used to determine the chemical structure of the monomers and polymers. Measurements were performed in solid powder form at room temperature using universal ATR sampling accessory within the wavelengths of 4000–650 cm$^{-1}$. UV–Vis spectroscopy, was used to study the electronic transition in the

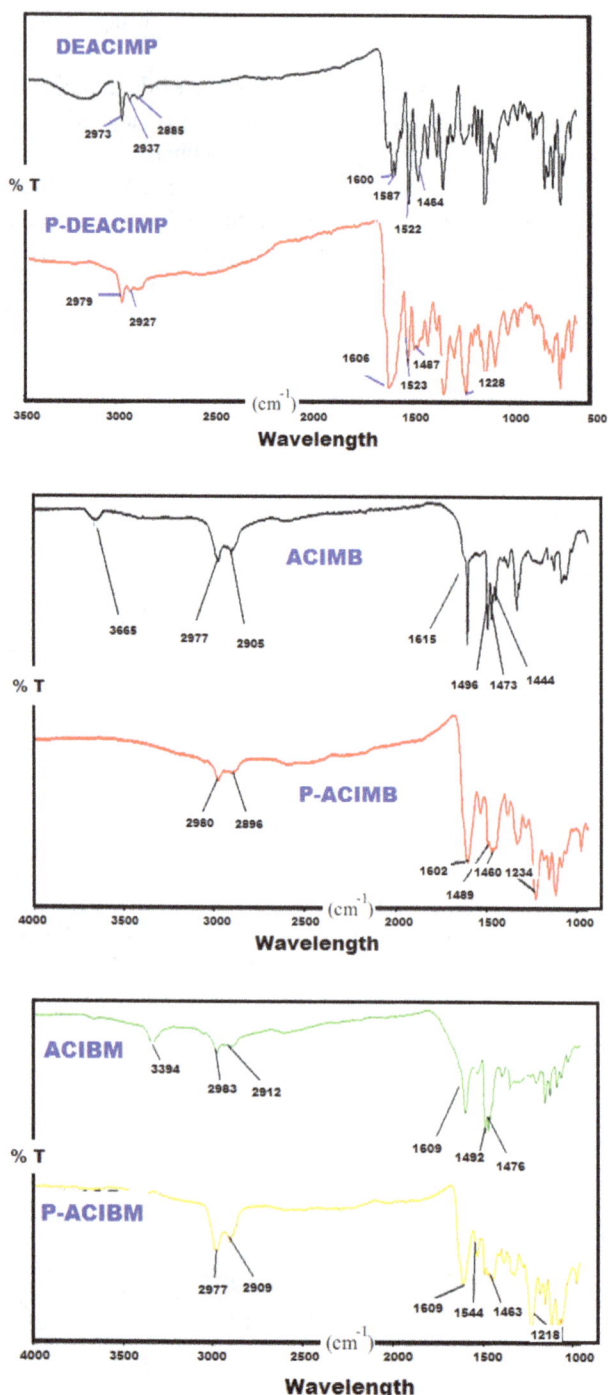

Fig. 1 FT-IR spectra of the monomers and polymers

**Table 1** FT-IR spectral data of monomers and polymers

| Compounds | Wave number (cm$^{-1}$) | | | | | |
|---|---|---|---|---|---|---|
| | –OH | C–H (aromatic) | C–H (aliphatic) | –C=N | C=C (aromatic) | C–O |
| DEACIMP | 3300 | 2973 | 2936 | 1600 | 1587, 1522 | – |
| P-DEACIMP | – | 2978 | 2927 | 1604 | 1532, 1486 | 1227 |
| ACIMB | 3665 | 2976 | 2905 | 1615 | 1495, 1473 | – |
| P-ACIMB | – | 2979 | 2895 | 1602 | 1489, 1460 | 1223 |
| ACIBM | 3393 | 2983 | 2911 | 1608 | 1492, 1476 | – |
| P-ACIBM | – | 2976 | 2908 | 1608 | 1544, 1463 | 1217 |

UV–Vis region of the all compounds. Measurements were performed by an AnalytikJena Specord 210 Plus with wavelength range of 190–900 nm by ethyl alcohol solvent at 25 °C. $^1$H and $^{13}$C NMR spectra (Bruker AV400 FT-NMR spectrometer operating at 400.1 and 100.6 MHz, respectively) were also recorded in DMSO-$d_6$ at 25 °C. The tetramethylsilane was used as internal standard. Thermal data were obtained by using a PerkinElmer diamond thermal analysis system. TGA–DTA measurements were performed between 20 and 1000 °C (in $N_2$, rate 10 °C min$^{-1}$). DSC analyses of polymers were carried out by a PerkinElmer Pyris sapphire DSC. DSC measurements were conducted between 25 and 450 °C (in $N_2$, rate 10 °C min$^{-1}$). The number average molecular weight ($M_n$), weight average molecular weight ($M_w$) and polydispersity index (PDI) were determined by gel permeation chromatography-light scattering (GPC-LS) device by Malvern Viscotek GPC Dual 270 max. For GPC investigations a medium 300 × 8.00 mm dual column light scattering detector (LS) and a refractive index detector (RID) were used to analyze the products at 55 °C. LiBr (40 mM) was added to the DMF mobile phase to dissociate molecular aggregates of polymer during GPC analysis. Surface morphology of the polymers was determined by scanning electron microscope (SEM) (JEOL, JSM-7100 model). Cyclic voltammetry (CV) measurements were carried out with a CH 660 C Electrochemical Analyzer (CH Instruments, Texas, USA) at a potential scan rate of 20 mV/s. A Shimadzu RF-5301PC spectrofluorophotometer was used in fluorescence measurements. Emission spectra of the synthesized polymers were obtained in different concentration of DMF solvent. Also, to obtain maximal emission intensity values of polymers were investigated in different concentrations of DMF solutions.

## Results and discussion

### Structural characterization of the monomers and polymers

The structures of the monomers and polymers were confirmed by FT-IR, $^1$H and $^{13}$C-NMR spectra. FT-IR spectra

of the monomers and polymers are shown in Fig. 1. According to these spectra, the characteristic peaks such as etheric groups, aromatic and aliphatic C–H stretching peaks are observed, as expected. The peaks related to Ar–O bonds are in the range of 1217–1235 cm$^{-1}$ [14], indicating that the polymerization was achieved. The bands due to hydroxyl groups (–OH) are observed at around 3300–3670 cm$^{-1}$ (Fig. 1) [15].

Aromatic –CH peaks were observed at around 2980 cm$^{-1}$ for all the compounds. Furthermore, peaks at 1450–1650 cm$^{-1}$ were assigned to benzene ring and (C=C) moiety and those at 1522–1590 cm$^{-1}$ were attributed to (C–O) stretching [16] while peaks for aliphatic group were seen at 2900–2940 cm$^{-1}$.

The peak at 1600–1620 cm$^{-1}$ corresponds to –CH=N stretching vibration of imine moiety. The peaks of vibration bonding are also shown in Table 1. As seen in Fig. 1, peaks of polymers were broader than that of monomers after the polycondensation reaction owing to their polyconjugated structures. Other characteristic absorption bands are also presented in Table 1.

Furthermore, the edged peaks of P-DEACIMP, P-ACIMB and P-ACIBM were broader and decreased numerically due to the increase in molecular weight after polymerization reactions [17], confirming polymerization of DEACIMP, ACIMB and ACIBM.

$^1$H NMR and $^{13}$C NMR spectra of the monomers and polymers were recorded in DMSO-$d_6$. $^1$H data of the monomers and polymers are listed in Table 2 and the spectrum of a representative polymer P-DEACIMP and DEACIMP are shown in Fig. 2. The $^1$H and $^{13}$C NMR spectra of DEACIMP and P-DEACIMP data are given in Figs. 2 and 3. $^1$H and $^{13}$C NMR data results of the synthesized compounds are listed in Table 2 except for DEACIMP and P-DEACIMP. When phenol-based Schiff bases were polymerized by oxidative polycondensation, they were combined by C–C binding at *ortho* and/or *para* position of the ring or alternatively C–O–C binding through oxygen atom of –OH moiety (Scheme 1) [2, 18]. $^1$H-NMR spectra of the polymers contained three types of signals assigned as follows: the singlet at 8.05–8.98 ppm, –CH=N–; the multiple at 6.40–8.50 ppm, aromatic protons. The proton resonances

**Table 2** NMR spectra data of the monomers and polymers

| Compounds | NMR spectra data |
|---|---|
| **ACIMB** | $^1$H NMR (DMSO-$d_6$, ppm, δ): 10.25 and 14.08 (s, –OH), 1.30 (t, Ar–He), 4.43 (m, Ar–Hd), 6.48–8.20 (m, Ar–H), 8.25 (s, –CH=N) <br> $^{13}$C NMR (DMSO-$d_6$, ppm, δ): 118 (C1-H), 121 (C2-H and C3-H), 126 (C4-ipso), 133 (C5-ipso), 109 (C6-H), 107 (C7-ipso), 138 (C8-ipso), 38 (C9-H), 14 (C10-H), 112 (C11-H), 160 (C12-ipso), 162 (C13-H), 139 (C14-ipso), 156 (C15-H), 140 (C16-H), 161 (C17-ipso), 129 (C18-H), 115 (C19-H), 122 (C20-H and C21-H) |
| **P-ACIMB** | $^1$H NMR (DMSO-$d_6$, ppm, δ): 11.02 and 13.20 (s, –OH), 1.27 (t, Ar–He), 4.38 (q, Ar–Hd), 7.40–8.03 (m, Ar–H), 8.05 (s, –CH=N) <br> $^{13}$C NMR (DMSO-$d_6$, ppm, δ): 133 (C1-H), 110 (C2-H), 121 (C3-H), 118 (C4-H), 120 (C5-H), 126 (C6-ipso), 105 (C7-ipso), 135 (C8-ipso), 36 (C9-H), 14 (C10-H), 111 (C11-H), 160 (C12-ipso), 165 (C13-H), 118 (C14-ipso), 107 (C15-H), 126 (C16-ipso), 161 (C17-ipso), 123 (C18-ipso), 139 (C19-ipso), 125 (C20-H), 112 (C23-H) |
| **ACIBM** | $^1$H NMR (DMSO-$d_6$, ppm, δ): 10.24 and 14.07 (s, –OH), 1.31 (t, Ar–He), 4.45 (m, Ar–Hd), 6.48–8.20 (m, Ar–H), 8.96 (s, –CH=N) <br> $^{13}$C NMR (DMSO-$d_6$, ppm, δ): 121 (C1-H), 119 (C2-H), 122 (C3-H and C4-H), 126 (C5-ipso), 109 (C6-H), 102 (C7-ipso), 133 (C8-ipso), 39 (C9-H), 15 (C10-H), 112 (C11-H), 140 (C12-ipso), 160 (C13-H), 118 (C14-ipso), 162 (C15-ipso and C18-ipso), 107 (C16-H), 118 (C17-H), 109 (C19-H), 121 (C20-H) |
| **P-ACIBM** | $^1$H NMR (DMSO-$d_6$, ppm,) δ): 11.00 and 13.00 (s, –OH), 7.22–8.33 (m, Aromatic protons), 8.87 (s, –CH=N), 1.30 (t, Ar–He), 4.43 (m, Ar–Hd). $^{13}$C NMR (DMSO-$d_6$, ppm, δ): 126 (C1-ipso), 107 (C2-H), 120 (C3-H), 116 (C4-H), 120 (C5-H), 125 (C6-ipso), 103 (C7-H), 127 (C8-ipso), 40 (C9-H), 14 (C10-ipso), 112 (C11-H), 140 (C12-ipso), 160 (C13-H), 110 (C14-ipso), 129 (C15-H), 116 (C16-ipso), 164 (C17-ipso), 126 (C18-ipso), 156 (C19-ipso), 112 (C20-H), 125 (C21-H) |

of hydroxyl (–OH) groups were observed at 12.01, 10.25 and 14.08, 10.24 and 14.07 ppm for DEACIMP, ACIMB and ACIBM, respectively. The proton signal of hydroxyl (–OH) groups were observed at 12.85, 11.02 and 13.20, 11.00 and 13.00 ppm for P-DEACIMP, P-ACIMB and P-ACIBM, respectively. Meanwhile, the FT-IR spectra of polymers show bands in 1217, 1223 and 1227 cm$^{-1}$ is assignable to the phenolic C–O stretching vibration [2].

The azomethine protons are observed in 8.84, 8.25 and 8.96 ppm for DEACIMP, ACIMB and ACIBM monomers, 8.90, 8.05 and 8.87 ppm for P-DEACIMP, P-ACIMB and P-ACIBM polymers, respectively. The $^1$H NMR spectra of polymers shows broad signals for aromatic protons which confirm the participation of aromatic ring in polymerization [2, 19]. New signals were observed in the region of 169, 162 and 168 ppm in the $^{13}$C NMR spectra of

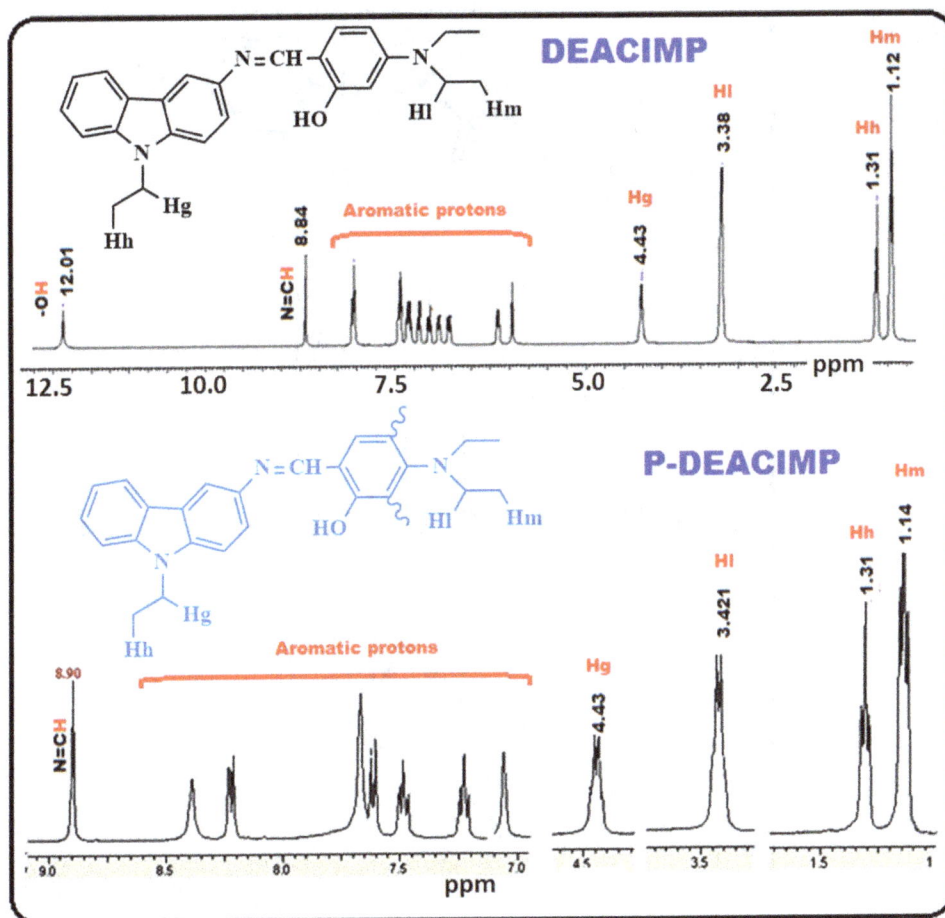

**Fig. 2** $^1$H NMR spectra of DEACIMP and P-DEACIMP

P-DEACIMP, P-ACIMB and P-ACIBM, respectively, due to –C–O–C– coupling, confirming polymerization occurred via –OH moiety [20].

Appearance of the new short resonances at 109,148 and 135,139 ppm obviously indicate polymerization of DEACIMP, ACIMB and ACIBM, respectively, i.e., –C–C binding occurs at ortho or para position of phenol by distribution of the phenoxy radical to the ring. The azomethine and aromatic carbon signals were observed at 159 and 97–163, 162 and 107–161, 160 and 102–162 ppm in the $^{13}$C NMR spectra of DEACIMP, ACIMB and ACIBM, respectively. Thus, we can easily conclude that the NMR data confirming the structures of the aimed products.

## GPC analysis of polymers

Gel permeation chromatography (GPC) analyses of compounds (P-DEACIMP, P-ACIBM and P-ACIMB) were performed at 55 °C by DMF as eluent at a flow rate of 1.0 mL min$^{-1}$. The number average molecular weight ($M_n$), weight average molecular weight ($M_w$) and polydispersity index (PDI, $M_w/M_n$) values of compounds were calculated according to a polystyrene standard calibration curve. The weight average molecular weight ($M_w$) and polydispersity index (PDI, $M_w/M_n$) values of P-DEACIMP, P-ACIBM and P-ACIMB were found to be 8350, 6400, and 9200 and 1.25, 1.22 and 1.27, respectively. These results also agree with the solubility tests. P-ACIBM had lower molecular weight than P-DEACIMP and P-ACIMB polymers. So it was a fine-soluble polymer in common organic solvents and the others had lower solubilities owing to their high-molecular weighted structures.

## Optical properties of compounds

UV–Vis spectra of the compounds recorded in ethyl alcohol at room temperature are given in Fig. 4. Their optical band gaps ($E_g$) were calculated as in the literature [21] and results are shown in Table 3.

$$E_g = 1242/\lambda_{onset}, \qquad (1)$$

where $\lambda_{onset}$ is the onset wavelength which may be determined by intersection of two tangents on the absorption edges. $\lambda_{onset}$ also indicates the electronic transition start wavelength. According to results, one can easily conclude

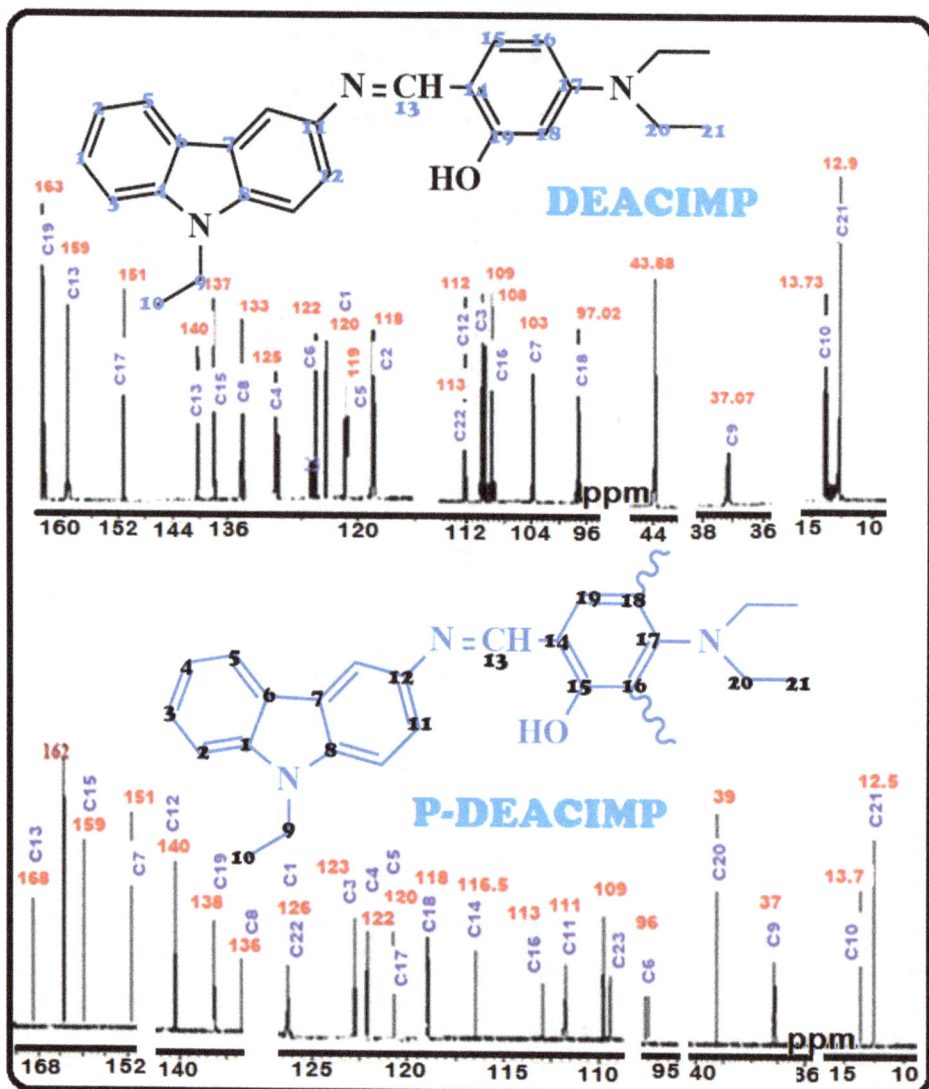

**Fig. 3** $^{13}$C NMR spectra of DEACIMP and P-DEACIMP

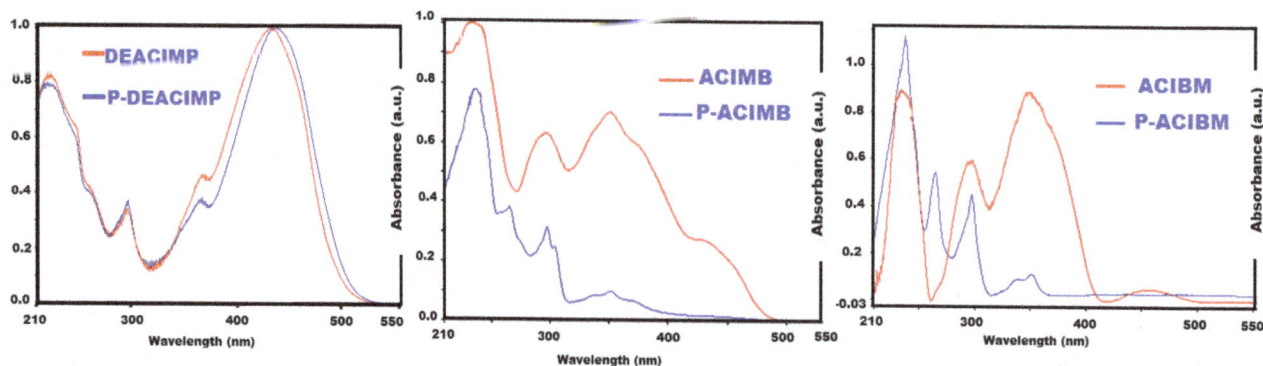

**Fig. 4** Absorption spectra of the monomers and polymers (conc.: 100 mg L$^{-1}$ in ethyl alcohol solvent)

that the polymers have low optical band gaps. Because of monomers and polymers compounds to be having the same functional groups, some UV–Vis absorption bands of their were observed as similar.

It is known that the solubility behavior is very advantageous for the processing of these materials for technological applications [4]. Thus, the solubility properties of the polymers and monomers were tested in

**Table 3** Optical electronic structure parameters of the monomers and polymers

| Compound | $E_g^a$ (eV) | $\lambda_{onset}$ (nm) | $E_{ox}^b$ (eV) | $E_{red}^c$ (eV) | HOMO$^d$ (eV) | LUMO$^e$ (eV) | $E_g'^f$ (eV) |
|---|---|---|---|---|---|---|---|
| DEACIMP | 2.43 | 510 | 1.44 | −1.73 | −5.83 | −2.66 | 3.17 |
| P-DEACIMP | 2.41 | 515 | 1.28 | −1.30 | −5.67 | −3.09 | 2.58 |
| ACIMB | 2.60 | 478 | 1.32 | −1.29 | −5.71 | −3.10 | 2.60 |
| P-ACIMB | 3.13 | 396 | 1.29 | −1.23 | −5.68 | −3.16 | 2.51 |
| ACIBM | 3.05 | 406 | 0.92 | −1.33 | −5.31 | −3.06 | 2.25 |
| P-ACIBM | 3.01 | 412 | 0.80 | −0.74 | −5.19 | −3.65 | 1.54 |

$^a$ Optical band gap

$^b$ Oxidation peak potential

$^c$ Reduced peak potential

$^d$ Highest occupied molecular orbital

$^e$ Lowest unoccupied molecular orbital

$^f$ Electrochemical band gap

**Table 4** Solubility characteristics of the monomers and polymers (1 mg mL$^{-1}$) at 25 °C

| Solvent | DEACIMP | ACIMB | ACIBM | P-DEACIMP | P-ACIMB | P-ACIBM |
|---|---|---|---|---|---|---|
| DMF | + | + | + | + | + | + |
| DMA | + | + | + | + | + | + |
| DMSO | + | + | + | + | + | + |
| THF | + | + | + | + | + | + |
| Ethyl acetate | + | + | + | ⊥ | − | − |
| Chloroform | + | + | + | ⊥ | ⊥ | ⊥ |
| Acetone | + | + | + | ⊥ | ⊥ | ⊥ |
| Ethanol | + | + | + | ⊥ | ⊥ | ⊥ |
| Acetonitrile | + | + | + | ⊥ | ⊥ | + |

(+), soluble; (⊥), partially soluble; (−), insoluble

different solvents by using 1 mg of compound in 1 mL solvent and shown in Table 4.

Fluorescence properties of the synthesized compounds are determined using DMF solutions at different concentrations [22]. Also, the optimization of the concentrations to obtain maximal emission intensity is investigated in DMF. Concentration effects on the fluorescence properties are shown in Fig. 5. Maximum emission intensity values of the compounds are also given in Table 5. When concentration of DEACIMP, P-DEACIMP, ACIMB were decreased, they showed bathochromic shift in the range 14–35 nm. On the other hand, a significant change in the emission wavelength of these polymers was not observed. As seen in Table 5, these results showed that P-ACIMB polymer has higher emission intensity values at $9.76 \times 10^{-6}$ and $1.95 \times 10^{-5}$ g mL$^{-1}$ than the other synthesized polymers and monomers. But P-DEACIMP and ACIBM have the lowest emission intensity as polymer and monomer, respectively.

Stokes shift ($\Delta\lambda_{ST}$) is the difference between positions of the band maxima of emission and excitation spectra of the same electronic transition. This knowledge offers significant conformational differences between ground state ($S_0$) and the first excited state ($S_1$). If the $\Delta\lambda_{ST}$ value is too small, the emission and excitation spectra will overlap more. So the emitted light will be self-absorbed and the photoluminescence efficiency will decrease. $\Delta\lambda_{ST}$ values are listed in Table 5. $\Delta\lambda_{ST}$ values of P-DEACIMP, P-ACIMB and P-ACIBM are calculated as 48, 31 and 61, respectively. According to $\Delta\lambda_{ST}$ values, the synthesized P-ACIBM can be used for the production of fluorescence sensor owing to high $\Delta\lambda_{ST}$ value [23]. It can be seen from Fig. 5 that with increase in the concentrations of the solutions, the absorption spectra of the compounds were observed to shift to the visible region.

**Thermal properties of the polymers**

Thermal degradation data (TGA–DTA and DTG) are listed in Table 6. TGA curves of the synthesized monomers and polymers are also shown in Fig. 6. According to the TGA results, the initial degradation temperatures ($T_{on}$) of the monomers are higher than their polymers, expect for P-ACIBM. This could be explained by the formation of C–

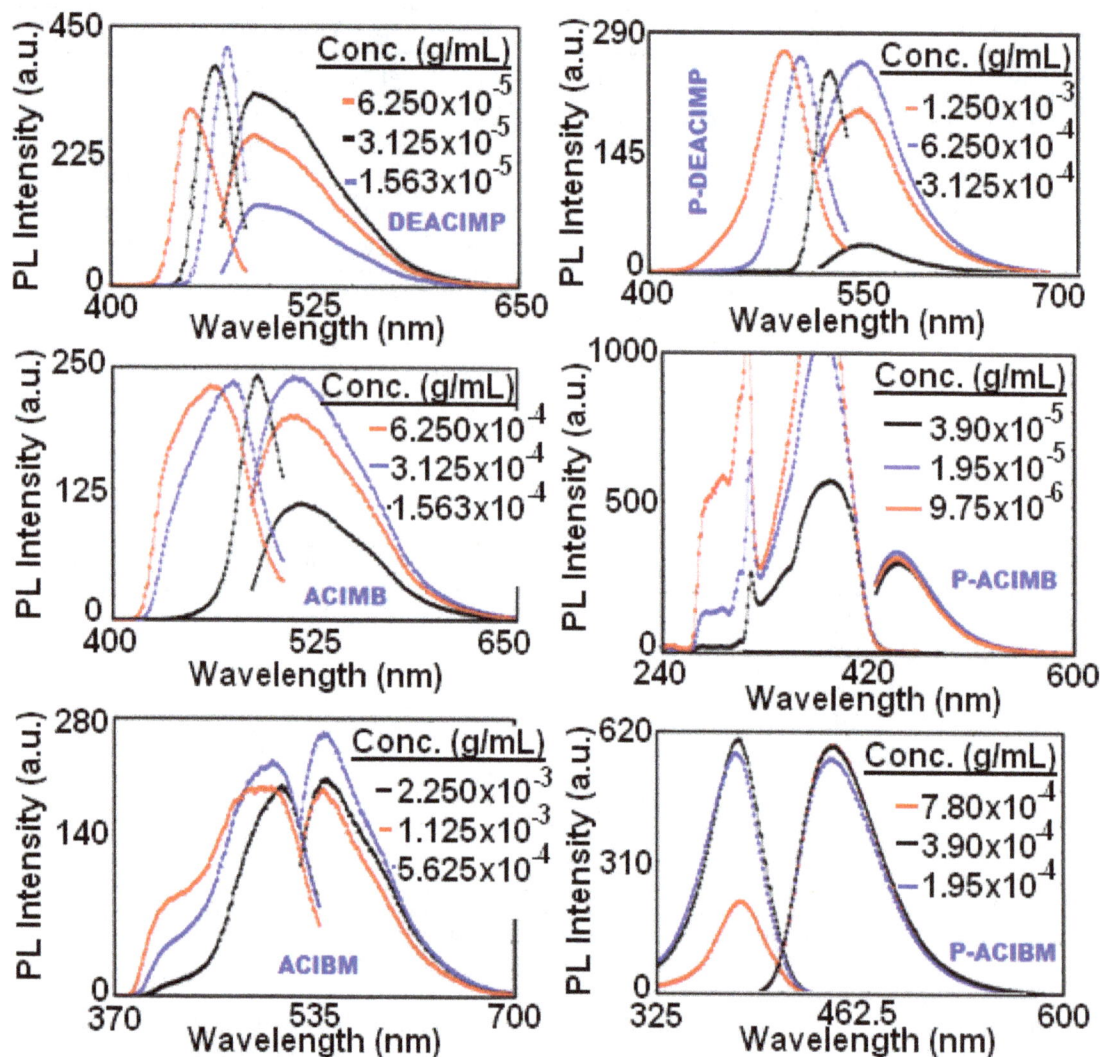

**Fig. 5** Emission spectra of the synthesized monomers and polymers in different concentrations (*PL intensity* photoluminescence intensity, slit width: 3 nm, in DMF solvent)

O etheric bond during polymerization. This weak bond is easily broken at mild temperatures and makes the polymer thermally unstable [21, 22]. Furthermore, when % char amounts are compared, % char of ACIMB was higher than other compounds at 1000 °C. In addition, Table 6 exhibits listed temperatures corresponding to 5, 10, 20 and 50% weight losses of the all compounds. $T_{on}$ values of monomers and polymers were found between 219–320 and 201–230 °C, respectively. According to TG curves of ACIMB, ACIBM, P-DEACIMP, P-ACIMB, and P-ACIBM, the losses of absorbed water were found to be 2, 1.7, 2.7, 5.1 and 3.9%, respectively, between 20 and 150 °C. DEACIMP monomer was degraded at the one step between 320 and 540 °C and its weight loss was 86.5% at this step. ACIMB and ACIBM monomers were degraded in two steps: between 220–385 and 219–308 for the first step; 385–690 and 308–665 °C for the second step and their weight losses

were 23, 16; 21 and 35% at these steps, respectively. P-DEACIMP, P-ACIMB and P-ACIBM polymers were degraded in two steps: between 230–370, 209–380 and 201–620 for the first step; 370–570, 380–600 and 620–1000 °C for the second step and their weight losses were 53, 37, 50; 25, 22 and 14% at these steps, respectively.

According to DSC measurements of polymers, the $T_g$ and $C_p$ values of P-DEACIMP, P-ACIMB, and P-ACIBM were found to be 123, 105 and 104 °C and 2.083, 0.086, and 0.075 J $g^{-1}$ $K^{-1}$, respectively. The results indicated that P-DEACIMP has the highest $T_g$.

**Morphologic properties**

Morphological properties of the polymers are obtained by SEM technique. SEM images of compounds were recorded using a Jeol JSM-7100F Schottky instrument in a powder

**Table 5** The maximum emission intensity values that obtained from fluorescence spectra of the monomers and polymers as a function of concentration

| Compounds | Concentration (mg mL$^{-1}$) | [a]$\lambda_{ex}$ (nm) | [b]$\lambda_{em}$ (nm) | [c]$\lambda_{max(ex)}$ (nm) | [d]$\lambda_{max(em)}$ (nm) | [e]$I_{ex}$ | [f]$I_{em}$ | [g]$\Delta\lambda_{ST}$ |
|---|---|---|---|---|---|---|---|---|
| DEACIMP | $6.250 \times 10^{-5}$ | 460 | 496 | 450 | 495 | 305 | 260 | 35 |
| | $3.125 \times 10^{-5}$ | 460 | 496 | 467 | 495 | 378 | 332 | 35 |
| | $1.563 \times 10^{-5}$ | 460 | 496 | 473 | 495 | 412 | 140 | 35 |
| P-DEACIMP | $1.250 \times 10^{-3}$ | 513 | 555 | 498 | 561 | 272 | 200 | 48 |
| | $6.250 \times 10^{-4}$ | 513 | 555 | 512 | 561 | 264 | 258 | 48 |
| | $3.125 \times 10^{-4}$ | 513 | 555 | 532 | 561 | 247 | 34 | 48 |
| ACIMB | $6.250 \times 10^{-4}$ | 475 | 514 | 460 | 515 | 243 | 117 | 60 |
| | $3.125 \times 10^{-4}$ | 475 | 514 | 474 | 515 | 237 | 241 | 60 |
| | $1.563 \times 10^{-4}$ | 475 | 514 | 484 | 515 | 233 | 204 | 60 |
| P-ACIMB | $3.90 \times 10^{-5}$ | 416 | 478 | 385 | 447 | 574 | 302 | 31 |
| | $1.95 \times 10^{-5}$ | 416 | 478 | 380 | 447 | >1000 | 338 | 31 |
| | $9.75 \times 10^{-6}$ | 416 | 478 | 375 | 447 | >1000 | 317 | 31 |
| ACIBM | $2.250 \times 10^{-3}$ | 528 | 568 | 526 | 561 | 233 | 204 | 33 |
| | $1.125 \times 10^{-3}$ | 528 | 568 | 519 | 561 | 237 | 117 | 33 |
| | $5.625 \times 10^{-4}$ | 528 | 568 | 518 | 561 | 243 | 241 | 33 |
| P-ACIBM | $7.80 \times 10^{-4}$ | 384 | 449 | 381 | 445 | 224 | 589 | 61 |
| | $3.90 \times 10^{-4}$ | 384 | 449 | 380 | 445 | 602 | 586 | 61 |
| | $1.95 \times 10^{-4}$ | 384 | 449 | 379 | 445 | 572 | 558 | 61 |

[a] Excitation wavelength for emission

[b] Emission wavelength for excitation

[c] Maximum emission wavelength

[d] Maximum excitation wavelength

[e] Maximum excitation intensity

[f] Maximum emission intensity

[g] Stokes shift

**Table 6** Thermal stabilities of the monomers and polymers

| TGA/DTG DTA | | | | | | | |
|---|---|---|---|---|---|---|---|
| Compounds | [a]$T_{on}$ | [b]$T_{max}$ | [c]$T_5$ (°C) | [d]$T_{10}$ (°C) | [e]$T_{20}$ (°C) | [f]$T_{50}$ (°C) | [g]% char | Endo (°C) |
| DEACIMP | 320 | 367 | 330 | 348 | 359 | 376 | 9.5 | 127, 373 |
| P-DEACIMP | 230 | 270, 425 | 236 | 248 | 269 | 345 | 20.0 | 129 |
| ACIMB | 220 | 302, 478 | 235 | 263 | 351 | – | 54.0 | – |
| P-ACIMB | 209 | 316, 432 | 216 | 247 | 298 | 424 | 23.0 | – |
| ACIBM | 219 | 244, 419 | 232 | 248 | 315 | 534 | 43.0 | 223 |
| P-ACIBM | 201 | 412, 862 | 234 | 298 | 380 | 542 | 35.0 | – |

[a] The onset temperature

[b] Temperature of the peak maxima

[c] Temperature corresponding to 5% weight loss

[d] Temperature corresponding to 10% weight loss

[e] Temperature corresponding to 20% weight loss

[f] Temperature corresponding to 50% weight loss

[g] % char at 1000 °C

form. Polymers were prepared by sprinkling on double-sided adhesive tape mounted on a carbon stub and then they coated with a thin gold/palladium film by a sputter coater.

SEM photographs of P-DEACIMP, P-ACIBM and P-ACIMB are given in Fig. 7. According to the SEM images, P-ACIBM consists of different, nano-sized particles, while P-ACIMB has sharp edges with the form of

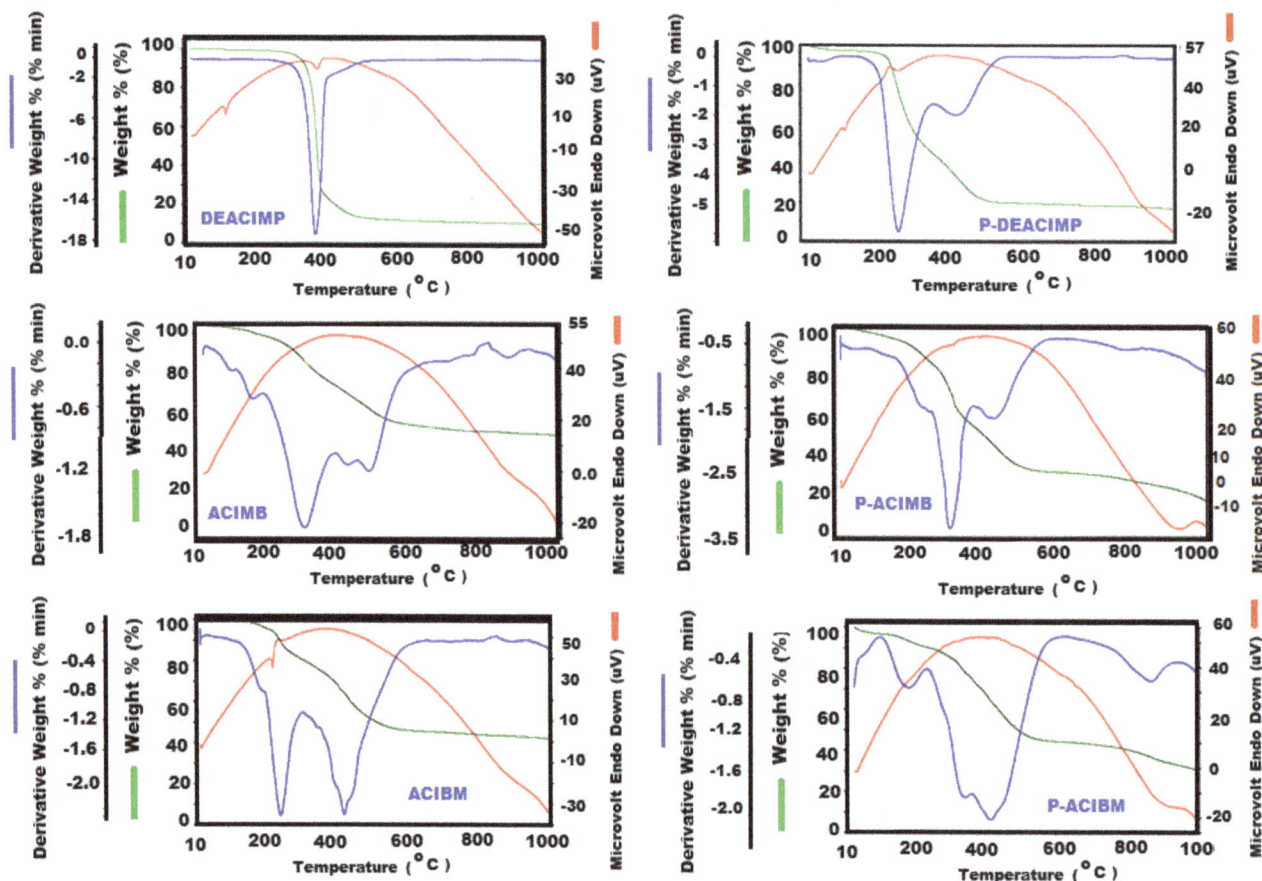

**Fig. 6** TGA–DTA–DTG curves of the monomers and polymers (heating rate: 10 C min$^{-1}$; N$_2$ atmosphere)

rods. Surface of P-DEACIMP has the folds in the form of brain.

### Electrochemical and conductivity properties

The voltammetric measurements were performed in acetonitrile. All the experiments were performed in a dry box filled with Ar at room temperature. The electrochemical potential of Ag was calibrated with respect to the ferrocene/ferrocenium (Fc/Fc$^+$) couple. The half-wave potential (E$^{1/2}$) of (Fc/Fc$^+$) measured in 0.1 M tetrabutylammoniumhexafluorophosphate (TBAPF$_6$) acetonitrile solution is 0.39 V with respect to Ag wire or 0.38 V with respect to saturated calomel electrolyte (SCE). The voltammetric measurements were carried out for all monomer compounds by acetonitrile and added to extra 1 mL DMF for polymers.

The values of electrochemical band gaps (E$_g$′) are given in Table 3. These data were estimated by using the oxidation onset (E$_{ox}$) and reduction onset (E$_{red}$) values, as given in Fig. 8 for the compounds where E$_{ox}$ is the oxidation peak potential and E$_{red}$ is the reduction peak

potential. The calculations were performed by using the following equations [24]:

$$E_{HOMO} = -(4.39 + E_{ox}) \qquad (2)$$

$$E_{LUMO} = -(4.39 + F_{red}) \qquad (3)$$

$$E_g' = E_{LUMO} - E_{HOMO}. \qquad (4)$$

To understand the electronic structure of conjugated polymers it is essential to establish the relative positions of the characteristic electronic energy levels such as the highest occupied molecular orbital (HOMO or π level), the lowest unoccupied molecular orbital (LUMO or π), and the associated energy parameters [25, 26].

The oxidation peaks in cyclic voltammograms probably correspond to the oxidation of hydroxyl groups to form phenoxy radicals. The reduction peaks were presumably due to the reduction of the azomethine linkages via protonation of azomethine nitrogen [23].

According to the Table 3, the order of the electrochemical band gap values of the polymer changes are as follows: P-ACIBM > P-ACIMB > P-DEACIMP. This was a result of the polyconjugated structures of the polymers,

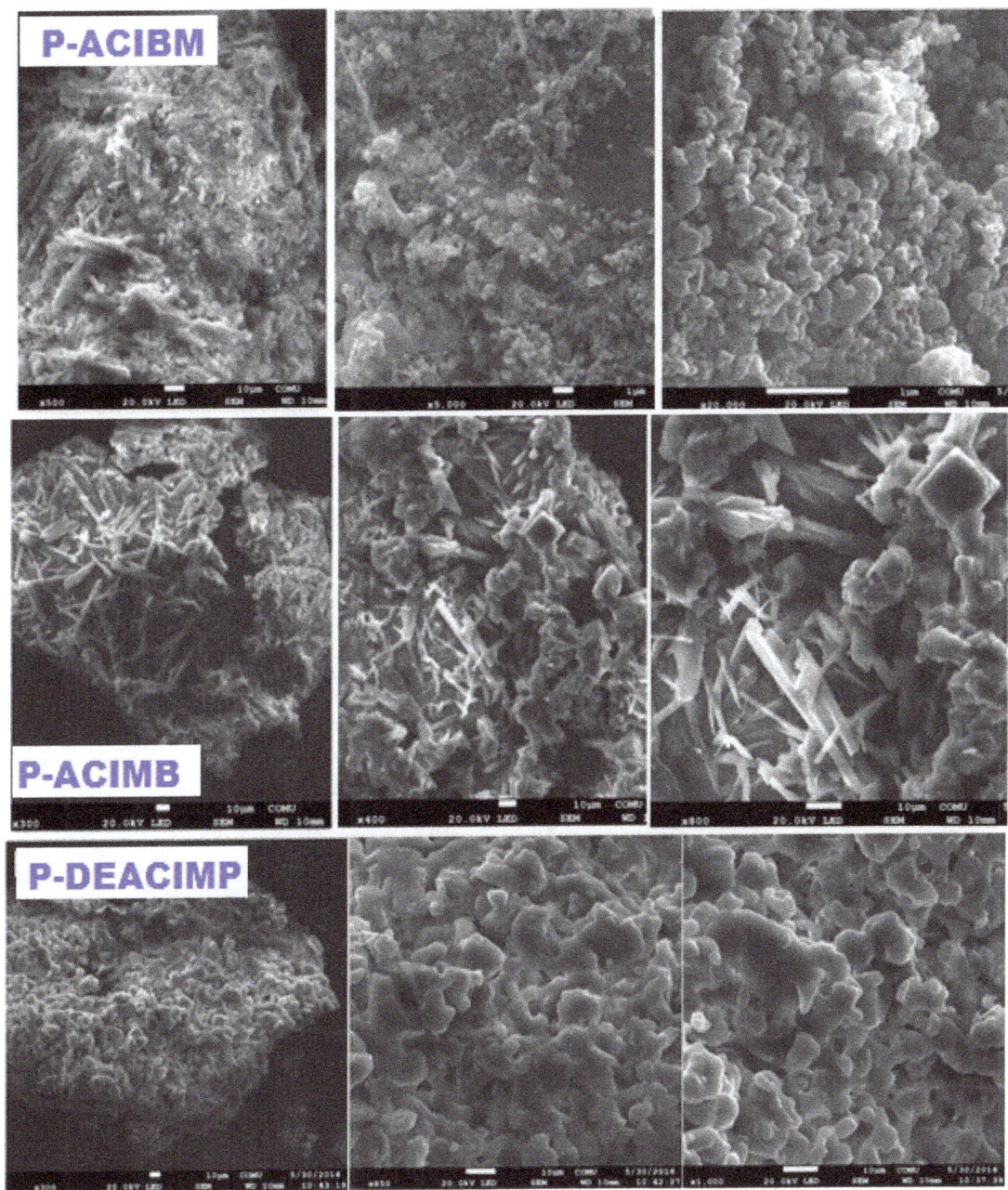

**Fig. 7** SEM images of polymers

which increase HOMO and decrease LUMO energy levels, resulting in lower electrochemical band gaps [23].

Conductivity was measured by a Keithley 2400 Electrometer (Keithley, Ohio, USA). The pellets were pressed on hydraulic press at 1687.2 kg/cm$^2$. Iodine doping was carried out by exposing the pellets to iodine vapor at atmospheric pressure and room temperature in a desiccator. Solid-state conductivities of the polymers measured under air atmosphere were shown in a graph plotted versus time. The measurements for the polymers were carried out in pure form and then polymers were exposed to iodine vapor

in a desiccator, and the change in their conductivities depending on time was measured at specific time intervals by doping. In the doping process, electron emitting amine nitrogen and electron pulling iodine coordinate, and the formation of radical cation (polaron) structure in polymer chain (on amine nitrogen) is enabled [27].

Electrical conductivities of the polymers and the changes of these values as a function of doping time with iodine were determined and shown in Fig. 9. Diaz et al. [28] suggested the doping mechanism of Schiff base polymers. According to doping mechanism, nitrogen, being a very

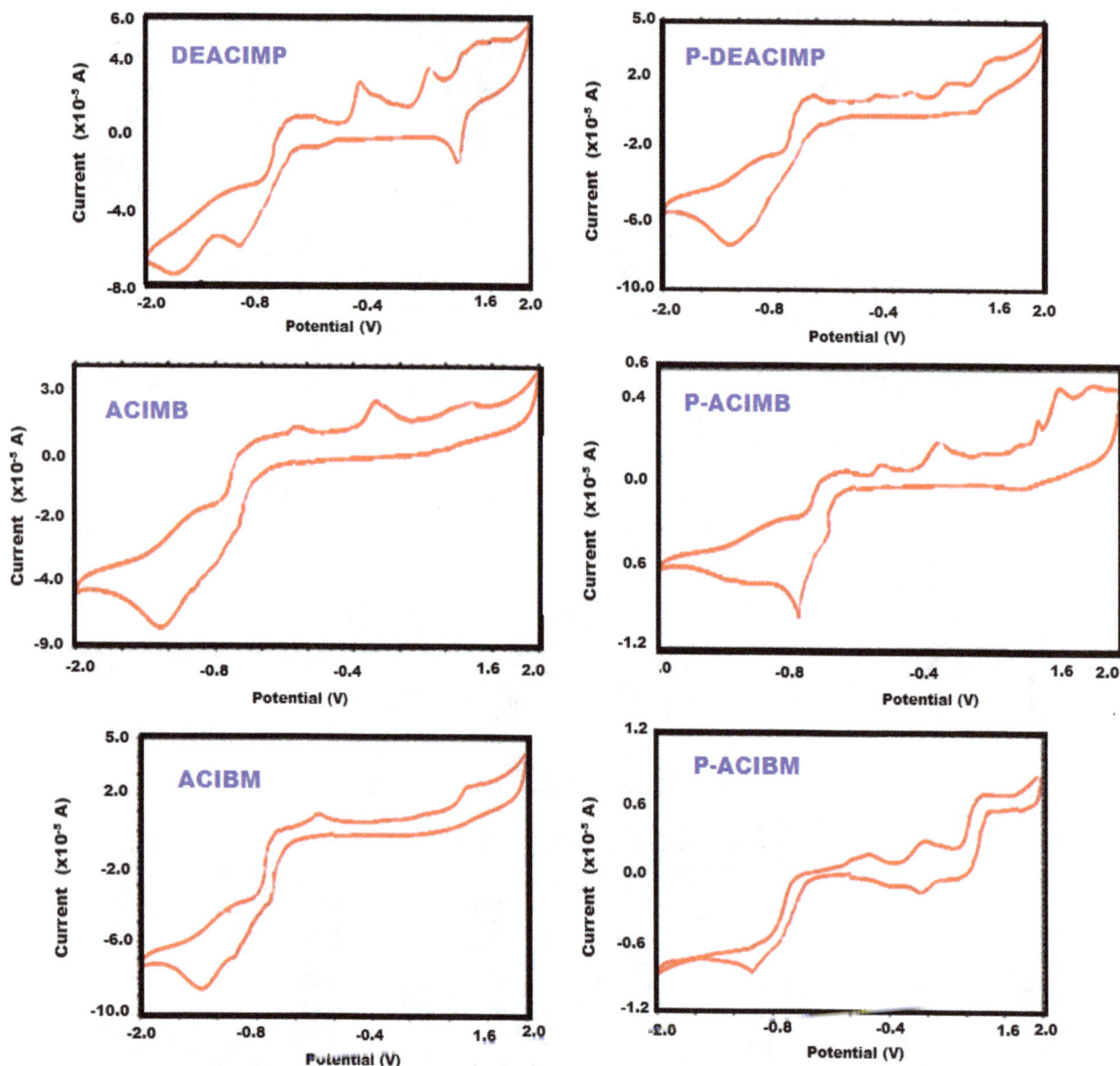

**Fig. 8** Cyclic voltammograms of the monomers and polymers

electronegative element, is capable of coordinating with an iodine molecule (Scheme 2). Consequently, a charge-transfer complex between imine compound and dopant iodine forms and thus conductivity increases [29].

The conductivity values of the undoped P-DEACIMP, P-ACIBM, P-ACIMB polymers were about $4.35 \times 10^{-8}$, $4.2 \times 10^{-7}$ and $5.3 \times 10^{-7}$ S cm$^{-1}$, respectively. After 120 h doping, the conductivity of P-DEACIMP, P-ACIBM, P-ACIMB polymers were found to be around $1.78 \times 10^{-5}$, $5.13 \times 10^{-5}$ and $2.65 \times 10^{-5}$ S cm$^{-1}$, respectively. According to these results, increasing of the conductivity values of polymers were observed as connected iodine doping time.

## Conclusions

Schiff base with different position of –OH group was oxidatively polymerized in aqueous alkaline medium by NaOCl as oxidant. The structures of the monomers and polymers were confirmed by UV–Vis, FT-IR, $^1$H, and $^{13}$C NMR spectroscopic techniques. According to $^1$H-NMR and $^{13}$C-NMR spectra, Schiff bases were polymerized by C–O–C or C–C binding. Iodine vapor-doped polymer P-DEA-CIMP gave the maximum electrical conductivity. The synthesized polymers have lower band gaps than the monomers because of polyconjugated structures. According to char % at 1000 °C, the thermal stability of the

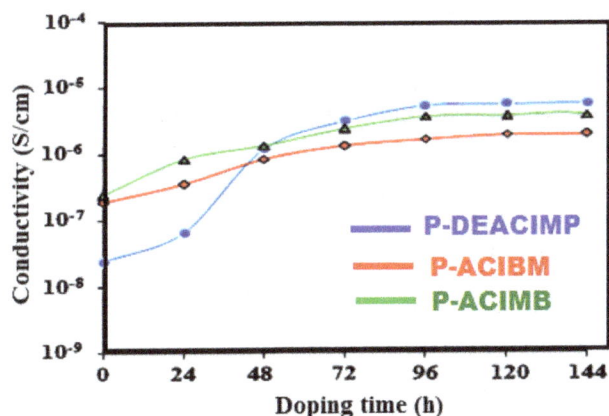

**Fig. 9** Electrical conductivity changes of the polymers $I_2$-doped at 25 °C

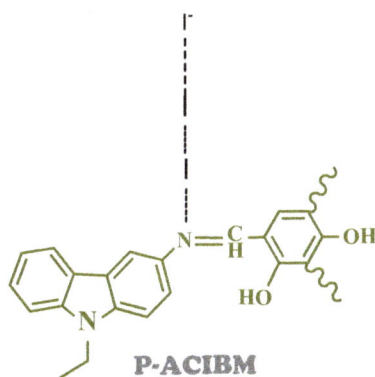

**Scheme 2** The nitrogen atom coordination of iodine to the P-ACIBM

polymers was in the order P-ACIBM > P-ACIMB > P-DEACIMP. The synthesized compounds were all soluble in common solvents such as DMF, THF and DMSO. Because of carbazole units in the structures of these polymers, both conductivity and thermal properties were better than that of other imine polymers in literature. Electrochemical band gap values of P-DEACIMP, P-ACIBM and P-ACIMB were found to be 2.58, 1.54 and 2.51 eV, respectively.

# References

1. Grigoras M, Antonoaia NC (2005) Synthesis and characterization of some carbazole-based imine polymers. Eur Polym J 41:1079–1089
2. Dineshkumar S, Muthusamy A (2016) Synthesis and spectral characterization of cross linked rigid structured Schiff base polymers: effect of substituent position changes on optical, electrical, and thermal properties. Polym-Plast Technol Eng 55:368–3788
3. El-Shekeil A, Al-Aghbari S (2004) DC electrical conductivity of some oligoazomethines. Polym Int 53:777–788
4. El-Shekeil A, Hamid S, Ali DA (1997) Synthesis and charac-

terization of some polyazomethine conducting polymers and oligomers. Polym Bull 39:1–7
5. Kaya İ, Bilici A (2006) Synthesis, characterization and thermal degradation of oligo-2-[(4-hydroxyphenyl) imino-methyl]-1-naphtol and oligomer-metal complexes. J Macromol Sci Part A 43:719–733
6. Kaya İ, Bilici A (2007) Synthesis, characterization, thermal analysis, and band gap of oligo-2-methoxy-6-[(4-methylphenyl) imino] methyl phenol. J Appl Polym Sci 104:3417–3426
7. Yao Y, Zhang QT, Tour JM (1998) Synthesis of imine ridged planar poly(pyridinethiophene). Combination of planarization and intramolecular charge transfer in conjugated polymers. Macromolecules 31:8600–8606
8. Kaya İ, Koça S (2004) Synthesis, characterization and optimum reaction conditions of oligo-2-amino-3-hydroxypyridine and its Schiff base oligomer. Polymer 45:1743–1753
9. Mart H, Yürük H, Saçak M, Muradoğlu V, Vilayetoğlu AR (2004) The synthesis, characterization and thermal stability of oligo-4-hydroxybenzaldehyde. Polym Degrad Stab 83:395–398
10. Cimrova V, Ulbricht C, Dzhabarov V, Výprachtický D, Egbe DM (2014) New electroluminescent carbazole-containing conjugated polymer: synthesis, photophysics, and electroluminescence. Polymer 55:6220–6226
11. Nishat N, Khan SA, Rasool R, Parveen S (2011) Synthesis, spectral characterization and biocidal activity of thermally stable polymeric Schiff base and its polymer metal complexes. J Inorg Organomet Polym 21:673–681
12. Özbülbül A, Mart H, Tunçel M, Serin S (2006) A new soluble Schiff base polymer with a double azomethine group synthesized by oxidative polycondensation. Des Monomers Polym 9:169–179
13. Kaya İ, Vilayetoglu AR, Mart H (2001) The synthesis and properties of oligosalicylaldehyde and its Schiff base oligomers. Polymer 42:4859–4865
14. Zhang Y, Shibatomi K, Yamamoto H (2005) Lewis acid catalyzed highly selective halogenation of aromatic compounds. Synlett 18:2837–2842
15. Kaya İ, Aydın A (2012) A new approach for synthesis of electroactive phenol based polymer: 4-(2,5-di(thiophen-2-yl)-1H-pyrrol-1-yl)phenol and its oxidative polymer. Prog Org Coat 73:239–249
16. Erdik E (2008) Spectroscopic methods in organic chemistry, 5th edn. Gazi Press, Ankara, p 531
17. Özbülbül A (2006) Synthesis and characterization of new type oligomer Schiff bases with based oligophenol. Master theses, Institute of Science and Technology, Çukurova University, Adana, p 108
18. Kaya İ, Aydın A (2011) Synthesis and characterization of the poly (amino phenol) derivatives containing thiophene in side chain: thermal degradation, electrical conductivity, optical-electrochemical, and fluorescent properties. J Appl Polym Sci 121:3028–3040
19. Kaya İ, Yıldırım M, Avcı A (2010) Synthesis and characterization of fluorescent polyphenol species derived from methyl substituted amino pyridine based Schiff bases: the effect of substituent position on optical, electrical, electrochemical, and fluorescence properties. Synth Met 160:911–920
20. Karakaplan M, Demetgül C, Serin S (2008) Synthesis and thermal properties of a novel Schiff base oligomer with a double azomethine group and its Co(II) and Mn(II) complexes. J Macromol Sci Part A 45:406–414
21. Colladet K, Nicolas M, Goris L, Lutsen L, Vanderzande D (2004) Low-band gap polymers for photovoltaic applications. Thin Solid Films 451:7–11
22. Kaya İ, Yıldırım M, Aydın A, Şenol D (2010) Synthesis and characterization of fluorescent graft fluorene-co-polyphenol derivatives: the effect of substituent on solubility, thermal sta-

bility, conductivity, optical and electrochemical properties. React Funct Polym 70:815–826

23. Kaya İ, Yılmaz T (2017) Preparation and Characterization of poly (azomethines) containing ether and methylene bridges: photophysical, electrochemical, conductivity and thermal properties. J Fluorescence. doi:10.1007/s10895-016-1966-1

24. Cervini R, Li XC, Spencer G, Holmes AB, Moratti SC, Friend RH (1997) Electrochemical and optical studies of PPV derivatives and poly (aromatic oxadiazoles). Synth Met 84:359–360

25. Yang C, Jenekhe S (1995) Conjugated aromatic polyimines. 2. Synthesis, structure, and properties of new aromatic polyazomethines. Macromolecules 28:1180–1196

26. Spiliopoulos IK, Mikroyannidis JA (1996) Soluble, rigid-rod polyamide, polyimides, and polyazomethine with phenyl pendent groups derived from 4,4″-diamino-3,5,3″,5″-tetraphenyl-*p*-terphenyl. Macromolecules 29:5313–5319

27. Kaya İ, Yıldırım M (2007) Synthesis, characterization, thermal stability and electrochemical properties of poly-4-[(2-methylphenyl)imino methyl]phenol. Eur Polym J 43:127–138

28. Diaz FR, Moreno J, Tagle LH, East GA, Radic D (1999) Synthesis, characterization and electrical properties of polyimines derived from selenophene. Synth Met 100:187–193

29. Kaya İ, Yıldırım M (2009) Synthesis and characterization of graft copolymers of melamine: thermal stability, electrical conductivity, and optical properties. Synth Met 159:1572–1582

# 13

# Effective removal of radioactive $^{90}$Sr by CuO NPs/Ag-clinoptilolite zeolite composite adsorbent from water sample: isotherm, kinetic and thermodynamic reactions study

Meysam Sadeghi[1] · Sina Yekta[2] · Hamed Ghaedi[3] · Esmaeil Babanezhad[2]

**Abstract** In this research, the natural zeolite named clinoptilolite (NCp) was considered as an applicable base structure for subsequent preparation and modification by pre-selected chemical agents to use for removal/adsorption goals in progress. To make Na-clinoptilolite form, the mentioned zeolite undergoes variations by treating with NaCl salt. Ion exchange procedure was utilized to imprint silver ions ($Ag^+$) from silver (I) nitrate solution as silver precursor on the modified clinoptilolite zeolite structure to attain Ag-clinoptilolite zeolite. Tenorite (CuO) NPs have been then dispersed and deposited on the external surface of the pre-prepared Ag-clinoptilolite zeolite through impregnation method to prepare the CuO NPs/Ag-clinoptilolite zeolite as a novel composite adsorbent. A series of analyses techniques including SEM-EDAX, XRD and FT-IR were applied for characterization and identification of crystal structure, morphology and elemental composition of the synthesized samples. The removal and adsorption process of radioactive strontium-90 ($^{90}$Sr) by CuO NPs/Ag-clinoptilolite zeolite composite adsorbent was exploited under various experimental conditions including pH, amount of adsorbent and the contact time at room temperature and the final optimized procedure used for removal of probable presence of $^{90}$Sr in water sample of

Bushehr city. Adsorption isotherm models including Langmuir, Freundlich, Temkin, D-R, H-J and Hasley have been applied and equilibrium adsorption data was better fitted to the Freundlich, Temkin, H-J and Halsey isotherms. The reaction kinetic information was studied by utilizing pseudo first and second orders, Elovich and Intra particle diffusion kinetic models. The adsorption kinetics is finely described by the pseudo second-order model. The energy of activation $E_a$ calculated using the Arrhenius equation was found to be 44.75 kJ/mol. Further, the evaluation of the thermodynamic parameters such as $\Delta G^0$, $\Delta H^0$ and $\Delta S^0$, denoted that adsorption process of $^{90}$Sr was spontaneous and illustrates a chemical adsorption properties and exothermic nature of the adsorption. The obtained results revealed that $^{90}$Sr ions were removed by CuO NPs/Ag-clinoptilolite zeolite composite under optimized conditions after 6 h with a yield more than 97 %.

**Keywords** Radioactive $^{90}$Sr · CuO NPs/Ag-clinoptilolite zeolite · Removal and adsorption · Isotherm

## Introduction

Liquid radioactive wastes (LRWs) are known as by-products of nuclear activities around the world including; nuclear power plants (NPPs) developments and activities, applications of nuclear fission or nuclear technology, nuclear weapon testing, and other world threatening researches. The entrance and existence of these radioactive wastes into the environment always has recognized as a serious threat for human being and other alive creatures. Radioactive contamination is seriously hazardous and if it comes about to an area, it can cause a real disaster not only for the main area but also for the other parts around its

✉ Hamed Ghaedi
 hamedghaedi@gmail.com

[1] Young Researchers and Elite Club, Ahvaz Branch, Islamic Azad University, Ahvaz, Iran

[2] Department of Chemistry, Faculty of Basic Sciences, Qaemshahr Branch, Islamic Azad University, Qaemshahr, Iran

[3] Faculty of Engineering, Bushehr Branch, Islamic Azad University, Bushehr, Iran

district depending on the amount of leaked or released radioactive wastes. Despite the other sorts of pollutants which can be removed or handled entirely in a certain period of time, the radioactive wastes cannot be handled in the same way. Because these radioactive wastes can spread widely in any possible way (air, soil, water, alive tissues) and large number of radioactive elements have years and years of half-life which means they are not going to be neutralized in easy way. Standing still and disintegration for years will go on till complete neutralization comes about. However, due to the wide spread of nuclear power plants and public raised concerns over nuclear safety, the new methods and regulations for monitoring and protection against these wastes are developing by researchers and governments.

For instance after the explosion took place in Fukushima in Japan, power plants accident, the enormous amounts of radioactive hazardous elements found their way to the natural environment and finding the best way to bring the best protection and monitoring service was of great concern [1]. For nuclear facilities, it can be very hazardous if such wastes are not well treated before discharged to the environment. Strontium-90 ($^{90}$Sr) is an important component of many nuclear wastes and is a high yield fusion product of uranium-235 ($^{235}$U) [2]. It is relatively short-lived with a half-life of 28.8 years. Its decay product that so called daughter of $^{90}$Sr, is yttrium-90 ($^{90}$Y) isotope which is $\beta^-$ emitter with half-life of 64 h and decay energy of 2.28 MeV distributed to an electron, an antineutrino and zirconium-90 ($^{90}$Zr) that is stable [3, 4]. Strontium isotope $^{90}$Sr is treated as one of the most dangerous products of nuclear fission for human beings [5, 6].

Radioactive strontium can be replaced instead of calcium in biosphere known as a bone seeker and it can also transfer to human body through food chain in which it has long retention time. $^{90}$Sr is taken up via gastrointestinal system and aggregate in the body turning to a part of the bone marrow tissue and hurting blood-producing cells [7]. Also, it can be a cause of leukemia or skeletal cancer. This is because of its chemical propinquity and alkaline earth metallic characteristics. For this reason, its characteristics and migration in the environment are widely studied. Drinking waters and fresh waters usually contain many natural radionuclides; Strontium, tritium, radon, radium and uranium isotopes, etc. In recent years, there has been an increase in the usage of zeolites in different compositions to remove and bury different radio-contaminations [8–12]. Zeolites are porous crystalline structurally—hydrated alumina silicates of group IA and IIA elements such as sodium, potassium, barium, magnesium and calcium. One of the most significant properties of zeolites is their ability to exchange cations. Clinoptilolite (Cp) is one of a very cheap, available and the most abundant natural

zeolites is in the chemical class of family of Heulandite (HEU-type), easily obtained from mines, appropriate as a sorbent due to its natural characteristics [13, 14].

The crystal structure of clinoptilolite has 3-dimensional aluminosilicate framework, which specific structure causes the developed system of micropores and channels occupied by water molecules and exchangeable cations. The combination of zeolites and metal oxide nanoparticles renders solid catalysts in which the high surface area of nanoparticles and the absorbent capacity provided by zeolites cooperate to increase the efficiency of the catalytic process [15]. As reported in previous researches, the natural clinoptilolite zeolite shows a high adsorbent capacity for the removal of non-radioactive strontium ions [16, 17]. The procedures for modifying zeolites are usually done by impregnation [18] and ion-exchange [17]. Also, the dispersion of metal oxide nanoparticles onto zeolite depends on the type of metal precursor used and its action during the preparation method. Among metal oxides, tenorite or copper oxide (CuO) nanoparticles actuated our attention due to its low cost and availability of the starting materials and also high certainty and purity compared with other applied metal oxides are the advantages of mentioned reagent [19–21].

Tenorite nanoparticles (NPs) as a p-type semiconductor exhibiting narrow band gap (Eg = 1.2 eV), have attracted a great scope of research interest in this decade. These nanoparticles are also utilized in a wide range of applications such nano devices such as degradation, bactericidal properties, etc. [22–24]. Herein, we will report the combination of Ag-clinoptilolite zeolite as host and CuO nanoparticles as guest materials to synthesize CuO NPs/Ag-clinoptilolite zeolite composite an adsorbent catalyst in which the high surface area of nanoparticles and the absorbent capacity provided by the zeolite cooperation to increase the efficiency of the removal process of radioactive $^{90}$Sr from water sample of Bushehr city. Ag$^+$ is the only noble mono-positive cation that forms mononuclear species with appreciable stability in aqueous solution. Besides, silver is known to have strong influence on the absorption properties of zeolites. To the best of our knowledge, there is no report on the application of CuO NPs/Ag-clinoptilolite zeolite composite catalyst used for the removal of radioactive $^{90}$Sr.

# Experimental

## Materials and reagents

The natural clinoptilolite (NCp) zeolite employed in our research was obtained from the region of the West Semnan, Central Alborz Mountains, Iran and its structural properties is as $(Na, K, Ca)_6(Si, Al)_{36}O_{72}.20H_2O$. Sodium chloride (NaCl), silver nitrate ($AgNO_3$), hydrochloric acid (HCl), copper nitrate trihydrate ($Cu(NO_3)_2.3H_2O$), acetone, sodium hydroxide (NaOH) and nitric acid ($HNO_3$) were purchased from Merck (Merck, Darmstadt, Germany). The high-capacity cocktail OptiPhase HiSafe-3 (Wallac Oy, Turku, Finland) and deionized water were used throughout the work.

## Instrumentation

The morphology, particle sizes and elemental composition of the prepared adsorbents were surveyed using a scanning electron microscope coupled with energy dispersive X-ray spectrometer (SEM-EDAX, HITACHI S-300 N). The powder X-ray diffraction (XRD) patterns were recorded using a Philips X'pert Pro diffractometer equipped with CuK$\alpha$ radiation at wavelength 1.54056 Å (30 mA and 40 kV) at room temperature. Data were collected over the range 4–80° in $2\theta$ with a scanning speed of 2° min$^{-1}$. The IR spectra were scanned on a PerkinElmer model 2000 FT-IR spectrometer (USA) in the wavelength range of 400–4000 cm$^{-1}$ using KBr pellets. An ultra low-level Quantulus 1220 liquid scintillation counter has been used for all measurements. A shaker Heidolph Vibramax 100 (Heidolph Co., Schwabach, Germany) was utilized for mixing of cocktail and sample. The samples and cocktail were mixed in 20 mL polyethylene vials, Polyvial (Zinsser Analytik Co., Frankfurt, Germany).

## Preparation of Na-clinoptilolite zeolite

First, 5 g of the clinoptilolite zeolite before processing was calcined at 300 °C for 2 h in a furnace for excluding moister and impurities from the surface. Then, to obtain the Na-clinoptilolite form, the calcined clinoptilolite was chemically treated with 250 ml of 1 M sodium chloride (NaCl) at 90 °C for overnight and was washed with deionized water several times until chloride ions were removed. Finally, treated clinoptilolite (sodium-clinoptilolite) was dried at 85 °C for 5 h [25].

## Preparation of Ag-clinoptilolite zeolite

Silver ions were loaded into the zeolite framework by ion exchange method. In a typical experimental procedure, 4.5 g of the prepared Na-clinoptilolite zeolite in the previous step was added to a 50 mL of a 0.1 M silver nitrate ($AgNO_3$) solution and the mixture was magnetically stirred at 60 °C for 5 h to perform ion exchange process in which $Ag^+$ ions were replaced with $Na^+$ ions. The synthesized product (Ag-clinoptilolite zeolite) was then filtered and washed with deionized water and 0.1 M HCl solution, to remove the excess and unreacted silver ions from the zeolite framework, sequentially and then dried at 110 °C for 16 h. At last, the clean and dry Ag-clinoptilolite zeolite was calcined at 400 °C for 4 h [26]. This process was repeated for three times to reach significant ion exchange.

## Preparation of CuO NPs/Ag-clinoptilolite zeolite composite

Preparation of CuO NPs/Ag-clinoptilolite zeolite composite has been achieved using impregnation method. Typically, 3.5 g of the Ag-clinoptilolite zeolite powder was added to a solution of 0.5 M of copper nitrate trihydrate ($Cu(NO_3)_2.3H_2O$) reagent in 250 mL deionized water, meanwhile, the suspension was vigorously stirred at room temperature for 6 h. When the reaction was completed, the green powders were filtered, washed with distilled water and dried overnight at 110 °C. Finally, the obtained powder was revealed as the CuO NPs/Ag-clinoptilolite zeolite composite after calcination at 500 °C in the air for 6 h [27]. On the other hand, pure CuO NPs was prepared but without the presence of zeolite under similar conditions.

## Removal of radioactive $^{90}$Sr from water sample by CuO NPs/Ag-clinoptilolite composite

The measuring of probable presence of radioactive $^{90}$Sr in water sample needs another independent chemical experiment and the polluted water sample can be examined and monitored before and after removal process via Ultra Low-Level Liquid Scintillation Counting (LSC) technique. To study the removal and adsorption of radioactive $^{90}$Sr, amounts of 0.5–3 g of the CuO NPs/Ag-clinoptilolite zeolite composite was added to 500 ml of the water sample. A tracer amount of $^{90}$Sr$^{2+}$ containing 533 μL amount of it (equal to 112.3 Bq(Becquerel)/L) as the optimized activity was added to the above solution samples. Next, the mixture was stirred in different pH ranges 2–12 and time intervals 1–12 h via varying the adsorption temperature at 298–323 K, respectively. After filtration of the mixture, 5 mL of supernatant solutions were analyzed by liquid scintillation spectrometry (LSC) instrument. In LSC, an aliquot of the sample is put into a vial and mixed homogeneously with 15 mL of scintillation cocktail. A shaker was utilized for mixing of cocktail and sample. The samples and cocktail were mixed together in 20 mL polyethylene vials, Poly vial. The outside of the vials was cleaned with acetone. In the next step, all the polyethylene vials were stored in a cool, dark shield (about 7 °C) for 2 h to eliminate the scintillation cocktail fluorescence. Finally, all samples were counted by LSC for 5 h. The relative error of radioactivity measurements did not exceed 2 %. The initial source activity was 210.8 Bq/L.

## Results and discussion

### SEM-EDAX analysis

To establish the morphology and crystalline size of the as-synthesized clinoptilolite zeolite, Ag-clinoptilolite zeolite, CuO NPs/Ag-clinoptilolite zeolite composite and pure CuO NPs, SEM analysis were utilized as depicted in Fig. 1. The SEM images explain homogenous morphology of the structures of clinoptilolite (1a) and Ag-clinoptilolite (1b and 1c) zeolites and quasi-spherical CuO nanoparticles dispersed and deposited on the external surface of Ag-clinoptilolite zeolite (1d and 1e) and also specify that these morphologies and the crystallinity of the structures are maintained with Ag ion exchange and CuO NPs loading processes which are indicated by SEM images in Fig. 1b–e. The average crystalline size of CuO NPs in the composite was illustrated to have nanometric dimensions (less than 100 nm). It also denote that CuO NPs loaded on the zeolite has lower crystalline size than that of pure CuO NPs. Figure 2 give the composition elements present in clinoptilolite, Ag-clinoptilolite zeolites, CuO NPs/Ag-

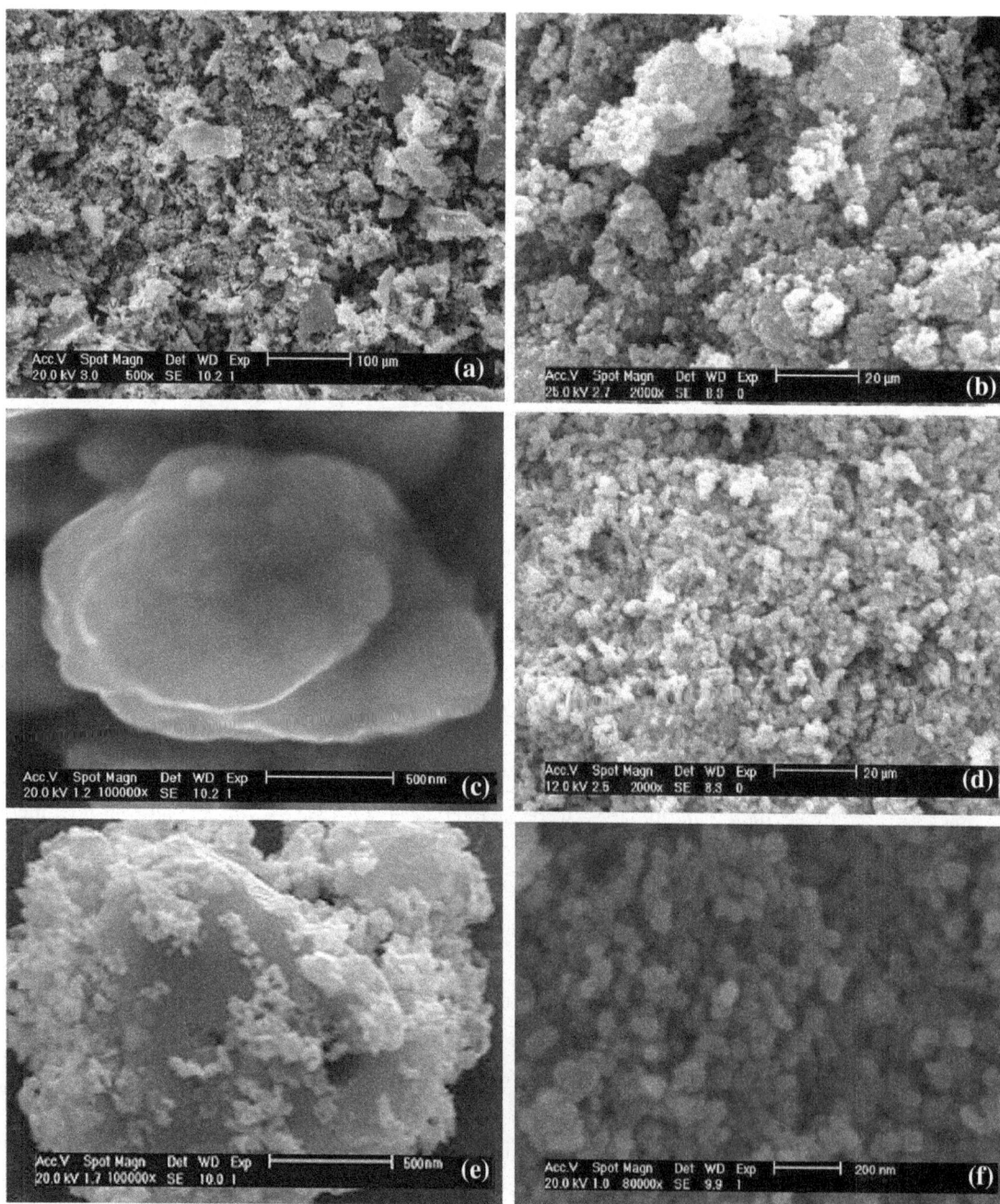

**Fig. 1** SEM images of the: **a** clinoptilolite, **b, c** Ag-clinoptilolite, **d, e** CuO NPs/Ag-clinoptilolite, and **f** pure CuO NPs

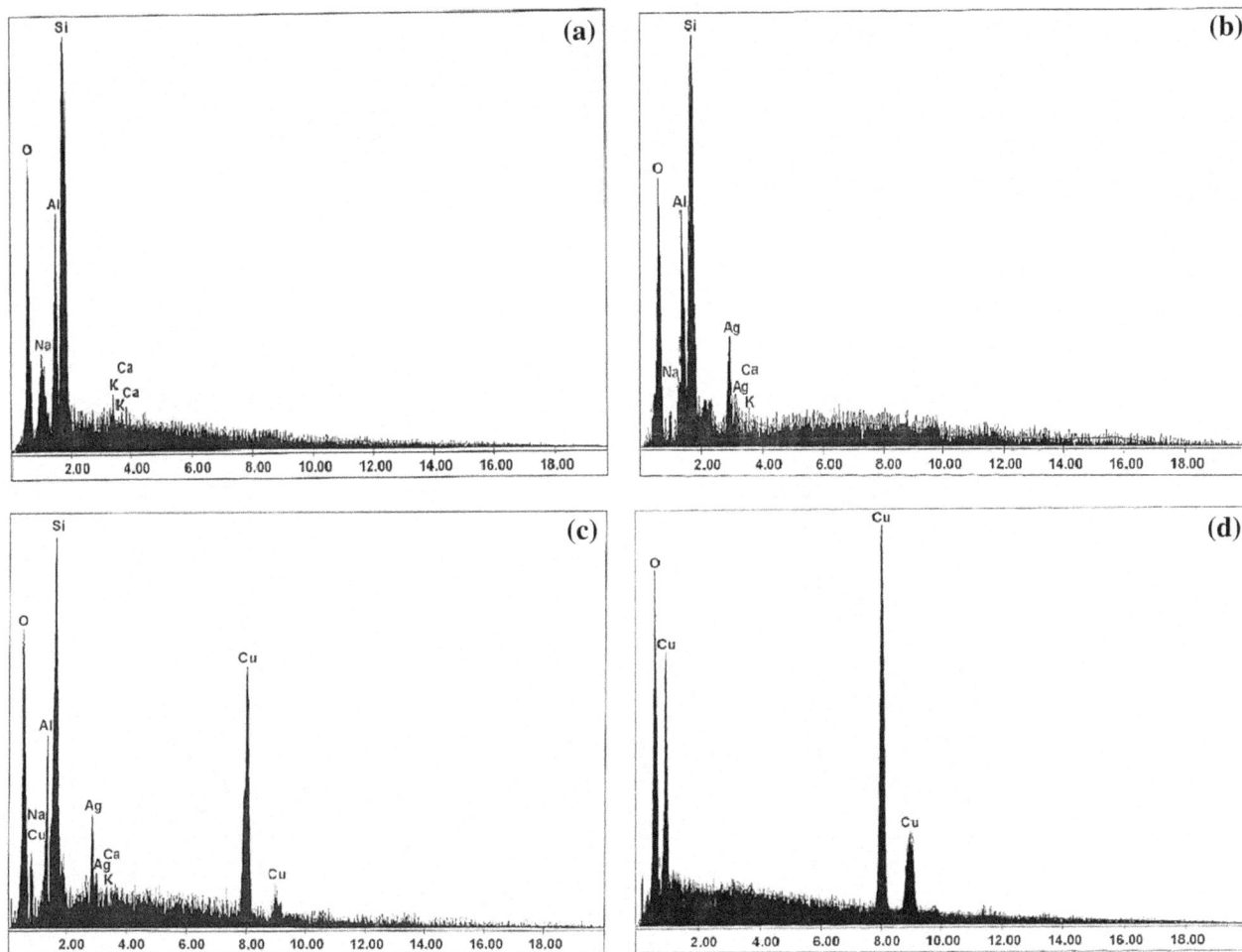

**Fig. 2** EDAX analysis of the **a** clinoptilolite, **b** Ag-clinoptilolite, **c** CuO NPs/Ag-clinoptilolite, and **d** pure CuO NPs

clinoptilolite zeolite and CuO NPs were investigated by energy dispersive X-rays (EDAX) analysis. In the EDAX spectra, the appeared peaks in the regions of approximately 0.55, 1.15, 1.50 and 1.75 are corresponded to the binding energies of oxygen (O), sodium (Na), aluminum (Al) and silicon (Si), respectively, that are related to the major elements of the clinoptilolite zeolite (Fig. 2a). On the other hand, in spectra (Fig. 2b, c), the appeared two peaks in the regions of 2.92 and 3.21 keV referred to the binding energies of silver (Ag) and three peaks in the regions of 0.93, 8.04 and 9.03 keV are related to the binding energies of copper (Cu) which reveals the presence of Cu in the composite. These results confirm coexistence of 6.4 and 19.7 wt% silver and copper in the prepared composite network, respectively.

**X-ray diffraction (XRD) patterns**

In Fig. 3, XRD patterns of the understudy clinoptilolite zeolite, CuO NPs/Ag-clinoptilolite zeolite composite and pure CuO NPs are given, respectively. As seen from the

patterns, the sharp peaks referring to clinoptilolite zeolite occurred at ($2\theta$) of 11.3993°–74.1895° (Fig. 3a) and are in good agreement with those of the clinoptilolite zeolite with Joint Committee on Powder Diffraction Standards: (JCPDS: 00-025-1349). Clinoptilolite zeolite structure was retained even after silver cation exchange in the Ag-clinoptilolite (Fig. 3b). Meanwhile, synthesized CuO NPs (as guest material) loaded as a 19.7 wt% of unit onto Ag-clinoptilolite zeolite as the host material, possesses a series of new peaks which were obtained at $2\theta$ of 41.5762°, 45.3174°, and 57.2960° corresponding to the diffraction planes of (002), (111) and (202), respectively [27–30]. No characteristic peaks related to the presence of impurities were observed in the patterns during CuO species loading. These peaks which are illustrated as purple points in Fig. 3b reveal that CuO NPs have been dispersed and deposited onto Ag-clinoptilolite and also indicate a host–guest interaction between Ag-clinoptilolite framework and CuO NPs. A definite line broadening of the scattering pattern in Fig. 3b is a demonstration upon which the synthesized CuO particles are in nanoscale range. However, a

**Fig. 3** XRD patterns of the catalyst samples:
**a** clinoptilolite, **b** CuO NPs/Ag-clinoptilolite, and **c** pure CuO NPs (*violet color filled triangle* indicates peak pattern depicting presence of CuO in the zeolite framework)

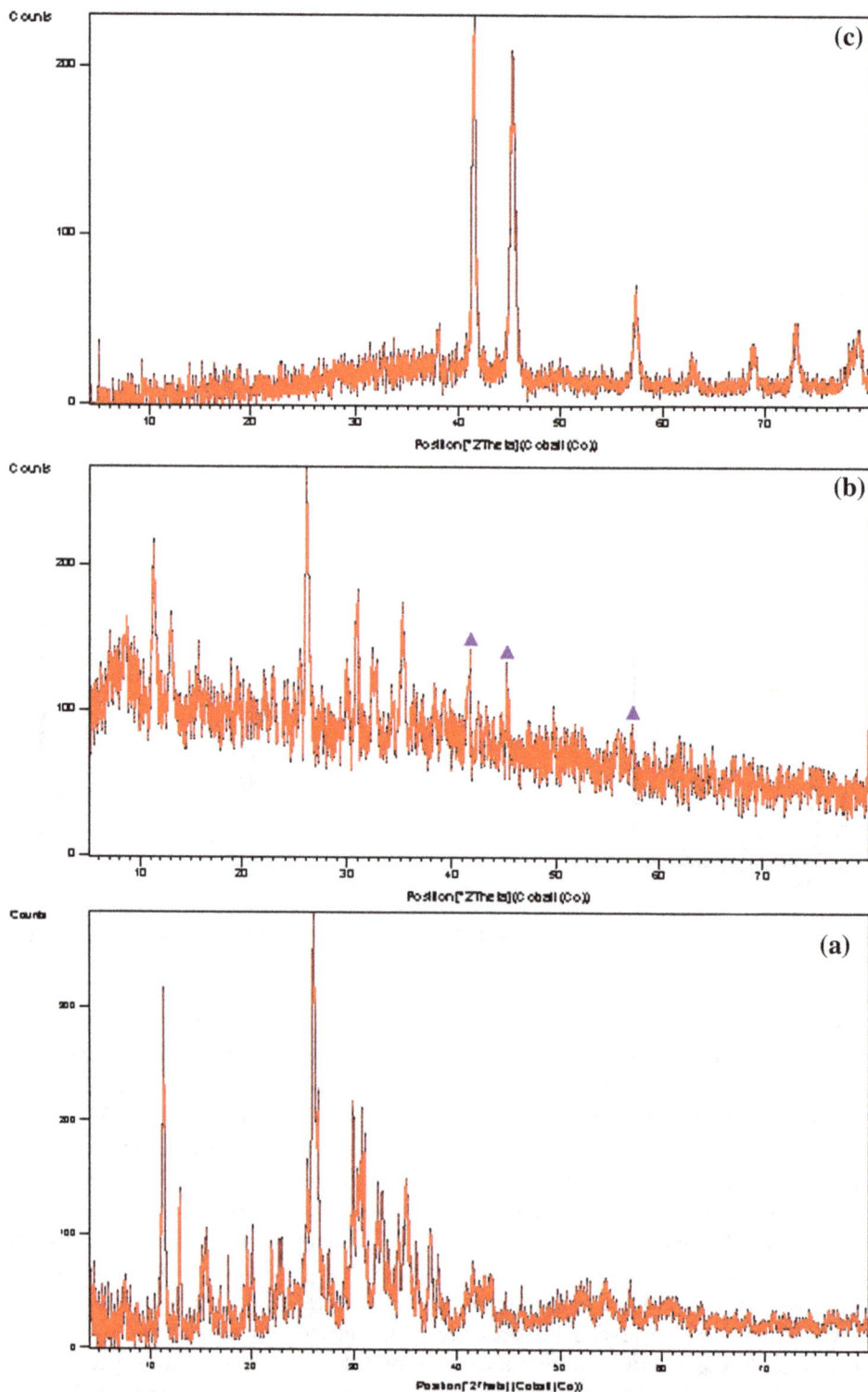

small loss of crystallinity is observed in Fig. 3b associated with the lower intensity of the peaks at $2\theta$ of 11.630° and 18.318°. This may be due to the dealumination process of Ag-clinoptilolite zeolite and CuO NPs/Ag-clinoptilolite zeolite composite and associated with the location of substituted silver and impregnated copper cations. The $Cu^{2+}$ ions within the zeolite framework can interact with the

aluminate sites more strongly than that of $Na^+$ or $Ag^+$ ions. Totally, it can be concluded that with silver ion exchange in clinoptilolite zeolite and subsequent loading of CuO NPs onto Ag-clinoptilolite, the structure of the zeolites did not change and can be stable after the above processing. On the other hand, the capacity of the clinoptilolite zeolite to keep the guest species is certainly limited. Consequently, the adsorption of the host cations (Si, Al and Na) will stop if the capacity is saturated. In contrast, the amount of the host species in the Ag-clinoptilolite increases as the copper oxide content increases. The introduced CuO NPs were dispersed and deposited on the external surface of Ag-clinoptilolite, however, due to the relative aggregation during processing of the composite, some particles are too large to perch inside the structure. Hence, high CuO NPs loading will cause structural damage to the zeolite framework. The size of the prepared CuO NPs deposited onto Ag-clinoptilolite was also investigated via XRD measurement and line broadening of the peak at $2\theta = 0°–80°$ using Debye–Scherrer Eq. (1) [31]:

$$d = \frac{0.94\lambda}{\beta \cos \theta}, \tag{1}$$

where $d$ is the crystal size, $\lambda$ is the wavelength of X-ray source, $\beta$ is the full width at half maximum (FWHM) and $\theta$ is Bragg diffraction angle. The peaks referring to the pure CuO NPs (Fig. 3c) occurred at scattering angles ($2\theta$) of 37.9379°, 41.5859°, 45.3296°, 57.3058°, 62.9655°, 68.9326°, 72.9516° and 78.8583° corresponding to diffraction planes of (100), (002), (111), (202), (113), (220), (311), and (222), respectively, that have been crystallized in the monoclinic phase and are in good agreement with those of CuO NPs with JCPDS = 01-072-0629. Using this equation, the average particle size for CuO NPs in the CuO NPs/Ag-clinoptilolite zeolite composite and pure CuO NPs are estimated to be 8.3 nm and 37.6 nm, respectively. The particle size obtained from XRD measurement is consistent with the results from the SEM study.

## FTIR study

The characterization of the prepared adsorbents along with the clinoptilolite zeolite precursors were further surveyed by FT-IR spectra as plotted in Fig. 4. Peak positions are nearly identical for three samples. All of the three as-synthesized typical samples, namely clinoptilolite zeolite and CuO NPs/Ag-clinoptilolite zeolite composite have peaks around 465 cm$^{-1}$ and 524 cm$^{-1}$ which are assigned to the bending vibrations of the insensitive internal TO$_4$ (T = Si or Al) tetrahedral units and double six rings (D6R) external linkage within the clinoptilolite zeolite structure, respectively. The peaks around 674 cm$^{-1}$ and 797 cm$^{-1}$ are attributed to the external linkage and internal tetrahedral symmetrical stretching vibrations, respectively. Furthermore, the peaks around 1034 cm$^{-1}$ are corresponded to the external linkage and internal tetrahedral asymmetrical stretching vibrations, and peaks around 1635 cm$^{-1}$, 3437 cm$^{-1}$ and 3623 cm$^{-1}$ are attributed to H–O–H bending O–H bonding (hydroxyl groups) vibrations and discrete water absorption bands of the clinoptilolite, respectively. Surveying Fig. 4a, b confirms

that no changes has occurred in the bands of Ag-clinoptilolite zeolite and CuO NPs/Ag-clinoptilolite zeolite composite compared with the original clinoptilolite zeolite, which tends to lend further support to the idea that the ion exchange modification of clinoptilolite zeolite by silver ion and copper oxide has a very little influence on the chemical structure of the zeolite framework. On the other hand, Fig. 4b illustrates two new peaks related to the synthesized loaded CuO NPs. The absorption bands in the 1467 cm$^{-1}$ region is referred to C–C bonding of probable trivial impurities existing in the applied stock materials. The absorption peak at 967 cm$^{-1}$ is also corresponded to Cu–O–Si and Cu–O–Al bonds and revealed the entrapped copper in the structure of zeolite [26–28]. Also, Fig. 4c reveals the FT-IR spectrum of pure CuO NPs. The broad at absorption peak around 3441 and 1632 cm$^{-1}$ were caused via the adsorbed water molecules. The absorption bands in the 1461 cm$^{-1}$ region is probably related to C–C bonding of trivial impurities existing in the applied stock materials as implied above. Three peaks at 487, 523 and 580 cm$^{-1}$ were related to the stretching vibrations of Cu–O bonding.

## Adsorption reaction isotherms

Several adsorption isotherm models were developed via analyzing solutions in contact with CuO NPs/Ag-clinoptilolite zeolite composite to find out the relation between the equilibrium concentrations before and after experimental in the liquid and solid phases. The Langmuir, Freundlich, Temkin, Dubinin–Radushkevich (D–R), Harkins–Jura (H–J) and Hasley models are used to describe equilibrium adsorption isotherms. The sorption isotherms were studied in water sample of Bushehr city as pH = 8.5, temperature (25 °C), and different initial solution concentrations related to the seven various activity of $^{90}$Sr from 112.3 Bq/L to 180.6 Bq/L. As can be seen in Fig. 5, by increasing the activity of solution, the removal efficiency of $^{90}$Sr is decreased. Thus, activity equal to 112.3 Bq/L as optimized activity for the adsorption reaction was chosen.

This model supposes that the adsorption takes place at a specific adsorption surface. The attraction between molecules decreases as they are getting further from the adsorption surface. The Langmuir adsorption isotherm is often utilized to describe the maximum adsorption capacity of an adsorbent and also show single coating layer on adsorption surface. Langmuir isotherm can be defined according to the following Eq. (2) [32].

$$\frac{1}{q_e} = \frac{1}{K_L q_m} \times \frac{1}{C_e} + \frac{1}{q_m} \tag{2}$$

where $q_m$ and $q_e$ are the maximum and equilibrium uptake of $^{90}$Sr per unit mass of adsorbent (mg/g), respectively. Also $K_L$ and $C_e$ refer to the Langmuir constant and the

**Fig. 4** FTIR spectra of the catalyst samples: **a** clinoptilolite, **b** CuO NPs/Ag-clinoptilolite, and **c** pure CuO NPs

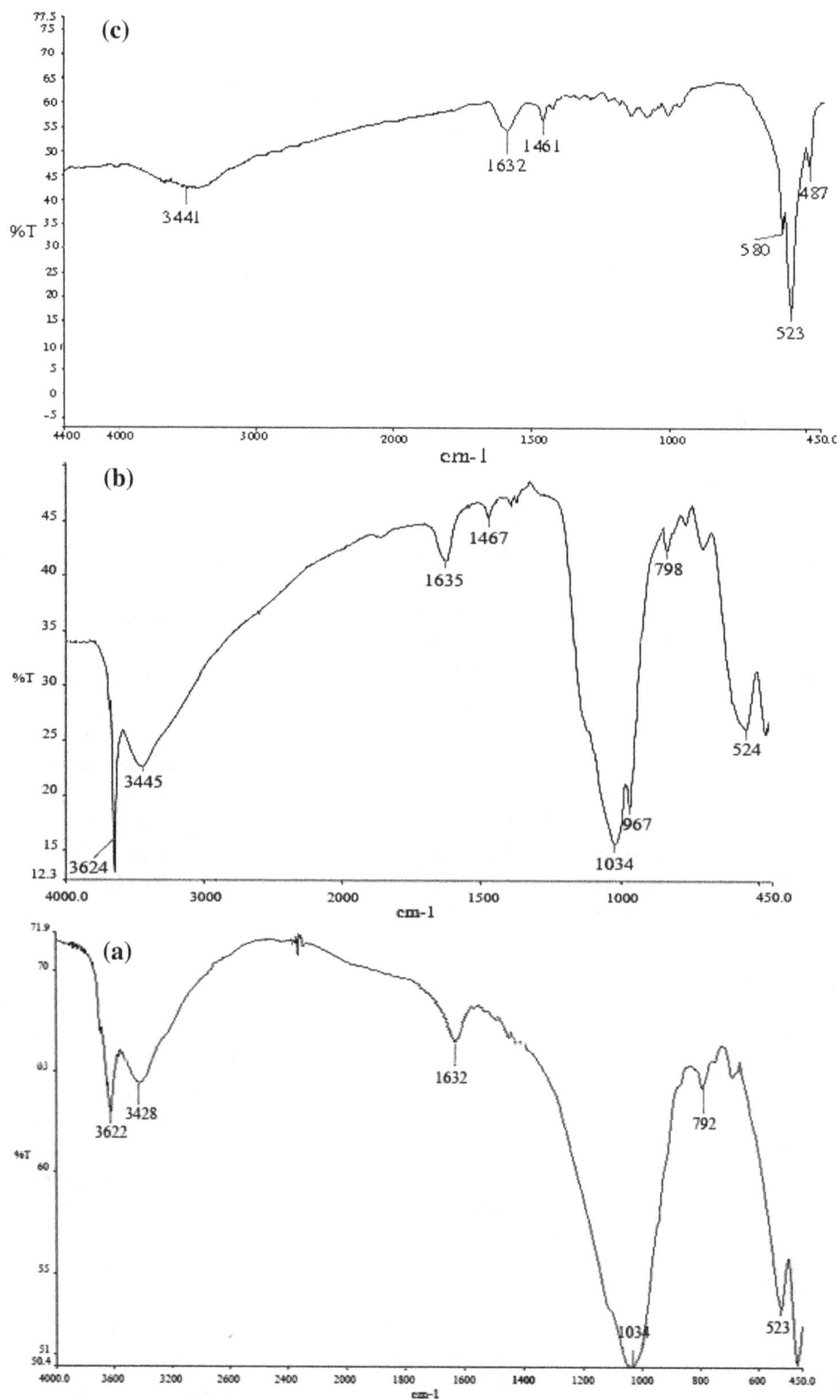

equilibrium concentration of the adsorbent, respectively. The plot corresponded to $\frac{1}{q_e}$ versus $\frac{1}{C_e}$ was depicted and all affiliated data applied in Table 1. Moreover, the plot of $q_e$ versus $C_e$ is shown in Fig. 6 in below.

The Freundlich isotherm is often used for modeling the multilayer adsorption on heterogeneous surfaces and also applied for the very low concentration. It suggests that sorption is not restricted to one specific class of the sites

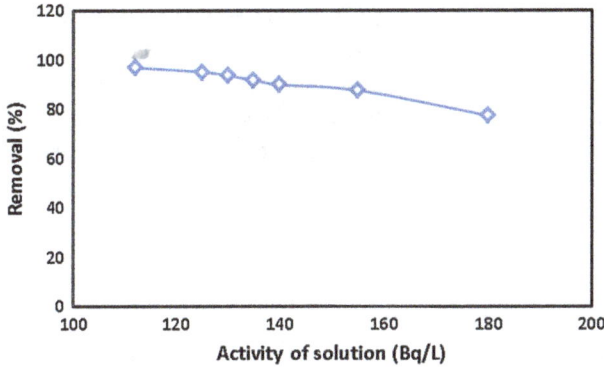

**Fig. 5** The effect of solution activity on the removal efficiency of $^{90}$Sr by CuO NPs/Ag-clinoptilolite zeolite ($T = 25$ °C, pH: 8.5, amount of adsorbent: 1.5 g, contact time: 6 h)

and assumes surface heterogeneity. This isotherm can be explained by the following Eqs. 3 and 4 [33]:

$$q_e = K_f C_e^{\frac{1}{n}} \tag{3}$$

$$\ln q_e = \ln K_f + \frac{1}{n} \ln C_e \tag{4}$$

where $K_f$ (mg/g) and n are Freundlich adsorption constants that were determined from the intercept and slope of the plot. The plot of this isotherm gives a straight line of slope $\ln q_e$ versus $\ln C_e$.

The Temkin isotherm suggests a linear decease of sorption energy and can be expressed via the following Eq. (5) [34]:

$$q_e = \beta \ln \alpha + \beta \ln C_e \tag{5}$$

Here $\alpha$ and $\beta$ are Temkin adsorption constants. The plot of this isotherm shows a straight line of slope $q_e$ versus $\ln C_e$.

The Dubinin–Radushkevich (D–R) can be simplified to the following equations below (6 and 7) [35]:

$$\ln q_e = \ln q_m - K\varepsilon^2 \tag{6}$$

$$\varepsilon = RT \ln\left(1 + \frac{1}{C_e}\right) \tag{7}$$

where $q_m$ and $K$ are Dubinin–Radushkevich adsorption constants that were calculated from the intercept and slope of the plot. $K$ is the adsorption energy. $q_m$ is the theoretical saturation capacity and $\varepsilon$ is the polanyi potential. The plot of this isotherm gives a straight line of slope $\ln q_e$ versus $\ln C_e$. The Harkins–Jura (H–J) isotherm can be defined to the following Eq. (8) [35]:

$$\frac{1}{q_e^2} = \left(\frac{B_{HJ}}{A_{HJ}}\right) - \left(\frac{1}{A_{HJ}}\right) \ln C_e \tag{8}$$

where $A_{HJ}$ is the Harkins–Jura isotherm parameter which accounts for multilayer adsorption and explains the existence of heterogeneous pore distribution in which $B_{HJ}$ is the isotherm constant. The plot of this isotherm gives a straight line of slope $\ln q_e$ versus $\ln C_e$.

The Hasley isotherm model can be utilized to evaluate the multilayer adsorption for the adsorption of $^{90}$Sr at a relatively large distance from the surface. Herein we discuss this isotherm model with equilibrium equation below (9) [36]:

$$\ln q_e = \left[\left(\frac{1}{n_H}\right) ln(K_H)\right] - \left(\frac{1}{n_H}\right) \ln\left(\frac{1}{C_e}\right) \tag{9}$$

$n_H$ and $K_H$ parameters are Hasley isotherm constants and were calculated from the slope and intercept of the linear plot based on $\ln q_e$ versus $\ln\left(\frac{1}{C_e}\right)$ respectively.

The conformity between experimental data and the model predicated values was expressed by the correlation coefficient ($R^2$). A relatively high $R^2$ value reveals that the model successfully describes the adsorption isotherm. It is obvious that the Langmuir and D–R isotherm models cannot correlate the experimental data well. Based on the

**Table 1** Various adsorption isotherm model parameters results for removal and adsorption of $^{90}$Sr by CuO NPs/Ag-clinoptilolite zeolite composite ($T = 25$ °C, pH: 8.5, amount of adsorbent: 1.5 g, contact time: 6 h)

| Isotherm type | Isotherm parameters | | Plot equation |
|---|---|---|---|
| Langmuir | $K_L = 6 \times 10^{12}$ (L/mg) $q_m = 0.1667 \times 10^{-5}$ (mg/g) | $R^2 = 0.8662$ | $1 \times 10^{-5}x + 6 \times 10^7$ |
| Freundlich | $K_f = 2.62 \times 10^6$ (mg/g) (L/mg) $n = 8.510$ | $R^2 = 0.9783$ | $0.1175 \times -14.781$ |
| Temkin | $\alpha = 1.586 \times 10^{15}$ (g/mg) $\beta = 2 \times 10^{-9}$ (mol$^2$/KJ$^2$) | $R^2 = 0.9737$ | $2 \times 10^{-9} \times +2 \times 10^{-8}$ |
| D–R | $K = 2 \times 10^{-10}$ (L/mg) $q_m = 2.87 \times 10^7$ (mg/g) | $R^2 = 0.8256$ | $2 \times 10^{-10} \times -17.174$ |
| H–J | $A_{HJ} = 1.11 \times 10^{-15}$ (L/mg) $B_{HJ} = 2220$ | $R^2 = 0.9759$ | $-9 \times 10^{14} \times -2 \times 10^{16}$ |
| Hasley | $n_H = 8.510$ $K_H = 4.26 \times 10^{54}$ (L/mg) | $R^2 = 0.9783$ | $-0.1175 \times -14.781$ |

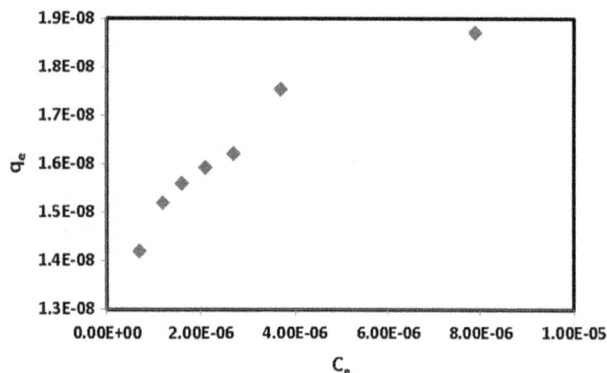

**Fig. 6** The plot for the adsorption of $^{90}$Sr by CuO NPs/Ag-clinoptilolite zeolite composite ($T = 25$ °C, pH: 8.5, amount of adsorbent: 1.5 g, contact time: 6 h)

$R^2$ value, high regression correlation coefficient was found in good straight linear with the Freundlich ($R^2 = 0.9783$), Hasley ($R^2 = 0.9783$), H–J ($R^2 = 0.9759$) and Temkin ($R^2 = 0.9737$) isotherms as compared to the Langmuir ($R^2 = 0.8662$) and D–R ($R^2 = 0.8256$) isotherm models. The obtained data from these models are summarized in Table 1.

## Removal and adsorptive properties study

Various radiometric methods such as gas flow GM (Geiger–Müller) counting and liquid scintillation counting are usually used for direct measurement of $^{90}$Sr and its daughter $^{90}$Y subsequently. Ultra low-level liquid scintillation counting (LSC) can be successfully utilized for counting alpha and beta activity derived from alpha, beta emitters to monitor the natural radioactivity, contamination related to nuclear fallouts, contaminants branched from nuclear power stations or fuel reprocessing plants. Reduced equipment requirements and relative readiness of radiochemical procedures make LSC an attractive technique which can be applied also by laboratories lacking specific radiochemistry facilities and experience. The determination of radiostrontium by this technique is based on the high counting efficiency for high-energy β-particles in aqueous solutions [37]. To evaluate the removal of radioactive $^{90}$Sr, the adsorption behavior of CuO NPs/Ag-clinoptilolite zeolite composite was assessed and those progresses were monitored by LSC instrument. The effects of several operational parameters such as pH, amount of adsorbent and contact time, and also kinetics and thermodynamic reactions were considered.

## Effect of pH

The effect of initial pH parameter on the removal and adsorption of radioactive Sr$^{2+}$ ions by CuO NPs/Ag-clinoptilolite zeolite composite has been investigated. The role of pH on the removal and adsorption yield of CuO

NPs/Ag-clinoptilolite zeolite adsorbent was surveyed via utilizing $^{90}$Sr solution of 112.3 Bq/L at optimized temperature (25 °C) for 6 h. As represented in Fig. 5, the adsorption characteristics of $^{90}$Sr were investigated in pH ranges 2–12 on the removal of $^{90}$Sr by CuO NPs/Ag-clinoptilolite zeolite composite. As pH was increased above pH 2, the percentage removal process of strontium increased in an approximately linear fashion up to a maximum value at about pH 8.5. Therefore, in the pH range of 2–8.5 strontium exists in the form of Sr$^{2+}$ and in high pHs (above 10) it can be found as Sr(OH)$^+$. Plus, when pH is low (pH 2 or less) no more affinity between the adsorbent and Sr$^{2+}$ ions can be observed. Besides, high acidity will lead to replacement of adsorbed Sr$^{2+}$ by H$^+$ from the solution and decrease the adsorption/removal efficiency as a result [38]. To gain the most selectivity and removal efficiency, pH of 8.5 was considered for the further modifications and more than 97 % adsorption yield. The solution pH was adjusted via 1 M solutions of NaOH and HNO$_3$. Moreover, the sorption equilibrium was achieved, the supernatant solution of $^{90}$Sr were brought out and introduced to the Ultra Low-Level Liquid Scintillation Counter (LSC). Subsequently, the removal and adsorption value percentage of $^{90}$Sr by composite adsorbent was calculated. The interaction of hydrogen ions with an oxygen radical of the CuO NPs/Ag-clinoptilolite zeolite composite body generates hydroxyl groups and lowers the charge of the matrix, which is accompanied by a decrease in the sorption ability of CuO NPs/Ag-clinoptilolite zeolite composite in relation to $^{90}$Sr. Besides, a higher sorption of the radioactive due to increasing pH shows that in the solution they are in an ionic state.

## Effect of amount of adsorbent

Selecting the desirable amounts of adsorbent is a key parameter which obviously affects the whole removal procedure. The adsorption characteristic of $^{90}$Sr was investigated at scope of 0.5–3 g of composite adsorbent to recognize the best and optimized amount of adsorbent for the removal of $^{90}$Sr. The obtained results revealed that the more the adsorbent amount, the better the removal efficiency, until the point after which no more significant variations is seen and the curve slope tend to a linear form which means constant values. Hence, the value of 1.5 g was chosen as the appropriate mass for CuO NPs/Ag-clinoptilolite zeolite composite to fulfill high yield removal and adsorption.

## Effect of contact time

To provide a perspicuous comparison between removal and adsorption capability of CuO NPs/Ag-clinoptilolite zeolite

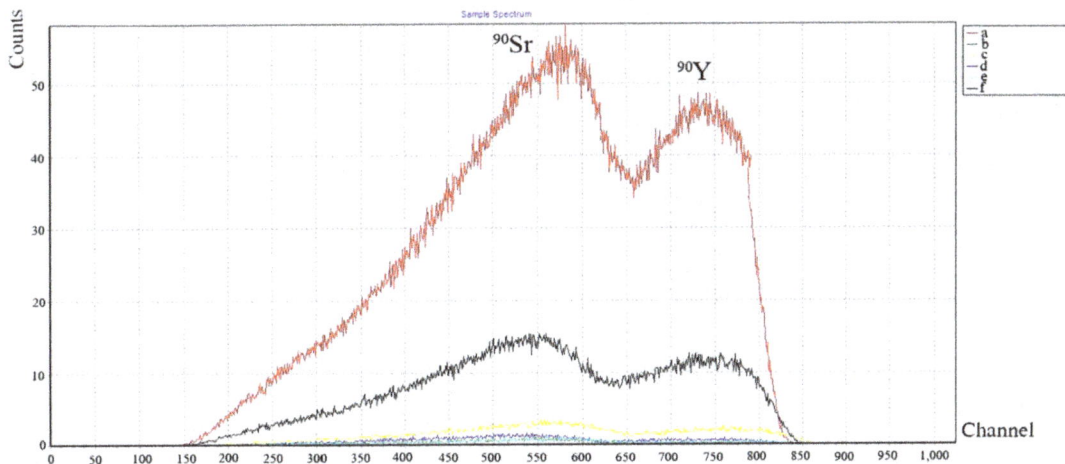

**Fig. 7** Liquid scintillation counting (LSC) spectra for removal of $^{90}$Sr (count versus channel); **a** before contacting with composite, **b** 1 h, **c** 3 h, **d** 6 h, **e** 9 h and **f** 12 h, under optimized conditions (pH = 8.5, temperature ($T = 25$ °C) and amount of adsorbent (1.5 g))

composite and reaction time, the effect of different contact time intervals on the adsorption process of $^{90}$Sr was accomplished. The spectra of $^{90}$Sr/$^{90}$Y and related results have been presented in Figs. 7, 8, and Table 2, respectively. The variation of adsorption value (%) with shaking time has been shown in Fig. 8. Figure 8 represents the reliability of adsorption yield of $^{90}$Sr on the composite adsorbent to the contact time. As the reaction time increases, the adsorption will increase scarcely. On the other hand, rate of counts per minutes (CPM) from 45,065 and 3959 decreases up to 1272 and 118 counts per minutes (CPM), respectively. The similar results are seen in previous researches [39, 40]. The adsorption time was investigated in the scope of 1–12 h, and LSC spectra analysis revealed that the removal first enhanced up to 6 h and then remained in constant value. Thereupon, to reach a shorter analysis period of time 6 h was considered as optimum value. The obtained results from designed experiment revealed that the sorption procedure was rapid and

equilibrium attained quickly after roiling the composite adsorbent with target containing solution. $^{90}$Sr uptake on the CuO NPs/Ag-clinoptilolite zeolite composite may be the cause of vicissitude of target metallic ion with the other ions presented on the adsorbent surface area. The Determination of the activity of $^{90}$Sr was accomplished using the double-energetic windows method. The energetic window A (150–760) contains all the $^{90}$Sr spectrum and low energy region of $^{90}$Y spectrum. The window B (760–940) contains the high-energy region of the $^{90}$Y spectrum. $^{90}$Sr analysis of natural water sample from the Bushehr city of Iran was considered. The removal efficiency was also computed using the following Eq. (10):

$$R(\%) = \left(\frac{A_0 - A_e}{A_e}\right) \times 100 \tag{10}$$

where $A_0$ is the initial radioactivity and $A_e$ is the radioactivity of $^{90}$Sr at equilibrium after sorption process. The minimum detectable activity (MDA) was evaluated using Currie formula (11) and (12) as can be seen in below [41]:

$$\text{MDA}\left(\frac{Bq}{\text{kg}}\right) = L_d(\varepsilon TQ)^{-1} \tag{11}$$

$$L_d(\text{counts}) = 2.71 + 4.65(BT)^{-1.2} \tag{12}$$

where $\varepsilon$ is the detection efficiency; $T$ is the counting time (s); $Q$ is the sample quantity (kg); $B$ is the background count rate (s$^{-1}$).

The following formula can be used for the conversation of strontium-90 activity it's mass in gram (13):

$$M = \frac{A \times T_{\frac{1}{2}} \times A_t}{\ln 2 \times A^0} \tag{13}$$

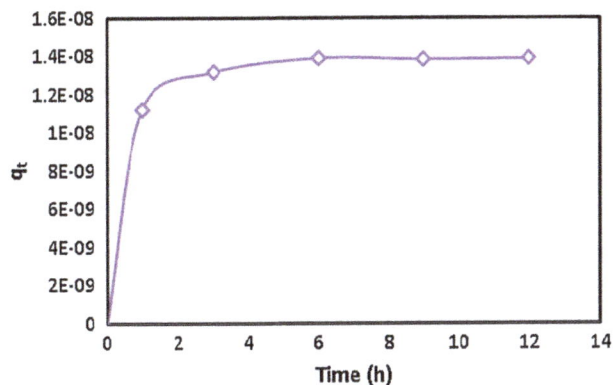

**Fig. 8** The effect of contact time on the removal efficiency of $^{90}$Sr by CuO NPs/Ag-clinoptilolite zeolite ($T = 25$ °C, pH: 8.5, amount of adsorbent: 1.5 g)

where $A$ is the $^{90}$Sr a activity ($Bq$), $T_{1.2}$ is the half-life (second), ln 2 is constant number, $A_t$ and $A^0$ are the mass

**Table 2** Liquid scintillation counting (LSC) results for removal and adsorption of $^{90}$Sr by CuO NPs/Ag-clinoptilolite zeolite composite under optimized conditions (pH = 8.5, temperature ($T$ = 25 °C) and amount of adsorbent (1.5 g))

| Time (h) | CPM (A) | CPM (B) | Activity (Bq/Sample) | Count time min. | MDA (mBq/sample) |
|----------|---------|---------|----------------------|-----------------|------------------|
| 0 | 45065.045 | 3959.91 | 110.5 | 60 | 6.92 |
| 1 | 9632.90 | 846.45 | 23.62 | | |
| 3 | 3474.69 | 317.15 | 8.52 | | |
| 6 | 1272.42 | 118.8 | 3.12 | | |
| 9 | 1488.57 | 130.8 | 3.65 | | |
| 12 | 1345.83 | 118.25 | 3.30 | | |

number of Sr ($A_t$ = 90) and Avogadro's number (g). $M$ is the mass of strontium-90 activity (g).

## Kinetics of adsorption reaction

The adsorption kinetics is a significant factor for designing adsorption systems and is required for selecting optimum operating conditions for adsorption reaction study. To investigate the adsorption kinetics of removal of $^{90}$Sr by CuO NPs/Ag-clinoptilolite zeolite composite, four different kinetics including models, pseudo first order, pseudo second order, Elovich and intra particle diffusion kinetic, were applied in this study. The pseudo first order Lagergren [42] model presumes that the rate of variation of solute uptake by reaction time is certainly related to versatility in glut concentration and solid uptake value via reaction time (14).

$$\log(q_e - q_t) = \log q_e - 2.303 k_1 t \tag{14}$$

where $q_e$ and $q_t$ parameters, are considered as the values of $^{90}$Sr which are adsorbed per mass unit of the composite adsorbent (mg·g$^{-1}$) at the equilibrium and time $t$, respectively. $k_1$ is recognized as the rate constant of the adsorption reaction (min$^{-1}$). $\log(q_e - q_t)$ was also plotted versus time interval, a straight line should be obtained with a slope of $k_1$, if the first order kinetics is credible. Ho and McKay [43] proposed a pseudo second order model for the adsorption of divalent metal ions onto sorbent particles that following below Eq. (15):

$$\frac{t}{q_t} = \frac{t}{q_t} + \frac{1}{k_2 q_{e^2}} \tag{15}$$

where $q_e$ and $q_t$ parameters, represent the amount of $^{90}$Sr (g·mg$^{-1}$) at equilibrium and other time intervals. $k_2$ is the rate constant of the pseudo second order equation (g·mg$^{-1}$ min$^{-1}$). When the second order model is a suitable expression, a pattern of $\frac{t}{q_t}$ against time ($t$) will gain a linear result with a slope of $\frac{1}{q}$ and an excise of $\frac{1}{k_2 q_{e^2}}$. The adsorbed amounts ($q$) of Sr$^{2+}$ were calculated using the following Eq. (16):

$$q = \frac{(C_0 - C_e)V}{m} \tag{16}$$

where $C_0$ and $C_e$ are the initial and equilibrium concentrations of Sr$^{2+}$ (g·mg$^{-1}$) in the liquid phase, respectively, $V$ is the volume of solution (L) and also $m$ is the mass of adsorbent (g). The rate constant of pseudo first order and pseudo second order of the adsorption and correlation coefficient ($R^2$) were determined from the pattern among $\log(q_e - q_t)$ versus time $t$ and the pattern of $t/q$ versus time $t$.

## Elovich kinetic model

The Elovich equation is represented as it can be observed below (17) [44]:

$$\frac{dq_t}{dt} = \alpha e^{\beta q_t^{-1}} \tag{17}$$

where $\alpha$ and $\beta$ are considered as the initial sorption rate and the desorption constant both (mg.g$^{-1}$), respectively. The Elovich equation can be simplified if it is presumed that $\alpha\beta t \gg 1$. At the boundary conditions $qt = 0$ at $t = 0$, the above mentioned equation changes to (18) [45]. Plot of Elovich kinetics is given.

$$q_t = \beta \log(\alpha\beta) + \beta \log t \tag{18}$$

## Intra particle diffusion model

Each adsorption procedure includes different surface diffusion followed by intra particle diffusion. Generally, the liquid phase mass transport managed the adsorption process. Also the mass transport rate can be imparted as a function of the square root of time ($t$). As clarified above, the intra particle diffusion model was stated by formula below (19) [45]:

$$q_t = k_i t^{1.2} + C \tag{19}$$

At the above mentioned formula, $q_t$ is the amount of the adsorbed $^{90}$Sr on the CuO NPs/Ag-clinoptilolite. Also t and C, are time and intra particle diffusion rate constant,

**Table 3** The different kinetics model rate constants for the removal and adsorption of $^{90}$Sr on the CuO NPs/Ag-clinoptilolite zeolite composite

| Kinetic model | $R^2$ | $k_1$ (min$^{-1}$) | $k_2$ (g.mg$^{-1}$.min$^{-1}$) | $k_i$ (mmol.mg$^{-1}$.min$^{-1}$) | $C$ | $\alpha$ | $\beta$ | Plot equation |
|---|---|---|---|---|---|---|---|---|
| First order | 0.7729 | 0.1103 | – | – | – | – | – | $-0.1103 \times -8.3238$ |
| Second order | 0.9996 | – | $7 \times 10^7$ | – | – | – | – | $7 \times 10^7 \times +9 \times 10^6$ |
| Elovich | 0.8864 | – | – | – | – | 1.1413 | $2 \times 10^{-6}$ | $2 \times 10^{-9} \times +1 \times 10^{-8}$ |
| Intra particle diffusion | 0.6854 | – | – | $4 \times 10^{-9}$ | $4 \times 10^{-9}$ | – | – | $4 \times 10^{-9} \times +4 \times 10^{-9}$ |

respectively. Plus, the amount of correlation coefficient ($R^2$) was calculated from the slope and intercept of the drawing of $q_t$ versus $t^{1.2}$. By drawing the plot of $q_t$ versus $t_{1/2}$, it can be obviously inferred that the Sr$^{2+}$ adsorption includes three main steps. First, the analyte ions diffuse among the particles in liquid phase and a rapid adsorption takes place which refers to the Sr$^{2+}$ transfers on the surface active sites of zeolite and then into the porous structure of adsorbent so-called intra particle diffusion. Afterwards, the gradual occupation of active sites slowed down the adsorption of analyte and maintained at a constant value when these active sites were thoroughly saturated. The slight indicated deviations of these lines from the origin specify that the intra particle transportations are not the only rate limiting operative.

The kinetic model along with upper correlation coefficient R$^2$ was considered as the most appropriate model. Table 3 shows the kinetic factors of the $^{90}$Sr adsorption on the CuO NPs/Ag-clinoptilolite. The obtained results illustrates that the $R^2$ value of pseudo second order kinetic compared to the $R^2$ value, other kinetics models is higher, thus the $^{90}$Sr adsorption on CuO NPs/Ag-clinoptilolite is followed via pseudo second order. The energy of activation $E_a$ calculated by the Arrhenius equation was found to be 44.75 kJ/mol and is expressed as below (20 and 21).

$$k = -A \exp\left(-\frac{E_a}{RT}\right) \tag{20}$$

$$\ln k = -\frac{E_a}{RT} + \text{Ln}A \tag{21}$$

where $k$ is the chemical reaction rate, $R$ is the gas constant (8.314 J mol$^{-1}$ K$^{-1}$), $T$ is the absolute temperature in kelvin (°K) and $A$ is the pre-exponential factor.

**Thermodynamic of removal reaction**

*Effect of temperature*

The temperature in which the experiment fulfills is an important factor that cannot be over looked. In this study, the removal and adsorption of $^{90}$Sr on the CuO NPs/Ag-clinoptilolite zeolite composite was surveyed in the temperature scope of 25–50 °C under certain optimized

conditions. The spectra of $^{90}$Sr/$^{90}$Y and its results are shown in Figs. 9 and 10. Figure 9 illustrates the effect of temperature on the removal of $^{90}$Sr on the composite adsorbent surface. As can be seen, the adsorption of $^{90}$Sr on the CuO NPs/Ag-clinoptilolite zeolite composite decreases as the temperature increases gradually. The removal reaction efficiency for the temperatures of 25, 30, 35, 40, 45 and 50 °C were 97.17, 90.24, 82.19, 73.72, 60.52 and 41.98 %, respectively. This is why in raised temperatures the formed bonds between $^{90}$Sr and active sites of nanoparticles adsorbent will be weaken and broken eventually. The behavior study of the adsorption of $^{90}$Sr ions by CuO NPs/Ag-clinoptilolite zeolite composite was investigated as a function of temperature. To determine the process of spontaneous reaction, both important energy and entropy factors should be considered. Moreover, the dependence of distribution ratios on the temperature was evaluated. The relationship between $K_d$ and Gibbs free energy $\Delta G^0$ variation in sorption has been shown below (22):

$$\Delta G = -RT\ln(55.5K_d) \tag{22}$$

$R$ is the universal gas constant (8.314 J mol$^{-1}$ K$^{-1}$), $T$ is the absolute temperature in kelvin (°K) and $K_d$ is the distribution coefficient. To gain a correct value of $\Delta G^0$, the $K_d$ value in Eq. (21) must be dimensionless. If the adsorption process was evaluated from an aqueous solution and $K_d$ is considered in dm$^3$ mol$^{-1}$, then the parameter $K_d$ can be simply recalculated as dimensionless by multiplying it by 55.5 which refers to the number of moles of water per liter of solution. Therefore, the correct value for $\Delta G^0$ can be gained from the Eq. (21). Thus, the term 55.5 $K_d$ (dm$^3$ mol$^{-1}$ mol dm$^{-3}$) is dimensionless [46]. The distribution coefficient was determined at equilibrium in the radioactive solution by the following Eq. (23) [47, 48]:

$$K_d = \frac{q_e}{C_e} = \left(\frac{A_0 - A_e}{A_e}\right) \times \frac{V}{m} \tag{23}$$

where $A_0$ (Bq/L) (or $C_0$ µmol/L) parameter is the initial strontium activity (or concentration) in solution, $A_e$ (Bq/L) (or $C_e$ µmol/L) parameter is the strontium activity (or concentration) in solution at equilibrium, $V$ (mL) is the solution volume, and $m$ (g) is the adsorbent mass [49].

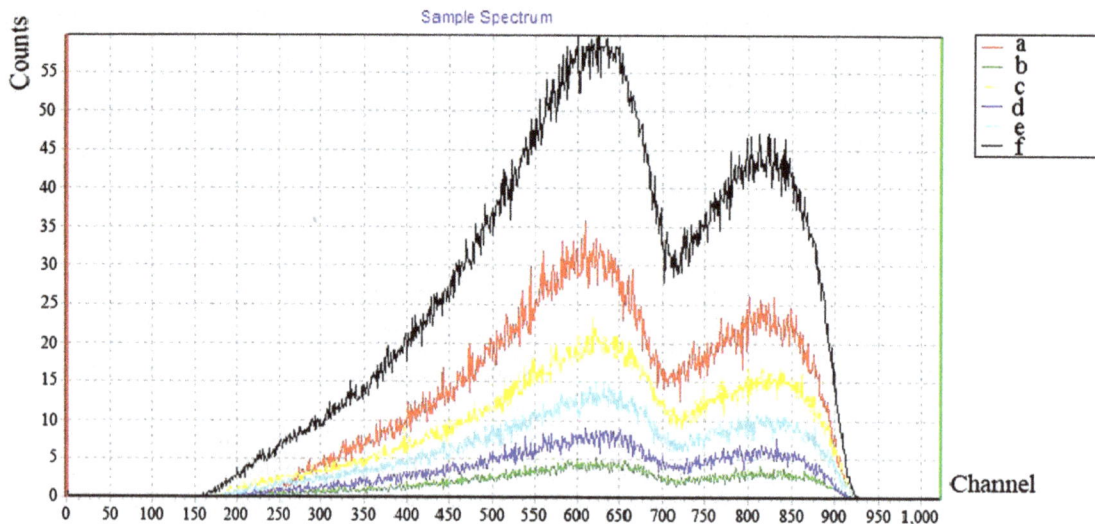

**Fig. 9** Liquid scintillation counting (LSC) spectra for removal of $^{90}$Sr (count versus channel) at different temperature; **a** 25 °C, **b** 50 °C, **c** 45 °C, **d** 40 °C, **e** 35 °C and **f** 30 °C, under optimized conditions (pH = 8.5, amount of adsorbent (1.5 g) and contact time (6 h))

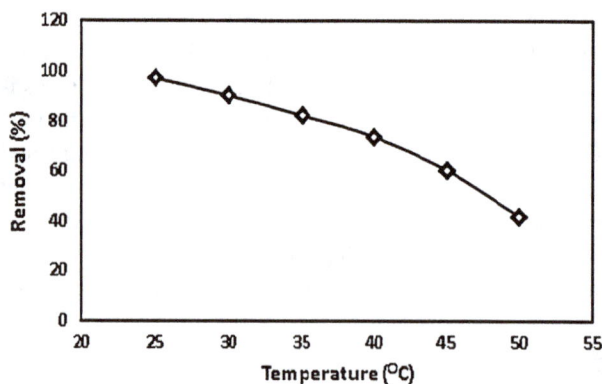

**Fig. 10** Plot of $^{90}$Sr removal % versus temperature (°C)

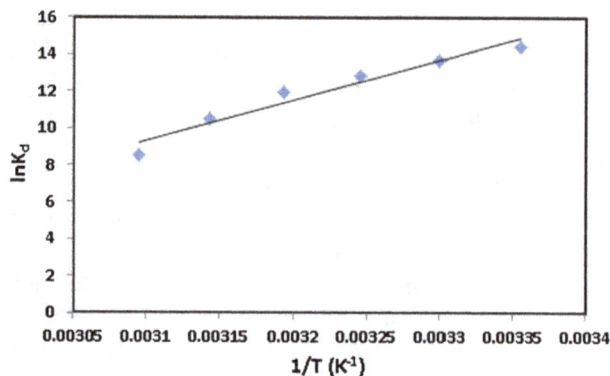

**Fig. 11** A plot of vans Hoff (ln $K_d$ versus $1/T$) for the removal and adsorption of $^{90}$Sr on the CuO NPs/Ag-clinoptilolite zeolite composite at different temperature

Gibbs free energy variation can also be introduced in terms of enthalpy variation, $\Delta H^0$, entropy variation, $\Delta S^0$, as stated in below (24):

$$\Delta G^0 = \Delta H^0 - T\Delta S^0 \qquad (24)$$

Besides, by mixing the two above mentioned Eqs. (21–23) a new exposition is attained as is seen in following (Vans Hoff Eq. (25)):

$$lnK_d = -\frac{H^0}{R} \times \frac{1}{T} + \frac{\Delta S^0}{R} \qquad (25)$$

The enthalpy ($\Delta H^0$) of adsorption and the entropy ($\Delta S^0$) of adsorption can be specified from the slope and the intercept of the linear fits which are gained by drawing ln $K_d$ against $\frac{1}{T}$ respectively. Also the negative amounts $\Delta G^0$ show that the adsorption procedure is spontaneous with affinity of $^{90}$Sr to the composite. Two significant types of adsorption process namely pure physical and pure chemical. During the pure physical adsorption, $\Delta H^0$ values are

from 2.1 to 20.9 kJ/mol, while $\Delta H^0$ values corresponded to the pure chemical adsorption are in a ranges 80–200 kJ/mol. $\Delta H^0$ values between 20.9 and 80 kJ/mol related to the physic-chemical adsorption process. The values are well under those related to chemical bond constitution, showing the chemical property of the adsorption process and means that the adsorption of $^{90}$Sr on the CuO NPs/Ag-clinoptilolite zeolite composite would be attributed to a chemical adsorption process ($\Delta H^0 = -182.550$ kJ/mol) comparing with above implied information and ranges. Forby, the enthalpy variation $\Delta H^0$ following adsorption is negative in all cases representing the exothermic nature of adsorption that is the removal of $^{90}$Sr is decreased as the temperature increases. The entropy variations $\Delta S^0$ of the system along with the adsorption of $^{90}$Sr ions on the CuO NPs/Ag-clinoptilolite zeolite composite is positive in all cases showing that more discovery

**Table 4** Thermodynamic function values for the removal and adsorption of $^{90}$Sr on the CuO NPs/Ag-clinoptilolite zeolite composite

| Temperature (°K) | $\Delta G^0$ (KJ/mol) | $\Delta H^0$ (KJ/mol) | $\Delta S^0$ (J/mol K) | $R^2$ | Plot equation |
|---|---|---|---|---|---|
| 298 | −35.816 | −182.550 | −488.514 | 0.9523 | 219578 × −58.758 |
| 303 | −34.553 | | | | |
| 308 | −32.895 | | | | |
| 313 | −31.061 | | | | |
| 318 | −27.803 | | | | |
| 323 | −22.896 | | | | |

is generated following adsorption and reflects that no specific change occurs in the internal structure of composite during adsorption of $^{90}$Sr. According to the plots data of ln $K_d$ versus $1/T$ in Fig. 11, the results as shown in Table 4.

## The mechanism of Sr$^{2+}$ adsorption on the CuO NPs/Ag-clinoptilolite zeolite composite

The properties such as porosity, presence of alkaline and earth alkaline metallic cations in zeolites structure along with high mechanical and chemical resistance provide good adsorptive, cation exchange, and catalytic characteristics and make them desirable for wide variant of analytical purposes [50, 51]. According to the results reported in this study, the adsorption mechanism of strontium ions on the pre-synthesized CuO NPs/Ag-clinoptilolite zeolite composite adsorbent can be demonstrated by three major possible theories including: (1) the formation of electrostatic or van-der-waals forces of Sr$^{2+}$ with negative charged oxygen and hydroxyl groups originated from zeolite structure on its surface area, (2) the ion exchange process between Sr$^{2+}$ and exchangeable cations (Ag$^+$, Cu$^{2+}$ and etc.) with negative charge balance of aluminum atoms and (3) the intra porosity diffusion of analyte into the zeolite structure which refers to the above mentioned binding potentials of Sr$^{2+}$ with active sites of zeolite.

## Conclusions

This research is focused on the preparation of CuO NPs/Ag-clinoptilolite zeolite composite adsorbent through two facile routes ion exchange and impregnation methods and applied for effective removal of radioactive $^{90}$Sr ions from water sample of Bushehr city. The prepared adsorbents were characterized by SEM-EDAX, XRD and FT-IR analyses and the removal process followed via Ultra Low-Level Liquid Scintillation Counting (LSC) as a rapid and suitable analytical technique. Also, different conditions such as pH, amount of adsorbent, the contact time and temperature were investigated and optimized to approach the highest adsorption/removal efficiency of strontium-90. Moreover, Adsorption isotherms including Langmuir, Freundlich, Temkin, D-R, H-J and Hasley have been analyzed to the equilibrium data. The Freundlich, Temkin, H-J and Hasley isotherms were found good to represent the measured adsorption data. The parameters including: pH = 8.5, amount of adsorbent (1.5 g), contact time (6 h) and temperature (25 °C) were considered as optimized conditions for this process. The experimental results denoted that CuO NPs/Ag-clinoptilolite zeolite composite leads to maximum removal and adsorption of $^{90}$Sr from water sample. On the other hand, the reaction kinetic information was surveyed utilizing pseudo first and second orders, Elovich and Intra particle diffusion kinetic models. Besides, the adsorption kinetics of $^{90}$Sr was matched nicely with the pseudo second order kinetic model. Then, thermodynamic study for the adsorption reactions was evaluated and the results showed that by increasing the temperature, efficiency reaction decreased. It was emphasized that CuO NPs/Ag-clinoptilolite zeolite composite has a high capacity and potential for the removal of radioactive $^{90}$Sr from water sample. It should be noted that the presented composite adsorbent as a potentially applicable adsorbent can be considered for further analytical procedures in future.

**Acknowledgments** The authors give their sincere thanks to the Islamic Azad University, Ahvaz, Iran, Islamic Azad University, Qaemshahr, Iran and Islamic Azad University, Bushehr, Iran for all their supports.

## References

1. Sachse A, Merceille A, Barre Y, Grandjean A, Fajula F, Galarneau A (2012) Macroporous LTA-monoliths for in-flow removal of radioactive Sr$^{2+}$ from aqueous effluents: application to the case of Fukushima. Macropo Mesopo Mater 164:251–258

2. Hari P, Bimala P, Inoue Katsutoshi, Ohto Keisuke, Kawakita Hidetaka, Kedar NG, Hiroyuki H, Shafiq A (2014) Adsorptive removal of strontium from water by using chemically modified orange juice residue. Separat Sci Technol 49:1244–1250

3. Mashkani SG, Ghazvini PTM (2009) Biotechnological potential of Azolla filiculoides for biosorption of $Cs^+$ and $Sr^{2+}$: application of micro-PIXE for measurement of biosorption. Biores Technol 100:1915–1921

4. Carboneau ML, Adams JP, Garcia RS (1994) National low-level waste management program radionuclide report series, vol. 7: Strontium-90. Idaho National Engineering Laboratory, Idaho Falls

5. Sebesta F, Motl John AJ (1993) Proceedings of International Conference on Nuclear Waste Management and Environmental Remediation, Prague, 3, p 871

6. Mardan A, Ajaz R, Mehmood A, Raza SM, Ghaffar A (1999) Preparation of silica potassium cobalt hexacyanoferrate composite ion exchanger and its uptake behaviour for cesium. Sep Purif Technol 16:147–158

7. Murthy ZVP, Parmar S (2011) Removal of $Sr^{2+}$ by electrocoagulation using stainless steel and aluminium electrode. Desalination 282:63–67

8. Bascetin E, Atun G (2010) Adsorptive removal of $Sr^{2+}$ by binary mineral mixtures of montmorillonite and zeolite. J Chem Eng Data 55:783–788

9. Chegrouche S, Mellah A, Barkat M (2009) Removal of $Sr^{2+}$ from aqueous solutions by adsorption onto activated carbon: kinetic and thermodynamic studies. Desalination 235:306–318

10. Ghaemi A, Mostaidi MT, Maragheh MG (2011) Charecterization of $Sr^{2+}$ and $Ba^{2+}$ adsorption from aqueous solutions using dolomite powder. J Hazard Mater 190:916–921

11. Hossein F, Mozhgan I, Mohammad M, Mohammad GM (2013) Preparation of a novel PAN–zeolite nanocomposites for removal of $Cs^+$ and $Sr^{2+}$ from aqueous solutions: kinetic equilibrium, and thermodynamic studies. Chem Eng J 222:41–48

12. Hossein F, Mohammad M, Alireza F, Mozhgan I (2013) Synthesis of a novel magnetic zeolite nanocomposites for removal of $Cs^+$ and $Sr^{2+}$ from aqueous solution: kinetic, equilibrium, and thermodynamic studies. J Colloid Interface Sci 393:445–451

13. Saheed SO, Modise SJ, Sipalma AM (2013) $TiO_2$ supported clinoptilolite characterization and optimization of operational parameters for methyl orange removal. Adv Mater Res 781–784:2249–2252

14. Mohammad A (2014) Khakizadeh, Hoda Keipour, Abolfazl Hosseini, Daryoush Zareyee, KF/clinoptilolite, an effective solid base in Ullmann ether synthesis catalyzed by CuO nanoparticles. New J Chem 38:42–45

15. Jahangirian H (2013) Synthesis and characterization of zeolite/$Fe_3O_4$ nanoparticle by green quick precipitation method. Dig J Nanomater Bios 8:1405–1413

16. Yusan S, Erenturk S (2011) Adsorption characterization of strontium on PAN/zeolite composite adsorbent. World J Nucl Sci Technol 1:6–12

17. Huang Y, Wang W, Feng Q, Dong F (2013) Preparation of magnetic clinoptilolite/$CoFe_2O_4$ composites for removal of $Sr^{2+}$ from aqueous solutions: kinetic, equilibrium, and thermodynamic studies. J Saudi Chem Soc 09:1–9

18. Dastafkan K, Sadeghi M, Obeydavi A (2014) Manganese dioxide nanoparticles-silver-Y zeolite as a nanocomposite catalyst for the decontamination reactions of O, S-diethyl methyl phosphonothiolate. Int J Environ Sci Technol. doi:10.1007/s13762-014-0701-1

19. Sadeghi M, Hosseini MH (2013) The detoxification of methamidophos as an organophosphorus insecticide on the magnetite ($Fe_3O_4$) nanoparticles/Ag-NaY faujasite molecular sieve zeolite (FMSZ) composite. Int J Bio-Inorg Hybd Nanomat 2:517–524

20. Buarod E, Pithakratanayothin S, Naknaka S, Chaiyasith P, Yotkaew T, Tosangthumb N, Tongsri R (2015) Facile synthesis and characterization of tenorite nanoparticles from gas-atomized Cu powder. Powder Technol 269:118–126

21. Engelbrekt C, Malcho P, Andersen J, Zhang L, Stahl Bin Li K, Hu J, Zhang JD (2014) Selective synthesis of clinoatcamite $Cu_2(OH)_2Cl$ and tenorite CuO nanoparticles by pH control. J Nanopart Res 16:2562–2564

22. Yin M, Wu CK, Lou Y, Burda C, Koberstein JT, Zhu Y, O'Brien S (2005) Copper oxide nanocrystals. J Am Chem Soc 127:9506–9511

23. Shaffiey SF, Shapoori M, Bozorgnia A, Ahmadi M (2014) Synthesis and evaluation of bactericidal properties of CuO nanoparticles against Aeromonas hydrophila. Nanomed J 1:198–204

24. Salavati-Niasari M, Davar F (2009) Synthesis of copper and copper (I) oxide nanoparticles by thermal decomposition of a new precursor. Mater Lett 63:441–443

25. Sadeghi M, Ghaedi H, Yekta S, Babanezhad E (2016) Decontamination of toxic chemical warfare sulfur mustard and nerve agent simulants by NiO NPs/Ag-clinoptilolite zeolite composite adsorbent. J Environ Chem Eng 4:2990–3000

26. Sadeghi M, Hosseini MH (2012) Preparation and application of $MnO_2$ nanoparticles/zeolite AgY composite catalyst by confined space synthesis (CSS) method for the desulfurization and elimination of SP and OPPJ. Nano Struct 2:439–453

27. Abdi MR, Shakur HR, Rezaee Ebrahim Saraee Kh, Sadeghi M (2014) Effective removal of uranium ions from drinking water using CuO/X zeolite based nanocomposites: effects of nano concentration and cation exchange. J Radioanal Nucl Chem 300(3):1217–1225. doi:10.1007/s10967-014-3092-3

28. Lou LL, Liu S (2005) CuO-containing MCM-48 as catalysts for phenol hydroxylation. Catal Commun 6:762–765

29. Li Z, Gao L (2003) Synthesis and characterization of MCM-41 decorated with CuO particles. J Phys Chem Solids 64:223–228

30. Hao XY, Zhang YQ, Wang JW, Zhou W, Zhang C, Liu S (2006) A novel approach to prepare MCM-41 supported CuO catalyst with high metal loading and dispersion. Microporous Mesoporous Mater 88:38–47

31. Debye P, Scherrer P (1916) The Scherrer equation versus the 'Debye-Scherrer equation'. Phys Z 17:277–283

32. Langmuir I (1916) The constitution and fundamental properties solids and liquids. J Am Chem Soc 38:2221–2225

33. Freundlich HMF (1906) User die adsorption in losungen. Z Phys Chem 57:385–470

34. Temkln MI (1941) Adsorption equilibrium and kinetics of processes on non-homogeneous surface and in the interaction between adsorbed molecules. Zh Fiz Chim 15:296–332

35. Foo KY, Hamed BH (2010) Insight into modeling of adsorption isotherm systems. Chem Eng J 156:2–10

36. Hasley GD (1952) The role of surface heterogeneity. Adv Catal 4:259–267

37. Abadzic SA, Ryan JN (2001) Particle release and permeability reduction in a natural zeolite (clinoptilolite) porous medium. Environ Sci Technol 35:4502–4508

38. Yavari R, Huang YD, Mostofizadeh A (2010) Sorption of strontium ions from aqueous solutions by oxidized multiwall carbon nanotubes. J Radioanal Nucl Chem 285:703–710

39. Talaie N, Aghabozorg HR, Alamdar Milani S (2012) Cs and Sr absorption study using synthesized titanosilicate nanoparticles, containing Nb and Ge, Proceedings of the 4th International Conference on Nanostructures (ICNS4) 12–14 March, Kish Island, IR Iran

40. Sadeghi M, Yekta S, Ghaedi H, Babanezhad E (2015) Synthesis and characterization of $\gamma$-$MnO_2$-AgA zeolite nanocomposite and its application for the removal of radioactive strontium-90 ($^{90}Sr$) Int. J Bio-Inorg Hybr Nanomater 4:101–111

41. Currie LLA (1968) Limits for qualitative detection and quantitative determination. Anal Chem 40:586–593
42. Lagergern S (1898) About the theory of so-called adsorption of soluble substances. K Sven Vetenskapsakad Handl 24:1–39
43. Mckay YSG (1998) The kinetics of sorption of basic dyes from aqueous solution by sphagnum moss peat. Can J Chem Eng 76:822–827
44. Liang S, Guo X, Feng N, Tian Q (2010) Isotherms, kinetics and thermodynamic studies of adsorption of $Cu^{2+}$ from aqueous solutions by $Mg^{2+}/K^+$ type orange peel adsorbents. J Hazard Mater 174:756–762
45. Annadurai G, Juang RS, Lee DJ (2002) Use of cellulose-based wastes for adsorption of dyes from aqueous solutions. J Hazard Mater 92:263–274
46. Milonjić SK (2007) A consideration of the correct calculation of thermodynamic parameters of adsorption. J Serb Chem Soc 72(12):1363–1367
47. Merceille A, Weinzaepfel E, Barre Y, Grandjean A (2012) The sorption behavior of synthetic sodium nonatitanate and zeolite A for removing radioactive strontium from aqueous wastes. Sep Purif Technol 96:81–88
48. Nishiyama Y, Hanafusa T, Yamashita J, Yamamoto Y, Ono T (2016) Adsorption and removal of strontium in aqueous solution by synthetic hydroxyapatite. J Radioanal Nucl Chem 307(2):1279–1285. doi:10.1007/s10967-015-4228-9
49. Lívia K, Lima S, Jean F, Silva L, Meuris G, da Silva C, Melissa G, Vieira A (2014) Lead biosorption by salvinia natans biomass: equilibrium study. Chem Eng Trans 38:97–102. doi:10.3303/CET1438017
50. Perić J, Trgo M, Vukojević Medvidović N (2004) Removal of zinc, copper and lead by natural zeolite-a comparison of adsorption isotherms. Water Res 38(7):1893–1899
51. Ersoy B, Çelik MS (2003) Effect of hydrocarbon chain length on adsorption of cationic surfactants onto clinoptilolite. Clays Clay Miner 51(2):172–180

# 14

# Influence of ZnO doping and calcination temperature of nanosized CuO/MgO system on the dehydrogenation reactions of methanol

S. A. El-Molla[1] · Sh. M. Ibrahim[1,2] · M. M. Ebrahim[1]

**Abstract**   Pure and doped catalysts were prepared with ZnO dopant concentration (0.04–0.08 mol) by wet impregnation method followed by calcination at 350–650 °C. The prepared solids were characterized by X-ray diffraction, $N_2$-adsorption at −196 °C and the methanol conversion as the catalytic probe reaction. The results revealed that the crystallite size and ordering of CuO phase decrease by ZnO-doping for samples calcined at 350–550 °C. Opposite trends were observed by increasing ZnO amount to 0.08 mol (i.e. 9.21 wt%). The specific surface area ($S_{BET}$) and the catalytic activity of pure catalyst increase by increasing the calcination temperature to 550 °C and/or increasing the amounts of ZnO up to certain extent reaching to a maximum at 7.07 wt% ZnO; above this concentration catalytic activity of doped samples decreases. But at calcination temperature above 550 °C, the catalytic activity and selectivity decrease. The prepared catalysts are selective towards formaldehyde and methyl formate formation.

**Keywords**   ZnO-doping · CuO/MgO system · Methanol dehydrogenation · Conversion · Selectivity

✉ Sh. M. Ibrahim
  shiemaa_332003@yahoo.com

[1]   Department of Chemistry, Faculty of Education,
     Ain Shams University, Roxy, Cairo 11757, Egypt

[2]   Present Address: Department of Chemistry,
     Faculty of Science, Al-Qassim University,
     Al-Qassim, Buraidah 51452, Saudi Arabia

## Introduction

Methanol has been considered as an important raw material in the synthesis of various chemicals. Dehydrogenation of methanol was expected as a promising process to synthesize formaldehyde and methyl formate [1]. Formaldehyde is an industrial chemical that is widely used to manufacture numerous household products [2]. Formaldehyde and methyl formate are produced in large quantity by the coal chemical industry and natural gas chemical industry [3], both of them are important organic chemical materials usually used as the raw material of medicines [4]. Methanol is also considered as a candidate for the chemical carrier to transport hydrogen [5]. Copper-based catalysts appeared to be active towards the dehydrogenation of methanol. The activity and selectivity of catalyst were strongly influenced by both reaction conditions and surface structure. On reduced copper surface, $CH_3O$ species decomposed near 370 K to form formaldehyde and methyl formate [6]. Over the past two decades, there has been considerable interest in the adsorption and decomposition mechanism of methanol over solid surfaces such as Cu, Pt, Ni, $TiO_2$ and other oxides [7–11].

Improving the catalytic activity and selectivity of metal oxides employed in some important industrial reactions is achieved using suitable support, exposure to radiation and doping with certain foreign cations [12, 13]. Doping of single or multicomponent metal or metal oxide system with certain foreign oxides was efficient approach for bringing about significant modifications in their thermal stability, electrical, optical, magnetic, surface and catalytic properties [14]. The addition of small amounts of certain foreign cations such as $Li^+$, $Zr^{4+}$ and $K^+$ to CuO/MgO has influenced the mutual solid–solid interaction between constituents [15, 16]. This influence may be accompanied by significant changes in the catalytic and physicochemical properties of the doped catalysts [14]. Doping CuO/MgO system with $K_2O$ decreased the degree of ordering of CuO and MgO phases [16]. $Li_2O$-doping of CuO/MgO solid enhanced its catalytic activity towards conversion of iso-propanol [15, 17]. ZnO is a well-known dopant for many catalytic systems such as NiO and $Co_3O_4$, in which ZnO brought about measurable changes in their specific surface areas and the crystallite sizes of their phases [18]. Doping $Ni–Al_2O_3$ system with ZnO affected on the range of reduction temperature of NiO due to formation of $ZnAl_2O_4$ spinel-liked structure [19]. It was reported that the reactivity of $Fe_2O_3$ to interact with MO (M = Mg, Ni, Co and Mn) yielding the corresponding $MFe_2O_4$ had been found to be much stimulated by doping with ZnO [20, 21]. $Zn^{2+}$ had a high activity to diffuse into $Co_3O_4$ and $Fe_2O_3$ solids at about 600 °C [22]. Doping $CuO/TiO_2$ with ZnO brought about a measurable decrease in the crystallite size of both anatase and CuO phases to an extent proportional to the amount of dopant added [23].

The present work aimed at studying the role of calcination temperature and ZnO doping of CuO/MgO system was prepared by impregnation method on its structural, surface characteristics, catalytic activity and selectivity. The techniques employed were X-ray-diffraction, $N_2$-adsorption at −196 °C.

## Experimental

### Materials

A known amount of magnesium hydroxide (previously prepared [24]) was impregnated with a solution containing a known amount of copper nitrate dissolved in the least amount of distilled water. The CuO content in all solid samples was fixed at 23 mol%. The solutions prepared contained different proportions of $Zn (NO_3)_2 \cdot 6H_2O$. The obtained pastes were dried at 110 °C and then calcined at 350–650 °C for 4 h. The amount of zinc oxide expressed in mole was 0.04, 0.05, 0.06 and 0.08. The undoped sample was nominated as CuMgO, but doped solids were determined and nominated as 0.04, 0.05, 0.06, 0.08 ZnO-doped CuMgO.

### Techniques

X-Ray diffractograms of various prepared solids were determined using a Bruker diffractometer (Bruker D8 Advance). The scanning rate was fixed at 8° in 2θ/min and 0.8° in 2θ/min for phase identification and line broadening analysis, respectively. The patterns were run with CuKα1 with secondly monochromator, $\lambda = 0.15005$ nm at 40 kV

and 40 mA. The crystallite size of the phases present was calculated using Scherrer equation [25].

The specific surface areas of the solid catalyst samples were determined from nitrogen adsorption desorption isotherms measured at −196 °C using a Quanta chrome NOVA 2000 automated gas-sorption apparatus model 7.11. All samples were degassed at 200 °C for 2 h under a reduced pressure of $10^{-5}$ Torr before undertaking such measurements.

The catalytic activities of pure and variously ZnO doped-CuMgO solid catalyst samples towards methyl alcohol dehydrogenation were determined at reaction temperatures 125–275 °C, the catalytic reaction being conducted in a flow reactor under atmospheric pressure. Thus, a 50 mg catalyst sample was held between two glass wool plugs in a Pyrex glass reactor tube 20 cm long and 1 cm internal diameter packed with quartz fragments 2–3 mm length. The temperature of the catalyst bed was regulated and controlled to within ±1 °C. Argon gas was used as the diluents and the methyl alcohol vapor was introduced into the reactor through an evaporator/saturator containing the liquid reactant at constant temperature 26 °C. The flow rate of the carrier gas was maintained at 15 ml/min. Before carrying out such catalytic activity measurements, each catalyst sample was activated by heating at 300 °C in a current of argon for 1 h then cooled to the catalytic reaction temperature. The injection time of the reaction products and the unreacted methyl alcohol was fixed after 15 min, and GC control at $t =$ constant was performed until achieving a steady state process. The reaction products in the gaseous phase were analyzed chromatographically using Perkin-Elmer Auto System XL Gas Chromatograph fitted with a flame ionization detector. Stainless steel chromatographic columns were used, 4 m length, packed with 10 % squalane supported on chromosorp. The reaction products were analyzed at a column temperature of 40 °C in all conversion runs. Detector temperature was kept at 250 °C.

## Results and discussion

### XRD investigation of pure ZnO-doped CuMgO solids calcined at different temperatures

X-Ray diffractograms of pure and variously ZnO-doped solids precalcined at 350–650 °C were determined and illustrated in Figs. 1 and 2. The values of crystallite sizes of different phases were calculated using Scherrer equation [25], are given in Table 1. Inspection of Figs. 1 and 2 and Table 1 revealed that: (1) all investigated solids precalcined at 350, 550 °C consisted of MgO as a major phase, CuO as a minor thereof together with $Cu_2O$ phase in nanocrystalline nature. The presence of $Cu_2O$ phase was evidenced from the brown color of the calcined solids beside presence of diffraction line at $d = 2.44$ Å (at $I/I > 10$ c/o). (2) Doping CuMgO system with 0.06 mol ZnO (i.e. 7.07 wt%) brought about a measurable decrease in both of ordering and crystallite sizes of MgO and CuO phases, further increase in ZnO amount to 0.08 mol (i.e.

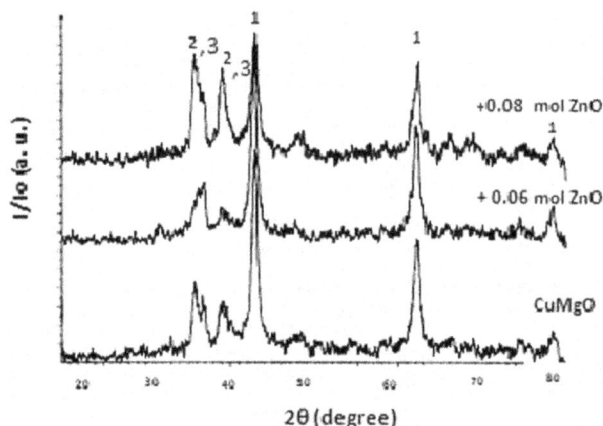

**Fig. 1** XRD diffractograms of pure and ZnO doped CuMgO solids precalcined at 550 °C. *Lines 1* refer to MgO, *lines 2* refer to CuO phases, *lines 3* refer to $Cu_2O$ phase

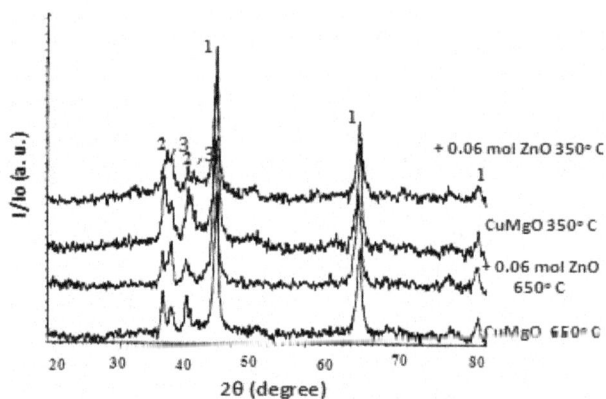

**Fig. 2** XRD diffractograms of pure and ZnO-doped CuMgO solids precalcined at 350 and 650 °C. *Lines 1* refer to MgO, *lines 2* refer to CuO phases, *lines 3* refer to $Cu_2O$ phase

9.21 wt%) resulted in an increase in the degree of ordering of CuO phase while opposite result was found in case of MgO phase. (3) Calcination at 350 and 550 °C of the doped solids resulted in the formation of nanocrystalline ZnO phase, its crystallite size increases with increasing ZnO content. (4) The rise in calcination temperature of pure and ZnO-doped solids to 650 °C causes decreasing the degree of ordering of CuO, MgO and ZnO phases, but led to increasing the crystallite size of CuO (4.7–25 nm).

These obtained results can be explained in the role of ZnO-doping in increasing the degree of dispersion of active phase [26, 27] via decreasing its crystallite sizes. The mechanism of this action had been tentatively attributed to coating the copper oxide crystallites by ZnO film that may hinder the particle adhesion of the doped oxide solids [26]. The re-increase of degree of crystallinity of CuO phase in the highly doped sample (0.08 mol) at 550 °C could be due to the presence of maximum limit for ZnO to hinder the grain growth of CuO. Increasing ZnO content to 0.06 mol at 650 °C led to dissolving ZnO in MgO and forms solid solution; this interaction led the finally divided CuO crystallites to grow on the top surface layer of MgO. This speculation was based on the effective increase in the crystallite size of CuO as shown in Table 1.

## Specific surface areas of pure and ZnO-doping CuMgO calcined at different temperatures

The surface characteristics of pure adsorbent sample and those treated with small amounts of ZnO precalcined in air at 550 and 650 °C were determined from nitrogen adsorption isotherms conducted at −196 °C. The dopant concentrations were 0.06 and 0.08 mol ZnO per mol MgO corresponding to 7.07 and 9.21 wt%, respectively. The surface characteristics, namely $S_{BET}$, $V_p$ "total pore volume" and $r^-$ "mean pore radius" calculated for various adsorbents are listed in Table 2. Inspection of the results in Table 2 shows that (1) addition of ZnO resulted in a progressive increase in the specific surface areas of the treated solids. The maximum increase in the $S_{BET}$ due to treatment with 0.06 and 0.08 mol ZnO attained about 33 and 52 %

**Table 1** Intensity count, crystallite sizes of various phases present in pure and ZnO-doped CuMgO solids calcined at 350, 550 and 650 °C

| Solid | Calcination temp. (°C) | Intensity count (a.u.) | | | Crystallite size (nm) | | | $I/Io$ |
|---|---|---|---|---|---|---|---|---|
| | | MgO | CuO | ZnO | MgO | CuO | ZnO | $Cu_2O$ |
| Pure solid | 350 | 131 | 52 | – | 17 | 11 | – | 27 |
| 0.06 ZnO | 350 | 81 | 36 | 36 | 13 | 4.7 | 4.7 | 39 |
| Pure solid | 550 | 114 | 43 | – | 12 | 14.6 | – | 31 |
| 0.06 ZnO | 550 | 109 | 30 | 30 | 12.4 | 8.7 | 8.7 | 30 |
| 0.08 ZnO | 550 | 73 | 61 | 45 | 8.6 | 16.5 | 13.6 | 55 |
| Pure solid | 650 | 93 | 32 | – | 14 | 21 | – | 23 |
| 0.06 ZnO | 650 | 94 | 25 | – | 14 | 25 | – | 33 |

| | Moles of ZnO | Calcination temp. (°C) | $S_{BET}$ (m²/g) | $V_p$ (cm³/g) | $r^-$ (Å) |
|---|---|---|---|---|---|
| **Table 2** The specific surface areas of pure and ZnO-doped 0.3CuO/MgO adsorbents precalcined at 550 and 650 °C | 0 | 550 | 21 | 0.108 | 103 |
| | 0.06 | 550 | 28 | 0.109 | 78 |
| | 0.08 | 550 | 32 | 0.102 | 64 |
| | 0 | 650 | 16 | 0.052 | 65 |
| | 0.06 | 650 | 22 | 0.048 | 44 |

for the solids calcined at 550 °C. The observed increase in the specific surface areas of ZnO-doped solids may be attributed to formation of new pores. The formation of these pores may be due to liberation of gaseous nitrogen oxides in the course of the heat treatment of zinc nitrate added. (2) Increasing the calcination temperature from 550 to 650 °C decreases the specific surface areas of the treated solids. The decrease in the $S_{BET}$ by increasing calcination temperature to 650 °C attained 24 and 21 % for pure and doped solid with 0.06 mol ZnO, respectively. This decrease in specific surface areas of the treated solids may be attributed to collapse of the pore structure and/or the particle adhesion process (grain growth). In fact, the BET-surface area of CuMgO treated with 0.06 mol ZnO conducted at 650 °C measured 22 m²/g while the ZnO-free sample calcined at the same temperature measured 16 m²/g. These findings might suggest that ZnO acted as convenient stabilizer against thermal sintering process of the treated solids.

The extension in the surface area due to ZnO-doping is expected to be accompanied by a corresponding improvement in their catalytic activity.

## Catalytic activity and selectivity of the pure and ZnO-doped CuMgO systems towards methanol dehydrogenation reactions

Doping with small amounts of ZnO affected both structural and surface properties of the investigated CuMgO system. The changes in structure and/or the surface can reflect the catalytic properties of these solid catalysts. The effect of ZnO–doping (0.04–0.08 mol) corresponding to (4.83–9.21 wt%) of CuMgO followed by calcinations at 350–650 °C on the catalytic activity and selectivity was investigated at different reaction temperatures (125–275 °C) as shown in Figs. 3 and 4 and Tables 3 and 4. The reaction proceeded via dehydrogenation to give both methyl formate and formaldehyde [28]. Examination of Figs. 3 and 4 and Tables 3 and 4 shows the following: (1) the catalytic activity of investigated solids (expressed as conversion of methanol) increases with increasing reaction temperature in the range of 125–175 °C. Further increase in the reaction temperature is followed by a small decrease in the catalytic activity (above this temperature the activity tends to be stable).

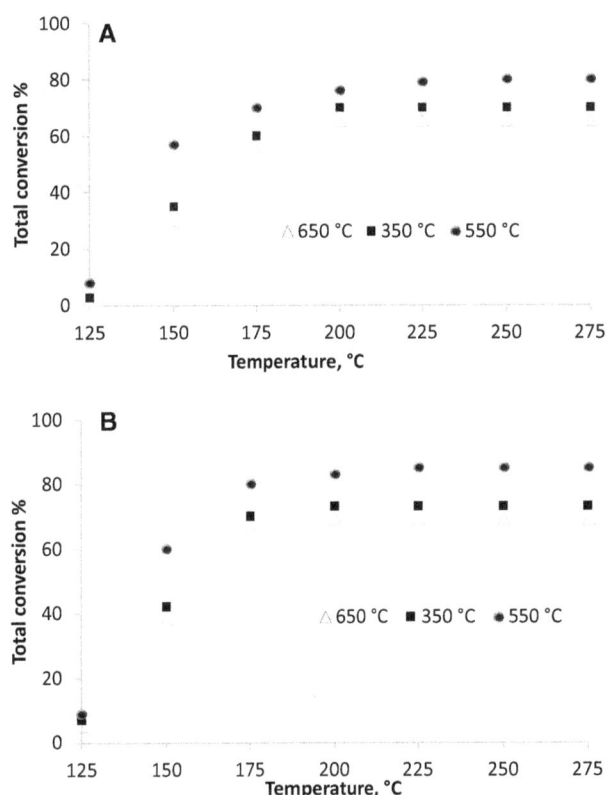

**Fig. 3** Total conversion of methanol as a function of reaction temperature over (**a**) CuMgO (**b**) 0.06 mol ZnO-doped CuMgO sample at different calcination temperatures

(2) Doping CuMgO solid with ZnO from 0.04 to 0.06 mol in the range of 125–175 °C led to increasing the conversion and selectivity of both formaldehyde and methyl formate. But increasing reaction temperature above 175 °C led to small increase in the conversion of methanol. (3) Further increase in ZnO content to (0.08 mol) decreases the conversion of methanol. (4) The catalytic activity and selectivity of pure and ZnO-doped solids increased with increasing the calcination temperature from 350 to 550 °C, but above this temperature the catalytic activity and selectivity decrease. (5)The investigated solids are selective to formaldehyde ($S_F$ %) at low temperature, this selectivity decreases with increasing reaction temperature from 125 to 275 °C. The selectivity towards methyl formate ($S_m$ %) was more pronounced at high reaction temperature.

**Fig. 4** Total conversion of methanol at reaction temperature 125 and 150 °C over pure CuMgO and those variously ZnO-doped samples calcined at 550 °C

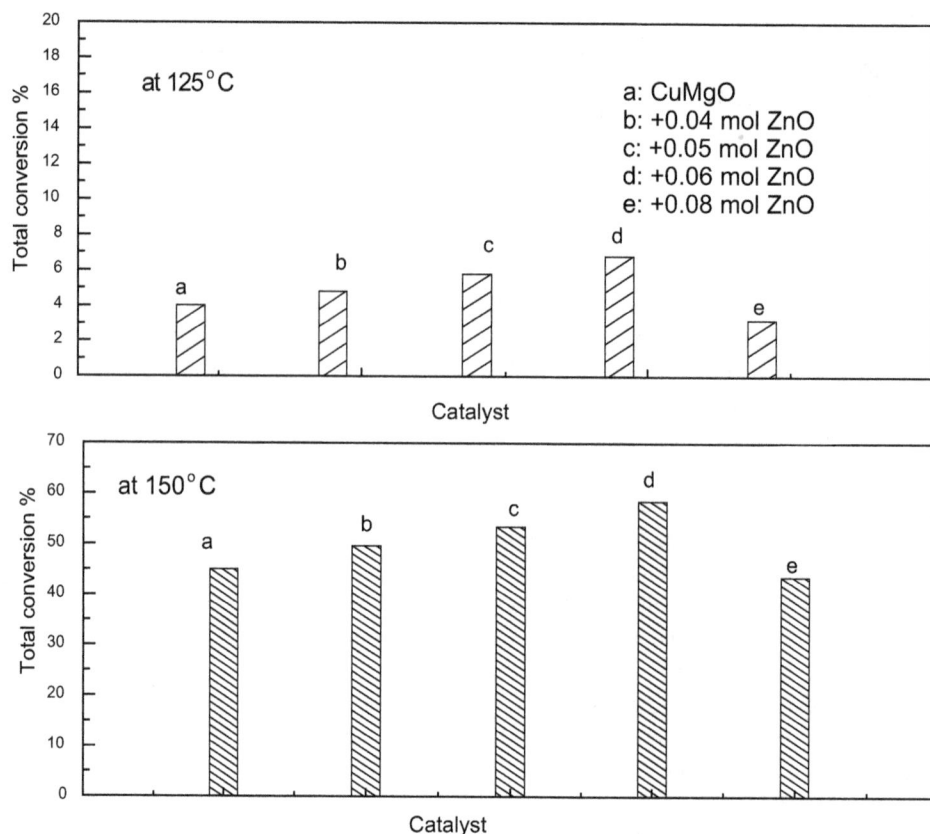

at 125°C

a: CuMgO
b: +0.04 mol ZnO
c: +0.05 mol ZnO
d: +0.06 mol ZnO
e: +0.08 mol ZnO

at 150°C

**Table 3** Effect of ZnO-doping on the selectivity of CuMgO solids calcined at 550 °C towards methanol dehydrogenation (with reference to formaldehyde ($S_f$ %) and methyl formate ($S_m$ %)

| The solids | Selectivities (%) | 150 °C | 175 °C | 200 °C | 225 °C | 250 °C | 275 °C |
|---|---|---|---|---|---|---|---|
| 0.3CuMgO | $S_m$ | 17 | 43 | 46 | 48 | 53 | 59 |
|  | $S_f$ | 83 | 57 | 54 | 52 | 47 | 41 |
| 0.04ZnO/0.3CuMgO | $S_m$ | 17 | 47 | 50 | 52 | 55 | 57 |
|  | $S_f$ | 83 | 53 | 50 | 48 | 45 | 43 |
| 0.05ZnO/0.3CuMgO | $S_m$ | 18 | 47 | 50 | 53 | 55 | 57 |
|  | $S_f$ | 82 | 53 | 50 | 47 | 45 | 43 |
| 0.06ZnO/0.3CuMgO | $S_m$ | 23 | 42 | 48 | 50 | 52 | 52 |
|  | $S_f$ | 77 | 58 | 52 | 50 | 48 | 48 |
| 0.08ZnO/0.3CuMgO | $S_m$ | 16 | 42 | 45 | 50 | 51 | 58 |
|  | $S_f$ | 84 | 58 | 55 | 51 | 49 | 42 |

The results mentioned above can be explained in the light of: (1) the investigated CuMgO system is dehydrogenation catalyst. The prepared catalysts are selective to formaldehyde formation due to the presence of dehydrogenation sites (copper species) [15, 29–31]. (2) The observed increase in the catalytic activity and selectivity of solid samples doped with ZnO (0.06 mol) 7.07 wt% may be attributed to an affective increase in the concentration of active sites involved in the catalytic reaction via decreasing the crystallite size of CuO and MgO phase (as shown in "XRD" section), beside increasing the $S_{BET}$ as shown in Table 2 and the presence of ZnO as dehydrogenation catalyst.

(3) Decreasing the catalytic activity and selectivity of doped solids above 0.06 mol% ZnO may be due to decreasing the ability of ZnO to hinder the grain growth of CuO and small amount of ZnO dissolves in MgO matrix. These effects yielded big crystallites of CuO (as shown in "XRD" section). (4) The observed increase in the catalytic activity by increasing the calcination temperature from 350 to 550 °C can be explained in the light of possible completeness of thermal decomposition of $Mg(OH)_2$ and also formation of new active sites responsible for increasing the catalytic activity, beside increasing the $S_{BET}$ as shown in Table 2. (5) Beside the possible dissolution of ZnO in both CuO and MgO

**Table 4** Effect of calcination temperature of pure 0.3CuMgO and ZnO-doped solids on their selectivities in the course of methanol dehydrogenation with respect to formaldehyde ($S_f$ %) and methyl formate ($S_m$ %)

| The Solids | Calcination temperature (°C) | Selectivities (%) | 125 °C | 150 °C | 175 °C | 200 °C | 225 °C | 250 °C | 275 °C |
|---|---|---|---|---|---|---|---|---|---|
| 0.3CuMgO | 350 | $S_m$ | 100 | 12 | 42 | 50 | 54 | 56 | 58 |
| | | $S_f$ | 0 | 88 | 58 | 50 | 46 | 45 | 42 |
| 0.06ZnO/0.3CuMgO | 350 | $S_m$ | 74 | 14 | 38 | 48 | 52 | 55 | 57 |
| | | $S_f$ | 26 | 86 | 62 | 52 | 48 | 45 | 43 |
| 0.3CuMgO | 650 | $S_m$ | 100 | 9 | 41 | 49 | 54 | 57 | 58 |
| | | $S_f$ | 0 | 92 | 59 | 51 | 46 | 44 | 42 |
| 0.06ZnO/0.3CuMgO | 650 | $S_m$ | 65 | 11 | 41 | 45 | 52 | 56 | 58 |
| | | $S_f$ | 35 | 90 | 59 | 55 | 48 | 44 | 42 |

lattice yielding various solid solutions [32], the obtained decrease in the catalytic activity as a result of increasing the calcination temperature above 550 °C may be attributed to an effective increase in the crystallite size of copper oxides in pure and doped solids (as shown in "XRD" section), which was evidenced also, by decreasing the $S_{BET}$ as shown in Table 2. The portion of CuO and/or ZnO involved in solid solution should have very small catalytic activity.

## Conclusions

In conclusion, the physicochemical and catalytic properties of CuMgO system are affected by ZnO-doping and calcination temperature. The results revealed that:

1.  The crystallite size and ordering of CuO phase decrease to (4 nm) by ZnO-doping (<0.08 mol) for samples calcined at 350–550 °C. Opposite trends was observed by increasing ZnO amount to 0.08 mol (i.e. 9.21 wt%).
2.  The BET surface area and catalytic activity of CuMgO catalyst increase by increasing the calcination temperature from 350 to 550 °C and/or by increasing the amounts of ZnO up to certain extent reaching to a maximum at 7.07 wt% ZnO.
3.  In the pure and doped samples, increasing the calcination temperature to 650 °C led to increasing the crystallite size of CuO, decreasing the catalytic activity and selectivity.
4.  The prepared catalysts were selective towards formaldehyde and methyl formate formation.

## References

1. Guerrero-Ruiz A, Rodriguez-Ramos I, Fierro JLG (1991) Dehydrogenation of methanol to methyl formate over supported copper catalysts. Appl Catal 72:119
2. Thorud S, Gjolstad M, Ellingsen DG, Molander P (2005) Air formaldehyde and solvent concentrations during surface coating with acid-curing lacquers and paints in the woodworking and furniture industry. J Environ Monit 7:586
3. Lee JS, Kim JC, Kim YG (1990) Methyl formate as a new building block in C₁ chemistry. Appl Catal 57:1
4. Lee YS, Kim JC, Lee JS, Kim YG (1993) Carbonylation of formaldehyde over ion exchange resin catalysts. Ind Eng Chem Res 32:253
5. Adamson K, Pearson P (2000) Hydrogen and methanol: a comparison of safety, economics, efficiencies and emissions. J Power Sources 86:548
6. Fu SS, Somorjai GA (1992) Roles of chemisorbed oxygen and zinc oxide islands on copper(110) surfaces for methanol decomposition. J Phys Chem 96:4542
7. Hsu WD, Ichihashi M, Kondow T, Sinnott SB (2007) Ab initio molecular dynamics study of methanol adsorption on copper clusters. J Phys Chem A 111:441
8. Spendelow JS, Goodpaster JD, Kenis PJA, Wieckowski A (2006) Methanol dehydrogenation and oxidation on Pt(111) in alkaline solutions. Langmuir 22:10457
9. Wang GC, Zhou YH, Morikawa Y, Nakamura J, Cai ZS, Zhao XZ (2005) Kinetic mechanism of methanol decomposition on Ni(111) surface: a theoretical study. J Phys Chem B 109:12431
10. Gong XQ, Selloni A, Vittadini A (2006) Density functional theory study of formic acid adsorption on anatase TiO₂(001): geometries, energetics, and effects of coverage, hydration, and reconstruction. J Phys Chem B 110:2804
11. Branda MM, Collins SE, Castellani NJ, Baltanas MA, Bonivardi AL (2006) Methanol adsorption on the beta-Ga₂O₃ surface with oxygen vacancies: theoretical and experimental approach. J Phys Chem B 110:11847
12. El-Molla SA, Ismail SA, Ibrahim MM (2011) Effects of γ-irradiation and ageing on Surface and catalytic properties of nano-sized CuO/MgO system. J Mex Chem Soc 55(3): 154–163
13. Ibrahim SM, Badawy AA, El-Shobaky GA, Mohamed HA (2014) Structural, surface and catalytic properties of pure and ZrO₂-doped nanosized cobalt–manganese mixed oxides. Can J Chem Eng 92:676–684
14. Rossignol S, Kappenstein C (2001) Effect of doping elements on the thermal stability of transition alumina. Int J Inorg Mater 3:51
15. El-Molla SA (2006) Dehydrogenation and condensation in catalytic conversion of iso-propanol over CuO/MgO system doped with Li₂O and ZrO₂. Appl Catal A 298:103
16. El-Molla SA, El-Shobaky GA, Amin NH, Hammed MN, Sultan SN (2013) Catalytic properties of pure and K-doped CuO/MgO system towards 2-propanol conversion. J Mex Chem Soc 57(1):36

17. Diez VK, Apesteguia CR, Dicosimo JI (2000) Acid–base properties and active site requirements for elimination reactions on alkali-promoted MgO catalysts. Catal Today 63:53

18. El-Shobaky GA, Ghozza AM (2004) Effect of ZnO doping on surface and catalytic properties of NiO and $Co_3O_4$ solids. Mater Lett 58:699

19. Chen J, Qiao Y, Li Y (2008) Promoting effects of doping ZnO into coprecipitated Ni–$Al_2O_3$ catalyst on methane decomposition to hydrogen and carbon nanofibers. Appl Catal A 337:148

20. El-Shobaky GA, Radwan NRE, Radwan FM (2001) Investigation of solid–solid interactions between pure and $Li_2O$-doped magnesium and ferric oxides. Thermochim Acta 380:27

21. Radwan NRE, El-Shobaky HG (2001) Solid–solid interactions between ferric and cobalt oxides as influenced by $Al_2O_3$-doping. Thermochim Acta 360:147

22. Zhou J-p, He H-c, Lin Y-h (2006) Effect of ZnO doping on the reaction between Co and Fe oxides. Mater Lett 60:1542

23. El-Shobaky HG, Ahmed AS, Radwan NRE (2006) Effect of $\gamma$-irradiation and ZnO-doping of $CuO/TiO_2$ system on its catalytic activity in ethanol and isopropanol conversion. Colloids Surf A 274:138

24. El-Molla SA, Abdel-all SM, Ibrahim MM (2009) Influence of precursor of MgO and preparation conditions on the catalytic dehydrogenation of iso-propanol over $CuO/MgO$ catalysts. J Alloys Compd 484:280

25. Klug HP, Alexander LE (1966) X-ray diffraction procedures for polycrystalline and amorphous materials. Wiley, New York, p 491

26. Dohiem MM, El-Boohy HA, Mokhater M, El-Shobaky GA (2001) Surface and catalytic properties of the $\gamma$-irradiated ZnO-treated $Co_3O_4/Al_2O_3$ system. Adsorpt Sci Technol 19:721

27. El-Shobaky GA, Amin NH, Deras NM, El-Molla SA (2001) Decomposition of $H_2O_2$ on pure and ZnO-treated $Co_3O_4/Al_2O_3$ solids. Adsorpt Sci Technol 19(1):45–58

28. Zaza P, Randall H, Dopper R, Renken A (1994) Dynamic kinetics of catalytic dehydrogenation of methanol to formaldehyde. Catal Today 20:325

29. Wojciechowska M, Haber J, Lomnicki S, Stoch J (1999) Structure and catalytic activity of double oxide system: Cu–Cr–O supported on $MgF_2$. J Mol Catal A 141:155

30. Pepe F, Angeletti C, Rossi S, Jacono ML (1985) Catalytic behavior and surface chemistry of copper/alumina catalysts for isopropanol decomposition. J Catal 91:69

31. Henrich VE, Cox PA (1994) The surface science of metal oxides. Cambridge University Press, Cambridge

32. El-Shobaky GA, Mostafa AA (2003) Solid–solid interactions in $Fe_2O_3/MgO$ system doped with aluminium and zinc oxides. Thermochim Acta 408:75

# 15

# Influence of synthesized nano-ZnO on cure and physico-mechanical properties of SBR/BR blends

Madhuchhanda Maiti[1] · Ganesh C. Basak[1] · Vivek K. Srivastava[1] · Raksh Vir Jasra[1]

**Abstract** This study focuses on the synthesis of zinc oxide (ZnO) nanoparticles by high temperature calcination as well as low-temperature hydrolysis methods and their efficiency as cure activator in styrene-butadiene rubber/polybutadiene rubber blend. The synthesized nano-ZnO samples were characterized by means of X-ray diffraction, BET surface area and transmission electron microscopy. The synthesized nano-ZnO samples had wurtzite structure and average particle size in the 'nm' range. ZnO nanoparticles, synthesized on sepiolite template, were of smallest particle size (maximum number of particles in the range of 7–12 nm) and highest surface area (104 $m^2 g^{-1}$). Polyethylene glycol (PEG)-6000 coated ZnO nanoparticles had rod-like structure; average diameter of the rods was 50 nm. In the case of PEG-coated ZnO containing compounds, optimum cure time of the blend was decreased by 5 min compared to that of standard rubber grade-ZnO containing compound (used as reference). Optimum cure time was lowered by 7–10 min in the case of synthesized nano-ZnO containing compounds compared to the reference ZnO based compound in presence of conventional filler, carbon black. It was also observed from ICP-OES analysis that the presence of very little amount of magnesium in one of the synthesized ZnO has noticeable impact on cure properties. PEG-coated ZnO increased the tensile strength of gum vulcanizates by 28% compared to the reference ZnO, acting as nanofiller at 3 phr loading. The study of curing behavior in dynamic condition was carried out using DSC. The results differ slightly from static curing except PEG modified nano-ZnO. Use of ZnO nanoparticles could provide faster crosslinking, better reinforcement at lower concentration compared to reference ZnO.

**Keywords** Nano-ZnO · SBR · BR · Curing · Cure properties

## Introduction

The rubber industries, specifically tire industries, contribute significantly to economy of a nation where automobile industry is growing at a very fast pace. Improvement in quality and safety of rubber products can have significant impact on this industry [1, 2]. Zinc oxide (ZnO) is primarily used as an activator for sulfur vulcanization of rubbers. Besides, inclusion of ZnO in the rubber compound brings other benefits viz., reduction in heat build-up, improvement of abrasion resistance and heat resistance of the vulcanizates. Furthermore, its high thermal conductivity helps to dissipate local heat concentrations in rubber products. Zinc oxide is a necessary ingredient in rubber compounds for bonding rubber to reinforcing steel cord, etc. Besides improving the properties of vulcanized rubbers, ZnO also assists in the processing of uncured rubbers. ZnO is added to rubber formulation to reduce shrinkage of molded rubber products and maintain the cleanliness of molds [3].

The road transport emission of zinc due to tire wear is the main sources of zinc pollution after iron and steel production and non-ferrous metals manufacture. This arises from the zinc content (1 wt%) of the tire-tread material [4, 5]. But some adverse environmental effects of zinc exposure have been reported. In view of the upcoming legislation and eco-labeling requirements for tires, it can be

✉ Ganesh C. Basak
ganesh.basak@ril.com

[1] Reliance Technology Group, Vadodara Manufacturing Division, Reliance Industries Ltd., Vadodara, Gujarat 391346, India

stated that it is desirable to keep the ZnO content in rubber compounds as low as possible.

In rubber industry, various kinds of vulcanization activators like CaO, MgO, CdO, CuO, PbO and NiO have been used in order replace conventional ZnO due to its toxic and fouling characteristics for aquatic flora and fauna. Although among the various activators studied, MgO shows most promising candidate in terms of activating properties in comparison to ZnO but maximum crosslinking can be achieved in the presence of ZnO only [6]. Moreover, few reports are also available that describe the effect of layered double hydroxide (LDH) on elastomeric materials in the place of ZnO. According to the literature reports, LDH material can be used as an alternative cure activator in place of ZnO and stearic acid combo in the conventional cure package for the preparation of rubber composites, and simultaneously can provide a strong platform for reduction of ZnO level in elastomer vulcanizate system [7].

In another approach, the concentration of ZnO can be minimized if the efficiency of ZnO during vulcanization can be enhanced by the maximization of the contact between the ZnO particles and the accelerators in the compound. This contact is dependent on the size, shape, specific surface area and dispersibility of the ZnO particles. Nano-sized ZnO particles have been paid more attention for their unique properties, even though there are limited open literatures available on nano-ZnO as cure activators. ZnO nanoparticles were studied as a cure activator and curing agent in natural rubber (NR), nitrile rubber (NBR), carboxylated nitrile rubber (XNBR) and chloroprene rubber (CR) by Bhowmick and his coworkers [8–10]. Similarly, it was used as cure activator in NR and CR by Joseph et al. [11, 12]. Nanostructured zinc oxide was used in crosslinking of hydrogenated butadiene-acrylonitrile elastomer and XNBR by Przybyszewska and Zaborski [13–15]. Guzman et al. synthesized mixed metal oxide nanoparticles of zinc and magnesium to reduce the ZnO levels in rubber compounds [16]. Heideman et al. studied the influence of nano-ZnO on the cure properties of solution styrene-butadiene rubber (SBR) and ethylene–propylene–diene rubber [17]. Kim et al. investigated the effect of nano-ZnO on the cure characteristics and mechanical properties of the silica-filled natural rubber/butadiene rubber compounds [18]. Jincheng and Yuehui studied the application of nano-ZnO master-batch in SBR [19].

In our previous work, we have studied the effect of nano-ZnO on the cure properties of polybutadiene rubber (BR) [20]. It was observed that the nano-ZnO reduces curing time and also enhances physico-mechanical as well thermal stability properties of butadiene rubber compound at lower concentration compared to the conventional micro-ZnO. However, to the best of our knowledge, the effect of nano-ZnO as cure activator has not yet been explored for SBR/BR rubber blend. In the open literature, it has already been reported that nano-zinc oxides are effective activators and reinforcing agents in rubber systems. The "little size effect," "surface effect" and "quantum effect" of nano-ZnO governs the properties of the composites [21]. Although considerable amount of work has been done so far on the use of nano-ZnO in place of conventional ZnO as a cure activator and for enhancing the mechanical properties of elastomer, the study on SBR/BR-nano-ZnO composites is scarcely available in the literature [22]. In the tire industry, SBR/BR blend is of considerable importance as it is widely used in passenger car tire-tread compound. Hence, investigation of nanocomposite based on SBR/BR blends and nano-ZnO would not only be providing valuable information but also have wide applications. Typically SBR/BR blend shows slower curing rate than other general purpose rubbers such as NR and BR [3]. Hence, it will be of interest to study the cure properties of this blend with nano-ZnO.

In this work, we have studied the influence of morphology, specific surface area and dispersibility of ZnO nanoparticles on the static and dynamic vulcanization of SBR/BR blends. We have studied the effect of sepiolite template and 'eco-friendly' metal oxide, magnesium oxide (MgO) on nano-ZnO in the crosslinking of the rubber blend. The influence of nano-ZnO on the properties of SBR/BR vulcanizates in the absence as well as in the presence of conventional filler was also evaluated.

## Experimental

### Materials

Zinc nitrate [$Zn(NO_3)_2 \cdot 6H_2O$] [molecular weight (M.W.) 297.48, 98% purity], ammonium carbonate [$(NH_4)_2CO_3$] (M.W. 157.13, 31% purity), acetone (M.W. 58.08, 99.5% purity), methanol (M.W. 32.04, 99.5% purity), sodium hydroxide pellets (M.W. 40.00, 98% purity), 1-octanol (M.W. 130.23, 99% purity), Stearic acid (M.W. 284.48, 98% purity), sulfur powder (M.W. 32.06, 99% purity), N-cyclohexyl-2-benzothiazole sulfenamide (CBS) (M.W. 264.42, 97% purity), microcrystalline wax, magnesium oxide (MgO) were procured from Labort Fine Chem. Pvt. Ltd., India. Standard rubber grade zinc oxide (ZnO), used as reference (designated as SZ), was supplied by Labort Fine Chem. Pvt. Ltd., India. Polyethylene glycol-6000 (PEG) was obtained from Alfa Biochem, Greece. Zinc acetate dihydrate [$Zn (CH_3COO)_2 \cdot 2H_2O$] (M.W. 219.50, 98.5% purity) and Oxalic acid (M.W. 126.07, 99.8% purity) were procured from S. D. Fine Chem. Ltd., India. N-(1,3-dimethyl butyl)-N'-phenyl-p-phenylenediamine (6PPD) was obtained from John Baker Inc., USA. Polybutadiene rubber (BR; Cisamer 01; $ML_{1+4}$ at 100 °C = 45; cis-content 96%) was collected from

Reliance Industries Ltd., India. Styrene-butadiene rubber (SBR 1502; $ML_{1+4}$ at 100 °C = 48) was supplied by Japan Synthetic Rubber, Japan. Sepiolite (Pangel S9) was generously supplied by Tolsa, Spain. Carbon black (N330) was procured from Philips Carbon Black Ltd., India.

## Preparation of nano-ZnO

Nano-ZnO was synthesized by high temperature calcination as well as low-temperature hydrolysis methods. The typical procedures are described below.

### Method-1 [9]

$Zn(NO_3)_2 \cdot 6H_2O$ and ammonium carbonate $(NH_4)_2CO_3$ were, respectively, dissolved in distilled water at a concentration of 1.0 M. Zinc nitrate solution was then slowly dropped into the vigorously stirred $(NH_4)_2CO_3$ solution with molar ratio of 2:1 to prepare the precursor. A white precipitate occurred immediately on mixing of the two solutions. Stirring was done for 3 h to have complete precipitation. The white precipitate thus obtained was filtered and repeatedly washed with distilled water to remove impurities and dried at 105 °C for 6 h. Calcination of the dried sample was carried out at 450 °C in a muffle furnace. The sample thus obtained is designated as Z1.

### Method-2 [23]

0.1 M aqueous solution of oxalic acid was added to 0.1 M aqueous solution of $Zn(CH_3COO)_2 \cdot 2H_2O$ and the solution was stirred for 4 h. The white precipitates thus obtained were filtered and washed with acetone and distilled water to remove impurities and dried at 120 °C for 6 h. The dried sample was calcined at 450 °C in a muffle furnace to remove CO and $CO_2$ from the compound. The sample is designated as Z2.

### Method- 3 [24]

The solution of $Zn (CH_3COO)_2 \cdot 2H_2O$ (0.1 M) was prepared in 50 ml methanol under stirring. 25 ml of NaOH (0.3 M) solution, prepared in methanol, was mixed with above solution under continuous stirring to get the pH of reactants between 8 and 11. These solutions were transferred into a Teflon lined sealed stainless steel autoclave and maintained at 150 °C for 12 h under autogenous pressure. It was then allowed to cool naturally to room temperature. After the reaction was complete, the resulting white solid product was washed with methanol, filtered and then dried in a laboratory oven at 100 °C. The sample is designated as Z3.

### Method-4 [20]

Equivalent volume of $Zn (CH_3COO)_2 \cdot 2H_2O$ (0.5 M) and sodium hydroxide (1.5 M) were mixed to obtain a solution A. 2.5 g of PEG-6000 was dissolved in 10 ml of water to obtain solution B. The solution B was then added into solution A to obtain solution C. 50 ml of 1-octanol was added to solution C under stirring at room temperature to obtain solution D. Then solution D was transferred to Teflon lined stainless steel autoclave which was then maintained at 180 °C for 4 h under autogenous pressure. The ZnO powder was obtained after filtering, washing and drying in oven at 120 °C. The sample is designated as Z4.

### Method-5

The solution of $Zn (CH_3COO)_2 \cdot 2H_2O$ (0.1 M) was prepared in 50 ml methanol under stirring. 25 ml of NaOH (0.3 M) solution, prepared in methanol, was mixed with above solution under continuous stirring to get the pH of reactants between 8 and 11, and then 4 g of sepiolite was added with vigorous stirring. It was then transferred into a Teflon lined sealed stainless steel autoclave and maintained at 150 °C for 12 h under autogenous pressure. Subsequently, it was allowed to cool naturally to room temperature. After the reaction was complete, the resulting white solid product was washed with methanol, filtered and then dried in a laboratory oven at 100 °C. The sample is designated as Z5.

### Method-6

0.1 M solution of $Zn(CH_3COO)_2 \cdot 2H_2O$ in 50 ml methanol was prepared. To this solution, 0.5 mol of MgO was added. 25 ml of NaOH (0.3 M) solution, prepared in methanol, was mixed with above solution under continuous stirring to get the pH of reactants between 8 and 11. After that it was transferred into Teflon lined sealed stainless steel autoclave and maintained at 150 °C for 12 h under autogenous pressure. It was then allowed to cool naturally to room temperature. After the completion of the reaction, resulting white solid products were washed with methanol, filtered and dried in a laboratory oven at 100 °C. The sample is designated as Z6.

## Characterization of zinc oxide particles

### X-ray diffraction (XRD)

X-ray diffraction analysis was done using X-ray diffractometer, Rigaku "Mini flex" model in the range of 10 to 80° (=2θ). The zinc oxide powder was deposited on the sample holder uniformly.

*Brunauer Emmet Teller (BET) surface area measurement*

BET surface area determination was done from $N_2$ adsorption data measured at 77.4 K using micromeritics-ASAP-2020 instrument. The samples were activated at 200 °C for 20 min under vacuum (10 mmHg) prior to measurements. Five point BET surface area and total pore volume were measured. The average of five reading is reported here.

*Transmission electron microscopy (TEM)*

Morphology of different ZnO samples was investigated by transmission electron microscopy (TEM) (JEOL 2010) having $LaB_6$ filament, operating at an accelerating voltage of 200 kV. ZnO powder samples were dispersed by ultrasonication in acetone for 30 min. A copper grid was immersed in and taken out of the suspension and dried at room temperature. Image analysis of the microphotographs was performed using UTHSCSA Image Tool for Windows Version 3.00. It was used to determine the particle size distribution.

*Differential scanning calorimetry (DSC)*

Cure-studies were done using differential scanning calorimetric analysis. It was carried out using modulated DSC (DSC 2910, TA Instruments, USA). The samples were heated from ambient temperature to 250 °C (at 5 °C min$^{-1}$ heating rate) in air. 5 mg of each sample was taken for the measurement. The error limit in the 'weighing measurements' was within ±5%.

*Elemental analysis*

Elemental analysis was done using a Perkin Elmer (Model: Optima 4300 DV) inductively coupled plasma-optical emission spectroscopy (ICP-OES).

*Scanning electron microscopy (SEM)*

SEM samples were fractured in liquid nitrogen immersion and mounted with carbon tape wrapping. The images were studied with a Nova NanoSEM 650 instrument, FEI, USA, operating at 1 and 10 kV for the micro and synthesized nano-ZnO samples, respectively.

**Preparation of rubber composites**

*Compounding and vulcanization*

ZnO was mixed with rubber by melt mixing method using Brabender Plasticorder (PL2000, Germany) internal mixer

(volume 50 cm³) for 3 min at 80 °C and 60 rpm. The formulation is given in Table 1. It was chosen as a typical tire-tread formulation. Amount of ZnO used (0.5, 1.5 and 3 phr) was lower than that used (5 phr) in the conventional formulations. The sample was then passed through a cold two roll open mixing mill at a friction ratio 1:1.2. The curing studies were followed with an Oscillating Disc Rheometer (ODR-2000, FLEXSYS) at 145 °C temperature and oscillating arc of 3° for 1 h. The samples were then compression molded at 145 °C at optimum cure time.

**Physico-mechanical properties of rubber composites**

*Tensile test*

Tensile test of the sample was carried out according to ASTM D412-98a on dumbbell shaped specimens using Instron 3367 universal testing machine at ambient temperature at a crosshead speed of 500 mm min$^{-1}$. Average of five samples is reported here.

*Hardness*

Hardness of each composition was obtained using Shore A Durometer tester as per ASTM D 2240-97.

*Volume fraction of rubber ($V_r$) and crosslink density*

The cured samples were immersed in toluene for 72 h at 25 °C temperature. The volume fraction of rubber in the swollen gel, at equilibrium swelling, was calculated using Eq. (1):

$$V_r = \frac{(D - FT)\rho_r^{-1}}{(D - FT)\rho_r^{-1} + A_0\rho_s^{-1}}, \tag{1}$$

where $D$ Deswollen weight, $F$ weight fraction of the insoluble component, $T$ initial weight of the test specimen, $\rho_r$ density of rubber, 0.89 g cm$^{-3}$, $\rho_s$ density of solvent, 0.86 g cm$^{-3}$, $A_0$ amount of solvent absorbed.

Further, the crosslink density, $\frac{1}{2M_C}$, in mol g$^{-1}$ of rubber hydrocarbon was calculated using the Flory–Rehner Eq. (2):

$$-[\ln(1 - V_r) + V_r + \chi V_r^2] = \frac{\rho_r V\left(V_r^{1/3} - V_r/2\right)}{2M_C}, \tag{2}$$

$\chi$ Flory–Huggins interaction parameter, 0.46 for BR-toluene system [25], $V$ molar volume of swelling solvent, toluene, $M_C$ number average molecular weight of the chain between two crosslinks.

**Table 1** Formulation and designation of different rubber compounds

| Ingredients | SBWZ | SBSZ | SBZ1 | SBZ2 | SBZ3 | SBZ4 | SBZ5 | SBZ6 | 0.5SBZ4 | 1.5SBZ4 | 0.5SBZ5 | 1.5SBZ5 | SBSZF | SBZ4F | SBZ5F |
|---|---|---|---|---|---|---|---|---|---|---|---|---|---|---|---|
| SBR | 70 | 70 | 70 | 70 | 70 | 70 | 70 | 70 | 70 | 70 | 70 | 70 | 70 | 70 | 70 |
| BR | 30 | 30 | 30 | 30 | 30 | 30 | 30 | 30 | 30 | 30 | 30 | 30 | 30 | 30 | 30 |
| Sulfur | 1.3 | 1.3 | 1.3 | 1.3 | 1.3 | 1.3 | 1.3 | 1.3 | 1.3 | 1.3 | 1.3 | 1.3 | 1.3 | 1.3 | 1.3 |
| CBS | 1.3 | 1.3 | 1.3 | 1.3 | 1.3 | 1.3 | 1.3 | 1.3 | 1.3 | 1.3 | 1.3 | 1.3 | 1.3 | 1.3 | 1.3 |
| ZnO | 0 | 3 | 3 | 3 | 3 | 3 | 3 | 3 | 0.5 | 1.5 | 0.5 | 1.5 | 3 | 3 | 3 |
| 6PPD | 2 | 2 | 2 | 2 | 2 | 2 | 2 | 2 | 2 | 2 | 2 | 2 | 2 | 2 | 2 |
| Wax | 1 | 1 | 1 | 1 | 1 | 1 | 1 | 1 | 1 | 1 | 1 | 1 | 1 | 1 | 1 |
| Stearic acid | 2 | 2 | 2 | 2 | 2 | 2 | 2 | 2 | 2 | 2 | 2 | 2 | 2 | 2 | 2 |
| Carbon black | 0 | 0 | 0 | 0 | 0 | 0 | 0 | 0 | 0 | 0 | 0 | 0 | 60 | 60 | 60 |
| Naphthenic oil | 0 | 0 | 0 | 0 | 0 | 0 | 0 | 0 | 0 | 0 | 0 | 0 | 8 | 8 | 8 |

Formulation of compounds is expressed in phr (parts per hundred rubber)

# Results and discussion

## Characterization of nano-ZnO

### XRD

Figure 1 shows the XRD patterns of different zinc oxide (ZnO) samples. The sharp intense peaks, confirming the good crystalline nature of synthesized ZnO, correspond to (100), (002), (101), (102), (110), (103), (200), (112), (201) and (004) planes. All of the indexed peaks in the obtained diffractograms match with that of the bulk ZnO (JCPDS card # 79-0207) which confirm that the synthesized samples are of wurtzite hexagonal structure [26]. Any other peak related to impurities was not detected in the diffractogram within the detection limit of the XRD. Absence of any extra peak in the diffractograms of final products indicates the purity of the products. In Z5, additional peaks can be observed, other than the earlier mentioned peaks for ZnO. These are for (060), (131) [at 20°], (260) [at 24°] and (080, 331) planes [at 27°, 28°] of sepiolite clay [27]. It proves that ZnO particles are formed on sepiolite without distorting the crystal structure of either material.

The average crystal size was calculated by Scherrer Eq. 28]:

$$L = \frac{K\lambda}{\beta \cos\theta},$$ (3)

where, $\beta$ is the full-width at half maximum (FWHM) of the peak corresponding to (100) plane, $K$ is a constant (0.89), $\lambda$ is the incident wavelength of $CuK_{\alpha}$ radiation ($\lambda = 0.154$ nm), $L$ is the crystallite size, and $\theta$ is the diffraction angle at a certain crystal plane.

The average crystallite size of Z1, Z2, Z3, Z4, Z5 and Z6 was calculated using Eq. (3) and was found to be 23, 20, 18, 27, 27 and 16 nm, respectively. It should be noted that crystallite size is assumed to be the size of a coherently diffracting domain. It is not necessarily the same as particle size [28].

### BET surface area

BET surface area of different prepared nano-ZnO is reported in Table 2. Surface area and pore volume both increase in the synthesized ZnO samples compared to the reference one. Highest surface area and pore volume can be observed in the case of Z5. This could be due to dispersion of ZnO particles on the fibrous sepiolite template surface. For the same sample, smallest particle size was also observed through TEM (Fig. 2e). So the surface area results corroborate well with the microscopic study. The sample Z4 shows minimum surface area among the synthesized samples, as it is coated with PEG. The organic coating of PEG resists the nitrogen to be absorbed on the surface of ZnO particles.

**Fig. 1** X-ray diffractogram of
the different zinc oxide samples

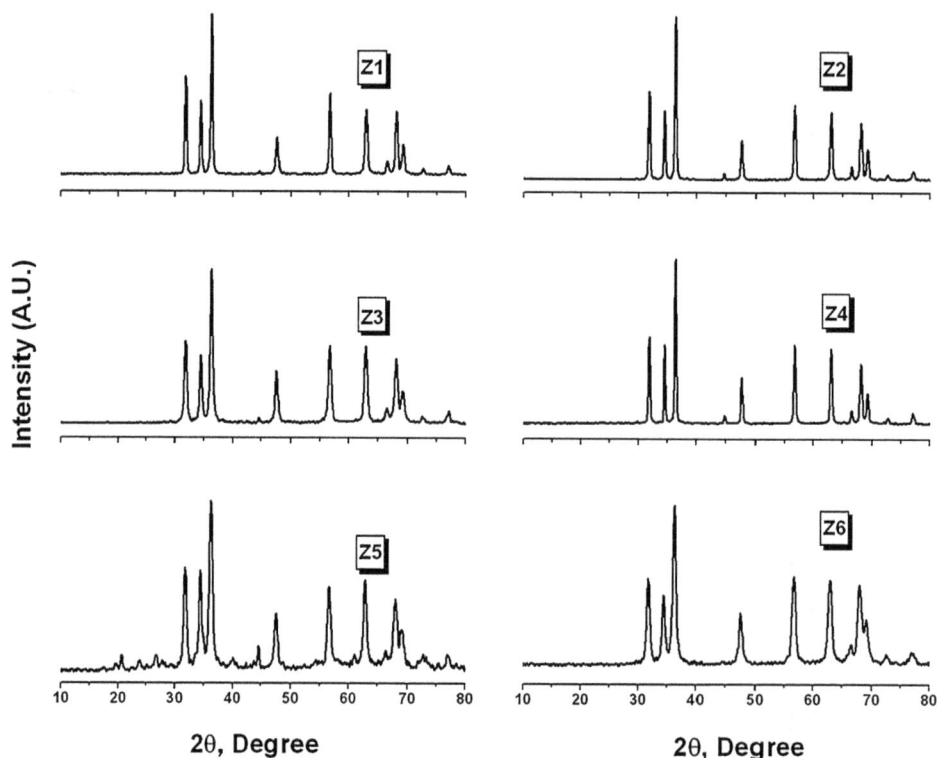

Table 2 Crystallite size of different zinc oxide samples

| Sample | FWHM of (100) plane (degree) | Average crystallite size (nm) |
|---|---|---|
| Z1 | 0.3444 | 23 |
| Z2 | 0.3936 | 20 |
| Z3 | 0.4428 | 18 |
| Z4 | 0.2952 | 27 |
| Z5 | 0.2952 | 27 |
| Z6 | 0.4920 | 16 |

*TEM*

Figure 2a–f portrays TEM photo-micrographs of different zinc oxide (ZnO) samples. TEM image exhibits the morphology of synthesized particles to be in nano region. Samples Z1, Z2, Z3, Z5 and Z6 show hexagonal structure. The particle size distribution curves for these samples are shown in Fig. 3. It shows that for Z1, maximum particles are in the range of 26–50 nm; for Z2, it is also in the range of 26–50 nm and for Z3, it is in the range of 15–28 nm. The sample Z4 evinces rod-like structures grown on PEG-sheets. The average rod diameter is ∼50 nm. These nano-rods are of 100–200 nm in length. Z5 exhibits an interesting morphology; it consists of smallest ZnO particles. Figure 2e infers that ZnO nanoparticles are grown on long bundles of sepiolite nanofibers. The ZnO particles are very small in size; most of the particles are in 7–12 nm range.

Sample Z6 has maximum particles in the range of 10–18 nm.

**Application of synthesized nano-ZnO**

*Cure properties*

The effect of synthesized nano-ZnO as cure activator has been studied on SBR/BR blend. A representative rheographic profile of SBR/BR blends at 145 °C is shown in Fig. 4 and cure time is tabulated in Table 3. In the absence of ZnO, the curing is extremely slow in the sample SBWZ and modulus is also lowest. The optimum cure time is faster with synthesized nano-ZnO samples by complex formation with acceleration compared to the reference one. ZnO helps in producing vulcanization precursor, hence faster curing can be observed in the presence of ZnO [29]. Due to decrease in particle size of ZnO, the area of contact increases which helps to react better with accelerator. This leads to the generation of vulcanization precursor quicker. It results in a faster curing rate and lower cure time. Fastest curing can be seen with the use of organo-coated ZnO, Z4, followed by Z5. Due to the presence of long-chain organic PEG molecules, it has more compatibility with elastomeric matrix leading to better dispersion. The curing reaction is not affected and slowed down in the presence of Z4 at lower dose, i.e., 0.5 and 1.5 phr. It indicates that dispersion of ZnO plays a major role in efficient vulcanization, rather

**Fig. 2** TEM image of the sample **a** Z1, **b** Z2, **c** Z3, **d** Z4, **e** Z5 and **f** Z6

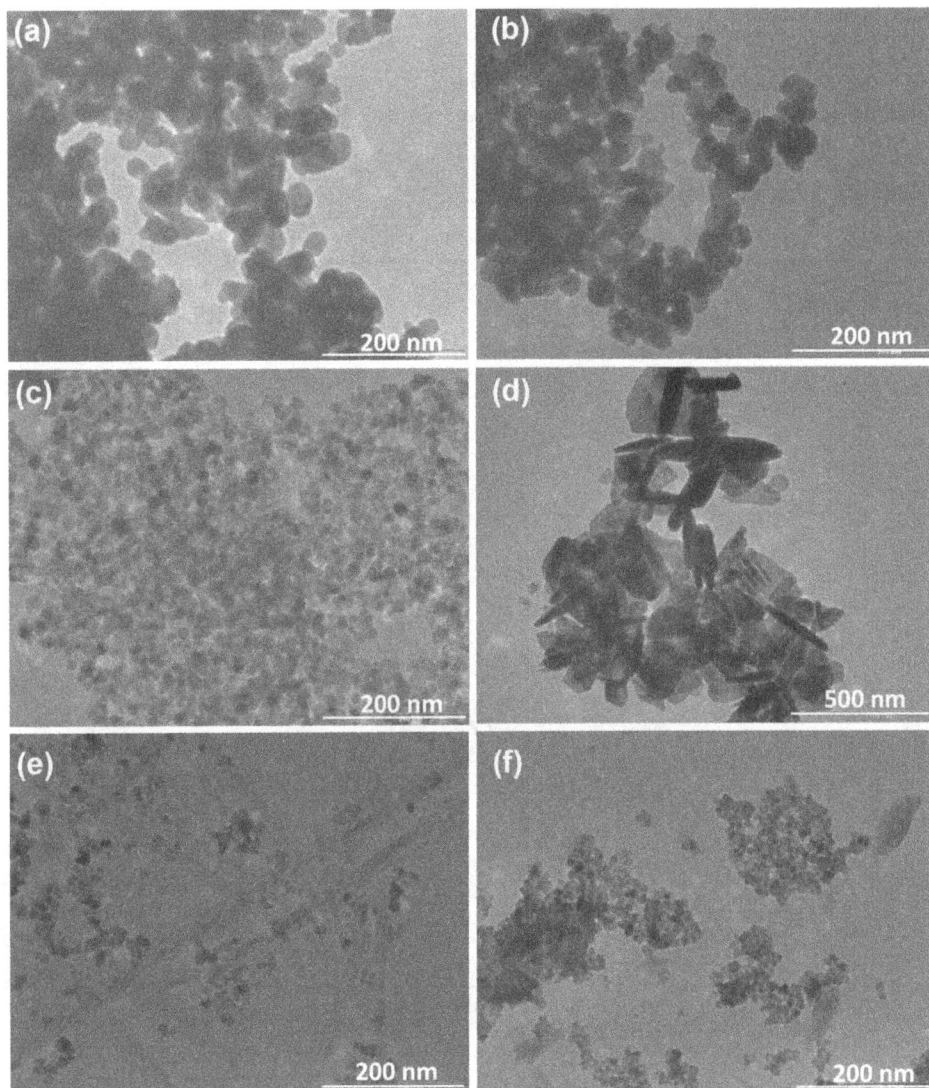

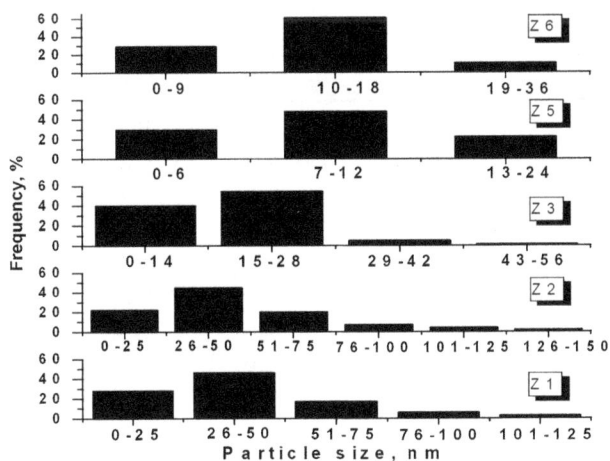

**Fig. 3** Particle size distribution curves of different ZnO samples

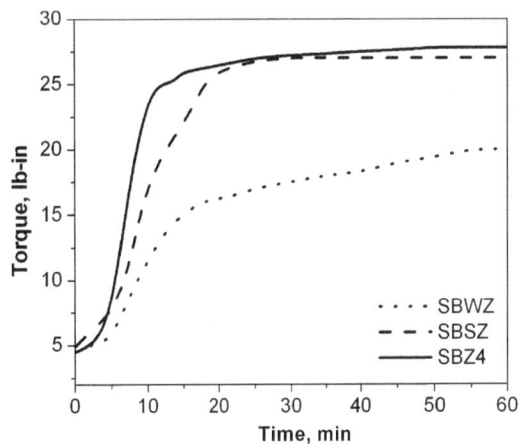

**Fig. 4** Representative rheographic profiles of SBR/BR blends containing different ZnO samples

than higher loading of ZnO. The difference in minimum ($M_L$) and maximum ($M_H$) torque value, $[\Delta S = (M_H - M_L)]$ has increased in SBR/BR blend with the use of synthesized nano-ZnO compared to that of reference ZnO

**Table 3** BET value of different nano-ZnO samples

| Sample | Surface area ($m^2\ g^{-1}$) | Pore volume ($cm^3\ g^{-1}$) |
|---|---|---|
| Reference ZnO, SZ | 5 | 0.025 |
| Z1 | 18 | 0.123 |
| Z2 | 11 | 0.084 |
| Z3 | 39 | 0.110 |
| Z4 | 5 | 0.039 |
| Z5 | 104 | 0.230 |
| Z6 | 55 | 0.110 |

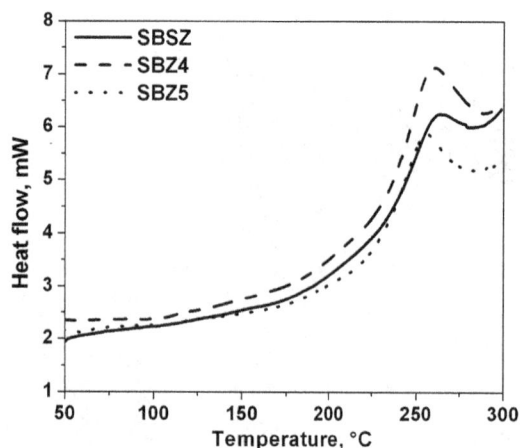

**Fig. 5** *DSC curves* of different ZnO containing SBR/BR compounds depicting dynamic curing

based sample. Increased $\Delta S$ indicates resistance to polymer chain mobility [30]. Due to the formation of increased crosslinks, chain mobility is restricted which will lead to higher crosslink density. In the case of Z5, the template, sepiolite which is fibrous clay with high aspect ratio, can help in better distribution of ZnO particles. As mentioned earlier, the area of contact increases in such case and helps in generation of vulcanizing precursors faster. These nano-ZnO samples impart better co-curing of both the rubbers in SBR/BR blend.

The presence of very little amount of magnesium in Z6 (Zn:Mg 90:1, as revealed from ICP-OES analysis) has visible impact on cure properties. It increases the optimum cure time but at the same time it also increases $\Delta S$ values. It produces maximum number of crosslinks but at a slower rate.

Most encouraging effect of synthesized nano-ZnO on curing can be observed in the presence of filler. Optimum cure time is lowered by 7–10 min in the case of synthesized nano-ZnO containing compounds compared to the reference ZnO based compound (which even contains higher dose of ZnO). Thus, ZnO nanoparticles can help in the reduction of the production cycle and also in minimizing the Zn-pollution due to lower dose.

Cure behavior in dynamic condition is shown in the representative plots (Fig. 5). The results differ slightly from static curing, though Z4 shows most efficient crosslinking activities ($\Delta H = 123.70$ Jg$^{-1}$) in dynamic curing, too (Table 4).

*Physico-mechanical properties of different rubber composites*

The nano-ZnO particles may also act as nano-fillers. Physico-mechanical properties of different rubber composites

**Table 4** Cure properties of different elastomeric compounds

| Sample | Cure time, $t_{90}$ (min) | $M_H$ (lb-in) | $M_L$ (lb-in) | $M_H - M_L$ (lb-in) |
|---|---|---|---|---|
| SBWZ | 38.2 (0.08)[a] | 20.28 (0.07) | 3.97 (0.14) | 16.31 |
| SBSZ | 18.5 (0.10) | 27.01 (0.10) | 4.87 (0.13) | 22.14 |
| SBZ1 | 17.0 (0.09) | 27.43 (0.15) | 3.62 (0.09) | 23.81 |
| SBZ2 | 16.9 (0.07) | 28.03 (0.20) | 3.72 (0.10) | 24.31 |
| SBZ3 | 16.4 (0.12) | 28.35 (0.04) | 4.12 (0.08) | 24.23 |
| SBZ4 | 12.9 (0.07) | 27.85 (0.13) | 4.05 (0.02) | 23.80 |
| SBZ5 | 13.0 (0.11) | 28.74 (0.09) | 4.48 (0.04) | 24.46 |
| SBZ6 | 19.2 (0.08) | 30.16 (0.07) | 5.03 (0.05) | 25.13 |
| 0.5SBZ4 | 12.2 (0.07) | 27.93 (0.13) | 4.42 (0.04) | 23.51 |
| 1.5SBZ4 | 12.5 (0.10) | 27.87 (0.14) | 4.43 (0.03) | 23.44 |
| SBSZF | 25.3 (0.07) | 35.83 (0.21) | 6.95 (0.12) | 28.88 |
| SBZ4F | 15.6 (0.09) | 43.76 (0.23) | 8.01 (0.14) | 35.75 |
| SBZ5F | 18.0 (0.04) | 39.88 (0.24) | 7.45 (0.11) | 32.43 |

[a] Values in parentheses are standard deviations

are reported in Table 5. Z2, Z3, Z4 and Z5 containing SBR/BR blend based nanocomposites show better tensile strength and hardness compared to the reference one. The sample containing organo-coated ZnO (Z4) displays highest tensile strength. This is ascribed to the better compatibility and in turn better dispersion of Z4 in the rubber matrix. This leads to better curing as observed in the previous experiments and higher value of volume fraction of rubber ($V_r$) as well as crosslink density. Z4 imparts highest reinforcement. From these results it can be concluded that dispersion of nanoparticles plays the major role in enhancement of properties.

Though lower filler loading maintains the cure properties unaltered (as observed in 0.5SBZ4 and 1.5SBZ4) but it does not provide the same amount of reinforcement as

3 phr loading. Tensile strength, 100% modulus and hardness decrease with decreasing ZnO loading.

Nano-ZnO containing compounds show slightly better properties than those of reference ZnO based compound. Z5 based compound exhibits highest overall properties. This may be due to some synergistic effect between nanofiller, sepiolite and conventional filler, carbon black [31]. The similar kind of effect has also been studied in our previous study using mesoporous silica as reinforcing filler in the poly butadiene rubber matrix in the presence of nanoclays, silica and carbon black [32].

## Morphology

The topographical images of SEM shown in Fig. 6 of different SBR/BR blend either having synthesized nano-ZnO or standard rubber grade-ZnO is studied to evaluate the extent of dispersion of ZnO within the rubber blend. The black phase implies the rubber matrix, whereas the white dot is the reflection of ZnO particles. In SEM images of SBSZ (Fig. 6a), some agglomeration of ZnO nanoparticles in the form of white dot can be seen. Figure 6b indicates that uniform dispersion of nano-ZnO occurs throughout the entire blend in comparison to standard rubber grade-ZnO (Fig. 5a). In case of SBZ4 (Fig. 6c), the rod-like structure as observed in TEM photo-micrographs (Fig. 2d) can also be seen. However, in the image of sepiolite based synthesized nano-ZnO (Fig. 5d), the homogenous distribution of ZnO with minimum particle

**Table 5** Dynamic cure properties of different elastomeric compounds

| Sample | $T_{max}$ (°C) | $\Delta H$ (J g$^{-1}$) |
|---|---|---|
| SBSZ | 261 | 63.66 |
| SBZ2 | 261 | 88.39 |
| SBZ3 | 262 | 81.06 |
| SBZ4 | 258 | 123.70 |
| SBZ5 | 254 | 86.58 |
| SBZ6 | 260 | 92.92 |

**Fig. 6** SEM images of **a** SBSZ, **b** SBZ3, **c** SBZ4 and **d** SBZ5

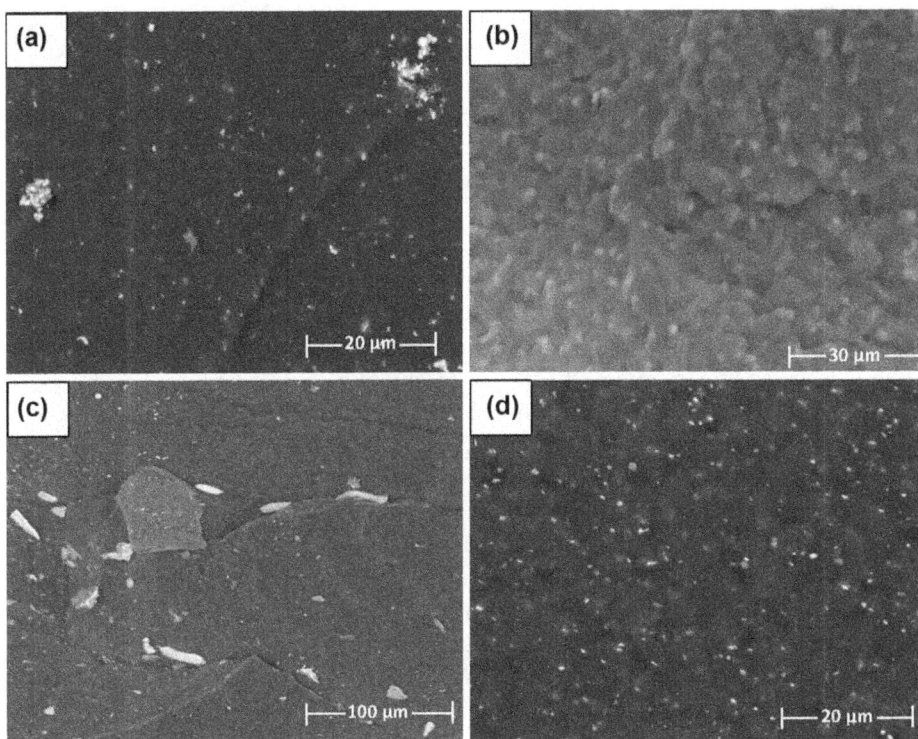

**Table 6** Physico-mechanical properties of different rubber composites

| Sample | Tensile strength (MPa) | Modulus at 100% elongation (MPa) | Elongation at break (%) | Hardness (shore A) | $V_r$ | Crosslink density $\times 10^{-5}$ mol cm$^{-3}$ |
|---|---|---|---|---|---|---|
| SBWZ | 1.73 (0.02)[a] | 0.70 (0.02) | 445 (10) | 37 (0.3) | 0.112 (0.003) | 2.50 (0.01) |
| SBSZ | 1.89 (0.02) | 0.84 (0.03) | 310 (12) | 41 (0.2) | 0.165 (0.002) | 6.31 (0.02) |
| SBZ1 | 1.82 (0.03) | 0.73 (0.03) | 320 (15) | 43 (0.3) | 0.164 (0.002) | 6.22 (0.03) |
| SBZ2 | 2.12 (0.03) | 0.78 (0.04) | 335 (18) | 43 (0.4) | 0.168 (0.003) | 6.60 (0.01) |
| SBZ3 | 2.07 (0.04) | 1.09 (0.03) | 265 (22) | 45 (0.3) | 0.164 (0.004) | 6.22 (0.01) |
| SBZ4 | 2.54 (0.01) | 1.05 (0.02) | 295 (14) | 44 (0.4) | 0.171 (0.002) | 6.90 (0.02) |
| SBZ5 | 2.05 (0.05) | 1.07 (0.02) | 250 (18) | 45 (0.2) | 0.167 (0.002) | 6.50 (0.03) |
| SBZ6 | 1.90 (0.03) | 0.90 (0.03) | 280 (12) | 46 (0.1) | 0.169 (0.003) | 6.70 (0.04) |
| 0.5SBZ4 | 1.61 (0.02) | 0.72 (0.02) | 335 (10) | 37 (0.3) | 0.165 (0.002) | 6.31 (0.02) |
| 1.5SBZ4 | 2.15 (0.01) | 0.94 (0.03) | 290 (17) | 41 (0.2) | 0.169 (0.004) | 6.70 (0.01) |
| SBSZF | 14.30 (0.5) | 1.62 (0.02) | 560 (09) | 61 (0.1) | 0.187 (0.005) | 8.65 (0.04) |
| SBZ4F | 14.84 (0.2) | 1.67 (0.01) | 565 (08) | 63 (0.2) | 0.191 (0.005) | 9.13 (0.03) |
| SBZ5F | 15.24 (0.3) | 1.82 (0.01) | 645 (10) | 63 (0.3) | 0.190 (0.007) | 9.01 (0.02) |

[a] Values in parentheses are standard deviations

size is observed (as seen in TEM images too) (Fig. 2e). As a result, the improvement of mechanical properties of the SBR/BR blend can be noticed (Table 5). From the SEM images, it can be distinguished that the distribution of ZnO in the SBR/BR blend is comparatively better for nano-ZnO which in turn is reflected in mechanical properties (Table 6).

## Conclusions

Six different nano-ZnO samples were synthesized by both high temperature calcination and low-temperature hydrolysis methods. All the samples had wurtzite structure and average particle size in the 'nm' range. ZnO, grown on sepiolite nanofiber, showed smallest particle size as well as highest surface area. PEG-coated ZnO nanoparticles were rod-like in structure. Effect of these nano-ZnO samples on cure properties of SBR/BR blends was studied by both static and dynamic curing methods. PEG-coated nano-ZnO sample exhibited maximum positive impact on cure properties. For PEG-coated ZnO, cure properties remained unaltered even at lower loadings (0.5 and 1.5 phr) of ZnO. From the observed results, it can be concluded that the cure properties are governed primarily by dispersion of cure-activator rather than its concentration and morphology. Nano-ZnO can act as nanofiller also. The sample containing organo-coated ZnO (Z4) displays highest tensile strength due to better compatibility and in turn better dispersion of Z4 in the rubber matrix. Dosing (of nano-ZnO) lower than 3 phr could not impart any reinforcement. Topographical images of SEM study indicates more

uniform dispersion of synthesized nano-ZnO over standard rubber grade-ZnO within rubber blend and this fact account for better mechanical properties. Nano-ZnO imparted faster curing even in the presence of conventional filler, carbon black compared with reference ZnO. Thus, the use of ZnO nanoparticles can provide faster curing, better reinforcement at lower dosing compared to standard ZnO, which can lead to shorter production cycles and less zinc pollution.

**Acknowledgements** The authors are highly thankful to Ms. Hetal Patel and Mr. Chirag S. Shah for their kind cooperation. Authors are grateful to Reliance Industries Ltd. for its consent to publish this work. Authors are also thankful to colleagues from catalyst, analytical and elastomer groups of RTG-VMD for their support.

## References

1. Frohlich J, Niedermeier W, Luginsland HD (2005) The effect of filler–filler and filler–elastomer interaction on rubber reinforcement. Compos A 36:449–460
2. Schuater RH (2001) The challenge a head-new polymer filler systems. Rubber World 224:24–28
3. Morton M (1959) Introduction to rubber technology. Reinhold Publishing Corporation, New York
4. Councell TB, Duckenfield KU, Landa ER, Callender E (2004) Tire-wear particles as a source of zinc to the environment. Environ Sci Technol 38:4206–4214
5. Smolders E, Degryse F (2002) Fate and effect of zinc from tire debris in soil. Environ Sci Technol 36:3706–3710
6. Heideman G, Noordermeer JWM, Datta RN, Baarle BV (2005) Effect of metal oxides as activator for sulphur vulcanisation in various rubbers. Kautschuk Gummi Kunststoffe 58:30–42
7. Basu D, Das A, Stockelhuber KW, Wagenknecht U, Heinrich G (2014) Advances in layered double hydroxide (LDH)-based elastomer composites. Prog Polym Sci 39:594–626
8. Sahoo S, Maiti M, Ganguly A, George JJ, Bhowmick AK (2007)

Effect of zinc oxide nanoparticles as cure activator on the properties of natural rubber and nitrile rubber. J Appl Polym Sci 105:2407–2415

9. Sahoo S, Bhowmick AK (2007) Influence of ZnO nanoparticles on the cure characteristics and mechanical properties of carboxylated nitrile rubber. J Appl Polym Sci 106:3077–3083

10. Sahoo S, Kar S, Ganguly A, Maiti M, Bhowmick AK (2008) Synthetic zinc oxide nanoparticles as curing agent for polychloroprene. Polym Polym Compos 16:193–198

11. Sabura Begum PM, Yusuff KKM, Joseph R (2008) Preparation and use of nano zinc oxide in neoprene rubber. Int J Polym Mater 57:1083–1094

12. Sabura Begum PM, Joseph R, Yusuff KKM (2008) Preparation of nano zinc oxide, its characterization and use in natural rubber. Prog Rubber Plast Recycl 24:141–148

13. Przybyszewska M, Zaborski M (2009) New coagents in crosslinking of hydrogenated butadiene-acrylonitrile elastomer based on nanostructured zinc oxide. Compos Interfaces 16:131–141

14. Przybyszewska M, Zaborski M (2010) Effect of ionic liquids and surfactants on zinc oxide nanoparticle activity in crosslinking of acrylonitrile butadiene elastomer. J Appl Polym Sci 116:155–164

15. Przybyszewska M, Zaborski M (2009) The effect of zinc oxide nanoparticle morphology on activity in crosslinking of carboxylated nitrile elastomer. Express Polym Lett 3:542–552

16. Guzman M, Reyes G, Agullo N, Borros S (2011) Synthesis of Zn/Mg oxide nanoparticles and its influence on sulfur vulcanization. J Appl Polym Sci 119:2048–2057

17. Heideman G, Datta RN, Noordermeer JWM, Van Baarle B (2005) Influence of zinc oxide during different stages of sulfur vulcanization. Elucidated by model compound studies. J Appl Polym Sci 95:1388–1404

18. Kim I, Kim W, Lee D, Kim W, Bae J (2010) Effect of nano zinc oxide on the cure characteristics and mechanical properties of the silica-filled natural rubber/butadiene rubber compounds. J Appl Polym Sci 117:1535–1543

19. Jincheng W, Yuehui CJ (2006) Application of nano-zinc oxide master batch in polybutadiene styrene rubber system. J Appl Polym Sci 101:922–930

20. Maiti M, Vaghasia A, Jasra RV (2012) Low-temperature synthesis of nano-to-submicron size organo-zinc oxide and its effect on properties of polybutadiene rubbe. J Appl Polym Sci 124:2857–2866

21. Wang J, Chen Y (2006) Application of nano-zinc oxide master batch in polybutadiene styrene rubber system. J Appl Polym Sci 101:922–930

22. Qi JY, Wu LX, Zhuo DX (2014) Preparation and properties of BR/SBR blends using surface-modified nano zinc oxide". Adv Mater Res 910:101–104

23. Sridevi D, Rajendran KV (2009) Synthesis and optical characteristics of ZnO nanocrystals. Bull Mater Sci 32:165–168

24. Aneesh PM, Jayaraj MK (2010) Red luminescence from hydrothermally synthesized Eu-doped ZnO nanoparticles under visible excitation. Bull Mater Sci 33:227–231

25. Gundert F, Wolf BA (1989) Solvents and non-solvents for polymers. In: Brandrup J, Immergut EM (eds) Polymer handbook, 3rd edn. Wiley, New York, VII, pp 173–182

26. Bhattacharyya S, Gedanken A (2008) A template-free, sonochemical route to porous ZnO nano-disks. Microporous Mesoporous Mater 110:553–559

27. Yalcin H, Bozkaya O (1995) Sepiolite-palygorskite from the Hekimhan region clay. Clay Miner 43:705–717

28. Monshi A, Foroughi MR, Monshi MR (2012) Modified Scherrer equation to estimate more accurately nano-crystallite size using XRD. World J Nano Sci Eng 2:154–160

29. Blow CM, Hepburn C (1982) Rubber technology and manufacture. Butterworth Scientific, London

30. Shamugharaj AM, Bae JH, Lee KY, Noh WH, Lee SH, Rye SH (2007) Physical and chemical characteristics of multiwalled carbon nanotubes functionalized with aminosilane and its influence on the properties of natural rubber composites. Comp Sci Technol 67:1813–1822

31. Maiti M, Sadhu S, Bhowmick AK (2005) Effect of carbon black on properties of rubber nanocomposites. J Appl Polym Sci 96:443–451

32. Maiti M, Basak GC, Srivastava VK, Jasra RV (2016) Mesoporous silica reinforced polybutadiene rubber hybrid composite. Int J Ind Chem 7:131–141

# 16

# Adsorption of methylene blue from aqueous solution using untreated and treated (*Metroxylon* spp.) waste adsorbent: equilibrium and kinetics studies

Jeminat O. Amode[1] · Jose H. Santos[1] · Zahangir Md. Alam[2] · Aminul H. Mirza[1] · Chan C. Mei[1]

**Abstract**

*Background* (*Metroxylon* spp.) waste is an inexpensive and abundantly available material with the characteristics of a good adsorbent for treating dye from wastewater. We studied the effectiveness of alkali and acid modification in enhancing the adsorption capacity of sago waste. The untreated and treated adsorbent was characterized by FTIR, elemental analysis and BET surface area. The capacity of each adsorbent to adsorb MB was evaluated at different pH values, adsorbent dosage and initial dye concentrations and contact time.

*Results* According to the results obtained, alkali treatment more than doubled the sorption capacity of sago waste by increasing the porosity, surface area and number of adsorption sites. The alkali-treated material also adsorbed significantly more than many known biosorbents. The effects of the initial concentration of methylene blue, solution pH and adsorbent dosage on methylene blue removal are reported. Equilibrium data were best represented by the Langmuir isotherm model with adsorption capacities of 83.5, 212.8 and 36.82 mg/g for untreated, potassium hydroxide-treated and phosphoric acid-treated sago wastes, respectively. The kinetics of adsorption were best described by a pseudo-second-order model ($R^2 = 0.999$).

*Conclusions* The alkali treatment of sago waste demonstrates the use of a low-cost agricultural waste and a simple modification process to produce an effective adsorbent for removing cationic dye from wastewater.

**Keywords** (*Metroxylon* spp.) waste · Methylene blue · Low-cost adsorbent · Adsorption · Alkali modification · Water treatment

## Introduction

Wastewater effluents from many industries, including paper, leather, textiles, rubber, plastics, printing, cosmetics, pharmaceuticals and food, contain several kinds of synthetic dyestuffs [1]. Dye-bearing wastewaters exhibit high chemical and biochemical oxygen demands [2]. The presence of even very low concentrations in discharge effluents to the environment is worrying for both toxicological and esthetic reasons [3, 4]. To reduce the negative effects of dye-contaminated wastewater on humans and the environment, the wastewater must be treated carefully before discharge into main streams [5]. Various physical, chemical and biological methods, including adsorption, biosorption coagulation and flocculation, advanced oxidation, ozonation, membrane filtration and liquid–liquid extraction, have been widely used for the treatment of dye-bearing wastewater [2, 6–8]. Adsorption is a very effective separation technique and is considered to be superior to other techniques for water treatment in terms of initial cost, simplicity of design, ease of operation and resilience to toxic substances [9, 10]. Although adsorption technologies are well established, a significant limitation is the cost of adsorbent materials. This has motivated the search for low-cost and renewable materials for use as sorbents and has led to a growing interest in the use of nonconventional and locally available materials such as natural materials and agricultural wastes.

✉ Jose H. Santos
 jose.santos@ubd.edu.bn

[1] Faculty of Science, Universiti Brunei Darussalam, Jalan Tungku Link, Gadong 1410, Negara Brunei Darussalam

[2] Bioenvironmental Engineering Research Unit (BERU), Faculty of Engineering, International Islamic University Malaysia (IIUM), Kuala Lumpur, Malaysia

Recently, a large number of low-cost adsorbents have been utilized to develop cheaper and effective adsorbents to remove dyes from wastewater, including cucumber peels [11], meranti sawdust [12], bagasse [13], durian leaf powder [14], watermelon seed hulls [11], grape pulp [15], chitosan [16, 17], kenaf core fibers [18], etc. The reported results showed that most of these readily available bio-adsorbents possess high efficiency in removing dyes from aqueous solutions [16, 19]. These lignocellulosic by-products possess various advantages, such as being eco-friendly, renewable, less expensive and abundantly available, as compared to commercial adsorbents [18, 20]. Studies have also shown that chemical modification of agricultural by-products significantly enhances their ion-binding properties, thereby providing greater flexibility in their applications to a wide range of dyes [21–24]. However, while agricultural by-products are often presented as low-cost adsorbents, their availability is often region specific. With the majority of costs in using biosorbents being associated with the transportation of materials [25], their viability may be limited to the region of origin. Agricultural waste is a low-cost and abundantly available material in Brunei Darussalam among other areas in the Asia–Pacific region. (*Metroxylon* spp.) waste also known as sago hampas is a by-product of starch extraction from *Metroxylon sagu* (sago palm). Sago palm is becoming an important socioeconomic crop in countries such as Papua New Guinea, Indonesia, Malaysia, Thailand and the Philippines [26]. This biomass has good chemical stability, high mechanical strength and a granular structure, making it a good adsorbent material for treating dye from wastewater. In the present study, agricultural sago waste was treated with alkaline and acid for the removal of methylene blue from aqueous solutions. The enhancement of unmodified *Metroxylon* spp. waste by alkali and acid treatment has been investigated here toward the development of a dye-removing adsorbent that is high in adsorption capacity, cost-effective and requires only simple processing. Batch studies were performed to evaluate the effects of various parameters such as pH, initial dye concentration and adsorbent dosage on the removal of a basic dye from an aqueous solution. The textural and physicochemical properties, adsorption isotherms and kinetic parameters have also been determined and discussed.

## Materials and methods

### Reagents

The following chemical reagents were of analytical grade (AR) and purchased from Merck and Sigma Aldrich companies: orthophosphoric acid, potassium hydroxide,

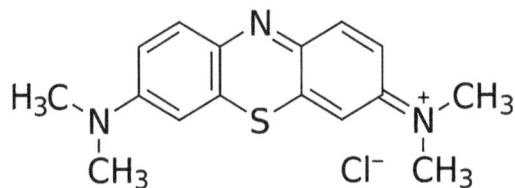

**Fig. 1** The structure of methylene blue chloride salt

sodium chloride, sodium hydroxide, hydrochloric acid and methylene blue (MB). Methylene blue was chosen in this study because of its known strong adsorption onto solids and its recognized usefulness in characterizing adsorptive material [27]. Methylene blue has a molecular weight of $319.85 \text{ g mol}^{-1}$, which corresponds to the hetero-cyclic aromatic chemical compound with the molecular formula $C_{16}H_{18}N_3SCl$. Methylene blue has a net positive charge and the structure of this dye is shown in Fig. 1.

Stock solutions were prepared by dissolving an accurately weighed $1.000 \pm 0.0005 \text{ g}$ of dye in 1 L of distilled water. Stock solutions were covered with aluminum foil and stored in a dark place to prevent UV degradation. The desired concentrations for batch adsorption tests were obtained by further dilution.

### Untreated adsorbent

The sago hampas waste used in this study was obtained from a sago processing plant in Ukong, Tutong District, Brunei Darussalam. When used on a dry basis, sago hampas contains 58 % starch, 23 % cellulose, 9.2 % hemicellulose and 4 % lignin [28]. Sago hampas is the starchy lignocellulosic by-product from the pith of *Metroxylon sagu* (sago palm) following the starch extraction process [26]. Sago waste collected from the processing plant was washed repeatedly with distilled water to remove soluble impurities, such as excess starch, and solid wastes from the industrial process including adhering dirt and debris. The sago waste was then dried to drain excess water and then further dried in an 80 °C oven for 24 h prior to storage. The dried biomass was ground using a laboratory blender to a fine powder and screened with a standard sieve size to ensure that particle sizes were not greater than 350 µm. The powdered biomass was stored in an airtight plastic container and used for batch adsorption tests and characterization. Sago waste samples used without chemical treatment are denoted here as SW.

### Modification of sago waste adsorbent

Alkali- and acid-treated sago waste adsorbent powders were prepared separately by mixing 10 g of the raw sample obtained after sieving with 100 ml of diluted $H_3PO_4$ (85 wt%) to a concentration of (20. 0 %) w/v in a beaker. Similarly, 10 g of sieved sample was mixed with 100 ml solution of dissolved KOH to a concentration of (1.0 M) in a beaker. The reaction mixtures were stirred with a magnetic stirrer for a period of 24 h prior to filtration to achieve good penetration of chemical into the interior of the precursor.

After that, the chemically treated samples were subjected to thorough washing with hot water (80 °C), mild acid (0.1 M HCl) and base (0.1 M NaOH) till the effluent water shows the neutral pH. The samples were dried at 70 °C overnight and used for batch adsorption tests and characterization. The treated sago wastes are denoted as SKOH and SHP for alkali- and acid-treated adsorbents, respectively.

## Characterization of adsorbent

The method for determining the point of zero charge, $pH_{PZC}$, for adsorbents is described elsewhere [29]. Briefly, 50 mL of 0.01 M $KNO_3$ solutions were placed in various Erlenmeyer flasks. The pH of the solutions was adjusted to values between 2 and 10 by the addition of 0.1 M HCl and NaOH solutions. For each solution, 0.2 g of adsorbent powder was added and the final pH recorded after 48 h. The $pH_{PZC}$ is the point where $pH_{final}$ and $pH_{initial}$ values are equal.

The Boehm titration method was applied to determine the amounts of acidic and basic surface functional groups. The main principle of this method is that surface oxygen groups are either acidic or basic moieties that are neutralized by bases and acids, respectively [30]. Prior to analysis, the adsorbent samples were dried in an oven at 110 °C for 3 h. For each adsorbent, 0.2 g amounts were placed in stoppered glass flasks and 30 mL of either 0.1 M NaOH or 0.1 M HCl was added. The bottles were sealed and shaken at 250 rpm for 48 h at 298 K to reach equilibrium. Suspensions at equilibrium were filtered and 20 mL of each filtrate was pipetted into 100 mL Erlenmeyer flasks. Filtrates containing excesses of NaOH and HCl were titrated with 0.1 M HCl and 0.1 M NaOH, respectively. The numbers of acidic and basic sites were calculated by determining the amounts of NaOH and HCl that reacted with the adsorbents [30]. The specific surface area was calculated from the amount of adsorbed methylene blue at maximum adsorption capacity according to:

$$\text{Specific Surface Area} = \frac{M_{MB}A_v A_{MB}}{319.85} \times \frac{1}{M_S}, \qquad (1)$$

where $M_{MB}$ is the mass (g) of methylene blue adsorbed at the point of maximum adsorption; $M_s$ is the mass (g) of the adsorbents; $A_v$ is Avogadro's number, $6.02 \times 10^{23}$; and $A_{MB}$ is the area covered by one methylene blue molecule (typically assumed to be $1.62 \times 10^{-18}$ m$^2$) [31]).

The physical properties such as specific surface area and pore volume distribution were measured by the nitrogen gas adsorption technique using a surface area analyzer (Quantachrome Corporation, USA) with liquid nitrogen at 77 K. The surface area was calculated using the Brunauer–Emmett–Teller (BET) method. Prior to the experiment, the samples were out-gassed at 393 K for 5 h. Field emission scanning electron microscopy (FESEM) images were obtained using a Superscan SS-550 field emission scanning electron microscope (Shimadzu Corporation; Kyoto, Japan) and used to investigate the surface morphologies of adsorbents. The effects of chemical treatment on the surface functionalization of adsorbents were evaluated by Fourier transform infrared spectroscopy (FTIR) with spectra recorded between 4000 and 400 cm$^{-1}$ (resolution of 4 cm$^{-1}$ and acquisition rate of 32 scan min$^{-1}$) using an IR Prestige-21/FTIR-8400S spectrometer (Shimadzu Corporation; Kyoto, Japan).

## Batch adsorption experiments

The effects of experimental parameters, such as MB concentration (75, 150 and 300 ppm), pH (2–12) and adsorbent dosage (1–10 g/L) on dye adsorption by treated and untreated sago wastes, were studied in batch adsorption experiments. The pH of the experimental solutions was adjusted by the addition of HCl and NaOH. All adsorption experiments were conducted in 250 mL conical flasks with 50 mL of dye solution (with the desired concentration and pH) for pH tests and 100 mL for adsorbent dosage tests added to SW, SKOH and SHP adsorbents. The solutions were mechanically agitated in a rotary shaker at 250 rpm and at a constant temperature of 298 K. For studies of pH and adsorbent dosage at initial dye concentrations of 300 mg/L, respectively, equilibrium was achieved within 2 h. The solutions were then filtered through Whatman No. 40 filter paper and the absorbance of filtrates was determined using a UV/visible spectrophotometer (PerkinElmer; USA) at a maximum wavelength of 665 nm. The MB concentrations were calculated from a calibration curve. The amounts of adsorbed dye per gram of adsorbent at equilibrium, $q_e$ (mg/g), and the percentage removal ($R_E$) were calculated by the following equations:

$$q_e = \frac{(C_0 - C_e)V}{W}, \qquad (2)$$

$$\%R_E = \frac{(C_0 - C_e)100}{C_0}, \qquad (3)$$

where $C_0$ and $C_e$ are the initial and equilibrium concentrations of MB, respectively (mg/L). $V$ is the volume of dye solution (L) and $W$ is the mass of the adsorbent used (g).

Adsorption isotherm experiments were carried out by agitating MB solutions of different concentrations

(10–1000 mg/L) with 0.1 g of adsorbent at a constant temperature of 298 K. To ensure full equilibration, a shaking time of 120 min was used for all concentrations of MB. To study the kinetics of biosorption, 0.1 g of different adsorbents was added to 50 mL of dye solutions at initial concentrations of 75, 150 and 300 mg/L. Aqueous samples were taken at different time intervals for the measurement of MB concentration.

## Results and discussion

### Characterization of adsorbents

The effect of chemical treatment on the surface structure of adsorbents was investigated by Fourier transform infrared spectroscopy (FTIR) (Fig. 2). Beyond the fingerprint region, a characteristic, broad band in all adsorbents occurs in the range of 3780–3000 cm$^{-1}$ corresponding to the stretching mode of free O–H groups, hydrogen-bonded O–H and chemisorbed water. A peak at 2920 cm$^{-1}$ for all samples can also be assigned to the C–H stretching of lignocellulosic components. While the composition of surface functional groups may differ between adsorbents, a notable feature of all samples is the presence of surface oxygen groups able to undergo protonation and deprotonation, and thus carry a surface charge.

The surface charge of adsorbents was further examined by comparing the pH$_{PZC}$ and pH of the adsorbent samples.

Surface charges arise from the presence of functional groups, such as surface oxygen complexes, and their interactions with the aqueous solution. The charge of each functional group contributes to the overall charge of the surface. When the solution pH is higher than the pH$_{PZC}$, the surface of the adsorbent has an overall negative charge, favoring the adsorption of cationic species. Alternatively, a solution pH below the pH$_{PZC}$ results in an overall positive surface charge and preferential adsorption of anionic species. The pH and pH$_{PZC}$ values for chemically treated adsorbents are listed in Table 1. For the untreated adsorbent, the pH$_{PZC}$ falls within a broad range. The pH of SW in solution is very similar to the pH$_{PZC}$ and so the untreated adsorbent is not considered to have an overall surface charge when added to aqueous solutions. A pH value above the pH$_{PZC}$ indicated an overall positive charge for the SHP samples. For SKOH, the pH was also above the pH$_{PZC}$ and a greater difference in these values indicated that the surface charge is more negatively charged compared with SW samples. The overall negative charge on SKOH adsorbents suggests a potential for adsorbing positively charged MB molecules.

The proportions of basic and acidic groups on the adsorbent surface obtained by Boehm titration are also shown in Table 1. All adsorbent samples indicated the presence of basic functional groups and thus the potential to act as MB adsorbents. The largest number of basic groups was found in SKOH and this is in good agreement with the overall surface charge indicated by comparison of

**Fig. 2** FTIR spectra of SW (*below*), SKOH (*middle*) and SHP (*top*)

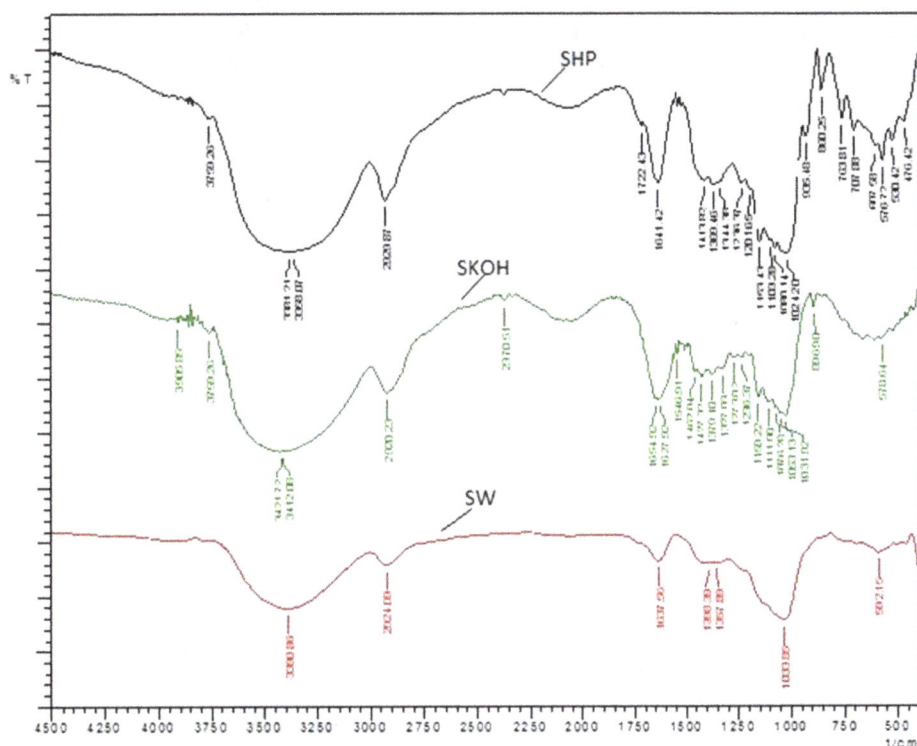

**Table 1** Physiochemical properties of the untreated and chemically treated *Metroxylon sagu* waste adsorbents

| Composition | pH | $pH_{PZC}$ | Surface basic group (meq) | Surface acidic group (meq) | Specific surface area ($S_{MB}$) $m^2$/g | Langmuir isotherm | | | |
|---|---|---|---|---|---|---|---|---|---|
| | | | | | | $R^2$ | $q_{max}$ (mg/g) | $q_{max}$ (mmol/g) | $R_L$ |
| SW | 5.61 | 6.75 | 0.79 | – | 246.4 | 0.9917 | 83.48 | 0.261 | 0.098 |
| SKOH | 7.29 | 7.17 | 2.1 | 1.3 | 549.4 | 0.9979 | 212.8 | 0.665 | 0.550 |
| SHP | 4.08 | 3.04 | 0.5 | 0.26 | 75.6 | 0.9488 | 36.82 | 0.115 | 0.101 |

pH and $pH_{PZC}$ values. However, the number of basic groups is lower for SHP than SW despite pH and $pH_{PZC}$ values indicating that SHP is more likely to have an overall positive charge than SW. The specific surface area was increased from 246.4 to 549.4 $m^2$/g for SW and SKOH and a significant decrease to 75.6 $m^2$/g for SHP is in agreement with the number of negatively charged functional groups available for adsorbing MB molecules.

The creation of the nitrogen adsorption–desorption curve provides qualitative information on the adsorption mechanism and porous structure of the materials. The $N_2$ adsorption–desorption isotherms and the pore size distribution by density functional theory (DFT) method of the untreated and treated sago waste was shown in Fig. 3. This adsorption behavior exhibits a combination of microporous–mesoporous structure. However, the adsorbents bear a resemblance to Type III isotherms which are generally obtained in case of nonporous adsorbents. The surface physical parameters obtained from the $N_2$ adsorption–desorption isotherms are summarized in Table 2. From the data, it is evident that the BET surface area, micropore surface area and total pore volume of SKOH were greatly improved after alkali treatment. The BET surface area of SKOH was obtained as 78.48 $m^2$/g, which may likely

supply more surface active sites, leading to an enhancement of adsorption performance. It is suggested that the pore structure of the adsorbent SKOH consists of mesopores and micropores. The total pore volume at $P/Po = 0.989$ was obtained as 0.157 $cm^3$/g, which indicated that SKOH has a mesoporous structure and makes it easy for methylene blue to penetrate into the mesopores of SKOH. Among the experimental samples, the alkali treatment (SKOH) had the highest surface area followed by untreated waste (SW = 32.01 $m^2$/g) and acid treatment (SHP = 30.07 $m^2$/g). However, the SHP sample and untreated samples showed only an insignificant pore volume. Earlier reports on rice straw fly ash (RSFA) found a higher surface (67.4 $m^2$/g) than our untreated samples [32]. This could be attributed to the quality of substrates where both the samples were of a fibrous nature.

It is evident that acid and alkali treatments affect the number of charged functional groups on the adsorbent surface. The correlation to the surface morphology was examined by field emission scanning electron microscopy (FESEM) of adsorbent samples (shown in Fig. 4). FESEM micrographs were obtained before and after treatment at an accelerating voltage of 10 kV and 1000× magnification. At such magnification, distinct differences in the surface

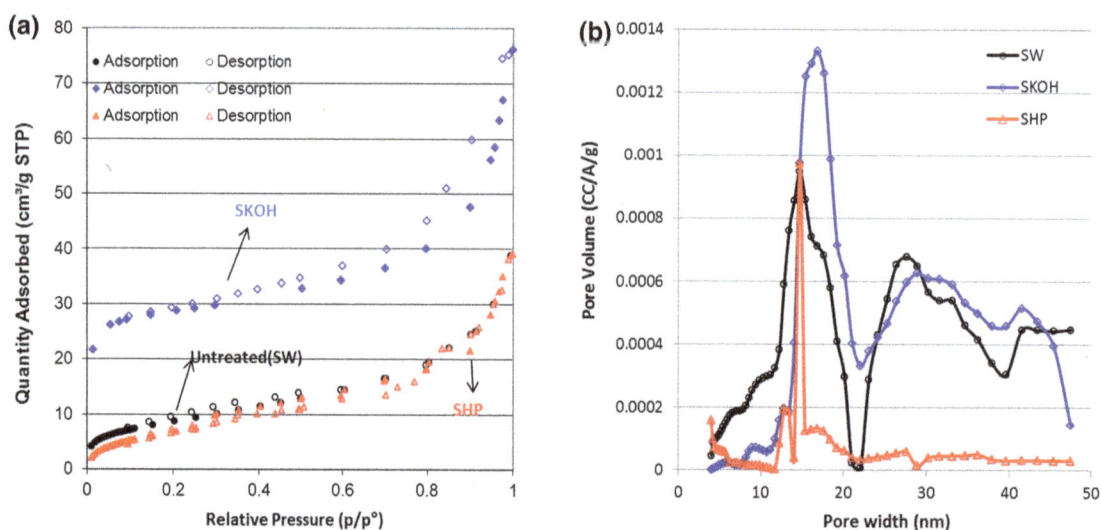

**Fig. 3** **a** Nitrogen adsorption/desorption isotherms; **b** density functional theory (DFT) pore volume distribution of the untreated and treated sago waste

**Table 2** Surface area of the untreated and treated samples

| Sample | Total surface area (BET) $(m^2\ g^{-1})$ | Micropore surface area $(m^2\ g^{-1})$ | Total pore volume $(cm^3\ g^{-1})$ | Micropore volume $(cm\ g^{-1})$ | Average pore diameter (Å) |
|---|---|---|---|---|---|
| SW | 32.01 | 39.48 | 0.015 | 0.014 | 188.1 |
| SKOH | 78.48 | 56.64 | 0.157 | 0.012 | 74.84 |
| SHP | 30.07 | 28.78 | 0.014 | 0.014 | 202.3 |

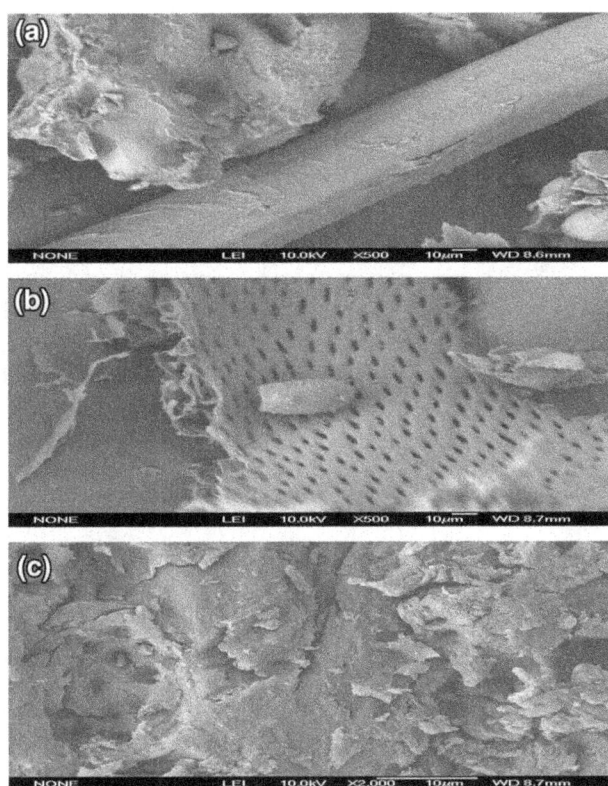

**Fig. 4** FESEM images of **a** untreated (SW), **b** KOH-treated (SKOH) and **c** $H_3PO_4$-treated (SHP) sago wastes

structure and porosity were clearly identifiable. After alkali treatment (Fig. 4b), the fibrous structure of the sago waste had clearly increased in surface roughness with the formation of pores throughout the structure. A previous study in which rice husk was exposed to NaOH demonstrated the dissolution of hemicellulose and lignin by alkali treatment and suggests that the increased porosity observed here is due to a similar dissolution of hemicellulose and lignin from the interfibrillar region of the sago hampas [33]. The increased porosity in SKOH not only increased the surface area of the adsorbent, but also the number of surface functional groups available as adsorption sites. This is reflected in the higher numbers of both basic and acidic functional groups listed in Table 1 for SKOH.

Acidification of sago waste with phosphoric acid, unlike alkali treatment, showed no indication of visible pores. Like the raw sago waste, SHP exhibited a rough surface morphology with a larva-like structure. However, comparison of FESEM images (Fig. 4a, c) also showed swelling or enlargement of particles in SHP and a decrease in surface roughness. The images suggest a decrease in both porosity and surface area. A lower surface area would also decrease the number of functional groups exposed and accounts for the lower numbers of basic and acidic groups observed for SHP compared with SW and SKOH.

### Effect of initial dye concentration and contact time

In all cases, an initially high rate of adsorption occurred because the MB concentration provided the driving force for the rapid attachment of MB onto the adsorbent surface. As adsorption proceeded, the ratio of MB molecules to available adsorption sites decreased, which resulted in a decrease in the adsorption rate until equilibrium was reached [34]. This behavior can be seen in Fig. 5a–c, for all adsorbents, and an increase in the initial concentration of MB also resulted in higher initial rates of adsorption. An increase in the initial concentration of MB corresponded to an increase in the ratio of MB molecules to available adsorption sites. This may have subsequently increased the initial driving force for the adsorption of MB by the adsorbent and led to a higher initial rate of adsorption. Figure 5a–c also shows that, in all cases, the time to equilibrium was completely reached within 30–60 min.

### Effect of pH

One of the important parameters in biosorption is the effect of pH. The effect of pH on the adsorption of MB onto untreated SW, treated SKOH and SHP was investigated at a pH range of 2–12 and is shown in Fig. 6. At ambient pH, under the conditions employed, 79.04 mg/g MB was adsorbed by SW (Fig. 6). As the pH increases from pH 4 to 12, there was a steady decrease in the amount of MB being adsorbed. The effect was the greatest at pH 12 where a reduction of 4.53 mg/g MB was observed. This effect was also reported for other low-cost biosorbents such as kenaf fiber char [35]. The high adsorption of SKOH between pH values of 6 and 10 can be attributed to electrostatic attraction between the negative charges of the adsorbent surface and the positive charge of the MB cation, since the amount of dye being removed was high (>140 mg/g) at

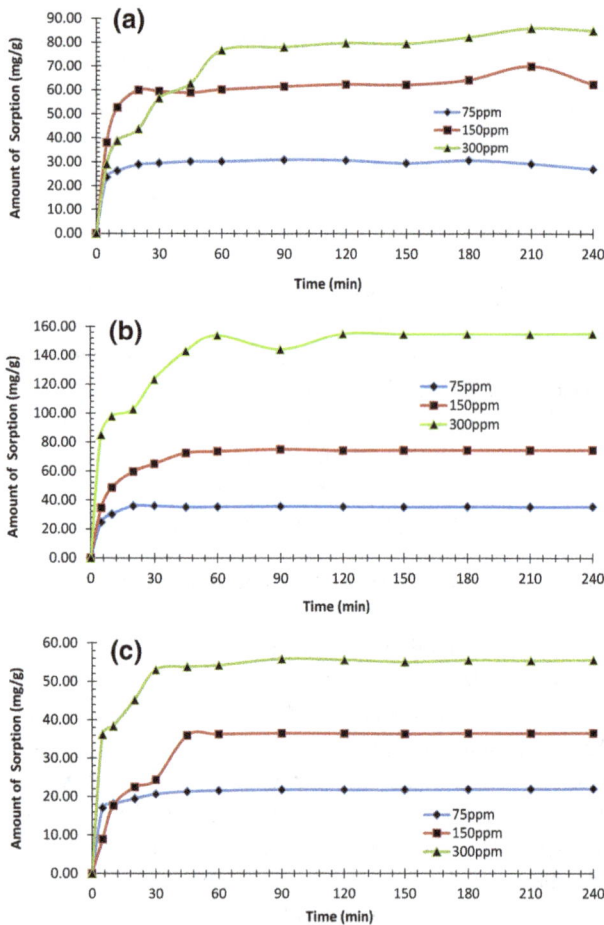

**Fig. 5** Effect of initial concentration and contact time on the sorptive capacity of MB on **a** SW, **b** SKOH and **c** SHP ($C_0 = 75$, 150 and 300 mg/L; dosage = 0.1 g; working volume = 50 mL; initial pH = ambient; agitation speed = 250 rpm)

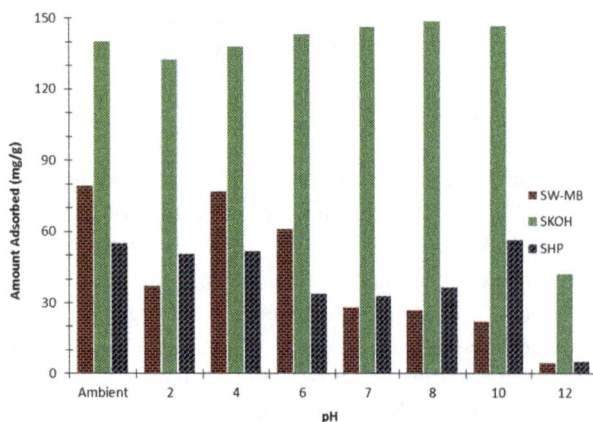

**Fig. 6** Effect of solution pH on the adsorption of MB on SW, SKOH and SHP ($C_0 = 300$ mg/L; dosage = 0.1 g; working volume = 50 mL; agitation speed = 250 rpm

ambient pH (MB = 4.6), as compared to that of other adsorbent pH values. The results indicate that removal of MB is less sensitive to the initial pH variation of the dye

solution and remains almost constant over the pH range 6–10. Further, an overall positive charge and repulsion of MB is expected when the solution pH drops below the $pH_{PZC}$ of 7.17. However, maximum sorptive capacity was maintained below the $pH_{PZC}$ until the pH had decreased to 2. Similar sorptive capacities within this pH range are likely the result of the number of adsorption sites being in excess of the number of MB molecules despite changes in pH potentially changing the ratio of positive and negative charges on the adsorbent surface. An excess of adsorption sites is consistent with the increase in specific surface area (shown in Tables 1, 2) for SKOH. The decrease in sorptive capacity at pH 2 is likely to result from the presence of excess $H^+$ ions competing with the cationic MB molecules and protonating negatively charged adsorbent sites, thereby reducing MB adsorption.

For SHP at pH conditions above the $pH_{PZC}$ of 3.04, there was a trend toward increasing sorptive capacities as pH increased to 10. As the excess hydroxyl ($-OH^-$) ions increased, surface functional groups were predominantly deprotonated resulting in an enhanced number of adsorption sites available for binding positively charged MB adsorbate [22, 35].

## Effect of adsorbent dosage

The adsorbent dosage is a major parameter because this concludes the capacity of a adsorbent for given initial concentration of the adsorbate at the operating conditions. Doses in the range of 1–10 g/L were used in adsorption experiments with an MB concentration of 300 mg/L to investigate the effect of adsorbent amount. The sorptive capacities at a dose of 1 g/L indicate that extremely low sorptive capacities would be observed at doses below 1 g/L for SW and SHP. Thus, doses below 1 g/L were not investigated. The results are shown in Fig. 7.

The percentage removal by individual sorbents increased with increasing dosage, although maximum sorptive capacities were reached at SW and SKOH doses of 5 g/L and then remained almost constant. In comparison, the percentage removal of MB onto SHP was poor and the optimum dosage was not reached at the range of doses investigated. The increased removal with dosage was expected because of the increased adsorbent surface area and availability of adsorption sites [12, 36]. This study demonstrates that optimum amounts of adsorbent may be established for specific dye concentrations.

## Evaluation of adsorption behavior of adsorbents

### Adsorption isotherm

Adsorption isotherms are often used to give an accurate description of adsorption behavior for design purposes. The

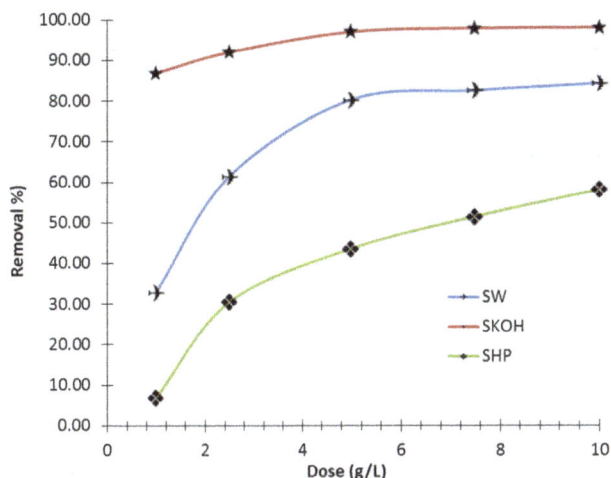

**Fig. 7** Effect of adsorbent dosage on the adsorption of MB on SW, SKOH and SHP ($C_0 = 300$ mg/L; working volume = 100 mL; initial pH = ambient; agitation speed = 250 rpm)

initial dye concentrations within the range from 0 to 1000 ppm were indicative of the completion of a monolayer by the leveling of the isotherm adsorption, thus preventing further adsorption leading to multilayer coverage in (Fig. 8). The equilibrium isotherms were analyzed by the following: Langmuir, Freundlich, Tempkin, Dubinin–Raduskevich (D–R) isotherm models. The nonlinear and linearized equations are represented by the expression in Table 3 and the results obtained from the best fitting isotherm models are shown in Table 1. The method of least squares was used for obtaining the isotherm constants, and error analysis using the following functions [37] confirmed the model of best fit: sum of squares of errors (SSE), sum of absolute errors (EABS), average relative error (ARE) and nonlinear Chi-test ($\chi$) as shown in Table 4. The Langmuir isotherm model could be used to describe the adsorption behavior of all adsorbents; all other models were a poor fit and so treatment of the adsorption behavior is shown in Table 1.

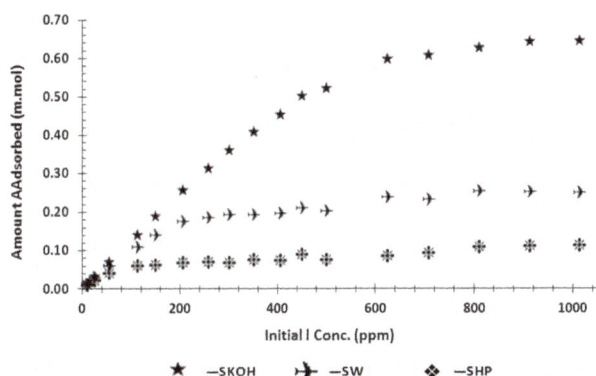

**Fig. 8** Variation of the extent of removal of MB by SW, SKOH and SHP as a function of initial dye concentration

A dimensionless constant separation factor or equilibrium parameter, $R_L$, is defined according to the following equation [38]:

$$R_L = \frac{1}{(1 + K_L C_0)}, \qquad (4)$$

where $C_0$ is the initial dye concentration (mg/L). The $R_L$ value indicates whether the isotherm is favorable $(0 < R_L < 1)$, unfavorable $(R_L > 1)$, linear $(R_L = 1)$ or irreversible $(R_L = 0)$. Coefficients of determination ($R^2$) values above 0.99 for all adsorbents are listed in Table 1 and indicate that the Langmuir isotherm fits the experimental data well. The Langmuir isotherm model suggests that monolayer adsorption onto a homogeneous surface is applicable for these adsorbents.

Of these four isotherm models represented in Table 5, error analyses show that the Langmuir model is the best to describe the biosorption of MB on SW, SKOH and SHP. All the error analyses for the Langmuir model give much reasonably close $R^2$ and lower errors indicating that it is overall a better fit. The biosorption capacity of MB on SKOH is determined to be 0.665 mmol g$^{-1}$ based on the Langmuir model, which is higher than the value obtained for the Freundlich, Temkin and Dubinin–Radushkevich model when compared with experimental data. Maximum sorption capacity of 36.8 mg g$^{-1}$ for SHP showed a decrease in adsorption capacity, and a significant improvement was observed for SKOH (212.8 mg g$^{-1}$) over SW (83.5 mg g$^{-1}$) in Table 1. Hence, Langmuir models can be used to describe the adsorption isotherm of MB on SW, SKOH and SHP, with the Langmuir model giving a better $R^2$ and lower errors in general. These results indicate that adsorption of MB on SKOH is stronger than that of SW and SHP on the different adsorbent. This could be due to the less bulky nature of SKOH than that of others, enabling more MB molecules to be adsorbed on the adsorbent.

A comparison of maximum adsorption capacities for MB with various biosorbents is summarized in Table 6. Only a few biosorbents had higher sorption capacities than SKOH; key considerations in which SKOH may prove more advantageous, however, are the availability and simple processing requirements of the adsorbent. Adsorbents may be considered low cost if they are abundant, require little processing and have low economic value [39]. As mentioned previously, sago hampas waste is readily available in the Asia–Pacific region as an agricultural waste that can be sourced at free cost. Furthermore, the chemical modification of the sago waste to SKOH was experimented in achieving high sorption capacities compared with previously reported biosorbents and was achieved here by a simple modification technique that requires few procedural steps. On the other hand, it can be considered as a good

**Table 3** Isotherm model equations

| Isotherm model | Linear | Non-linear | Plot |
|---|---|---|---|
| Langmuir | $\dfrac{C_e}{q_e} = \dfrac{1}{q_{max}K_L} + \dfrac{1}{q_{max}}C_e$ | $q_e = \dfrac{q_{max}K_L C_e}{1+K_L C_e}$ | $\dfrac{C_e}{q_e}$ vs $C_e$ |
| Freundlich | $\mathrm{Log}\,q_e = \mathrm{Log}\,K_F + \dfrac{1}{n}\log C_e$ | $q_e = K_F C_e^{1/n}$ | $\mathrm{Log}\,q_e$ vs $\ln C_e$ |
| Temkin | $q_e = B\ln K_T + B\ln C_e$ | | $q_e$ vs $\ln C_e$ |
| | | $q_e = \left(\dfrac{RT}{b}\right)\ln(K_T C_e)$ | |
| Dubinin–Radushkevich | $B = \dfrac{RT}{b}$ $\ln q_e = \ln q_D - B\varepsilon^2$ $\varepsilon = RT\ln\left[1+\dfrac{1}{C_e}\right]$ $E = \dfrac{1}{\sqrt{-2B}}$ | $q_e = (q_D)\exp(-k_{ad}\varepsilon^2)$ | $\ln q_e$ vs $\varepsilon^2$ |

$q_e$ is the amount of dye adsorbed, $C_e$ is the equilibrium concentration of the dye, $q_{max}$ is the maximum adsorption capacity (mg/g), $K_F$ (mg$^{1-1/n}$L$^{1/n}$ g$^{-1}$) is the Freundlich constant, n is the empirical parameter which is related to the biosorption intensity, $K_T$ is the equilibrium binding constant (L/mmol) corresponding to the maximum binding energy, constant $B$ is related to the heat of adsorption, $R$ is gas constant, $T$ is the absolute temperature, $\beta$ gives the mean free energy, $E$ is the sorption per molecule of sorbate

**Table 4** Error functions

| Error function | Abbreviation | Definition/expression |
|---|---|---|
| Sum squares errors | ERRSQ/SSE | $\sum_{i=1}^{n}(q_{e,calc} - q_{e,meas})_1^2$ |
| Average relative error | ARE | $\dfrac{100}{n}\sum_{i=1}^{n}\left|\dfrac{q_{e,meas}-q_{e,calc}}{q_{e,meas}}\right|$ |
| Sum of absolute error | EABS | $\sum_{i=1}^{n}\left|q_{e,meas} - q_{e,calc}\right|_i$ |
| Nonlinear chi-square test | $\chi^2$ | $\sum_{i=1}^{n}\dfrac{(q_{e,meas}-q_{e,calc})^2}{q_{e,meas}}$ |

biosorbent in the context of its higher qmax value as it is able to remove MB better than many other treated biosorbents, such as citric/EDTA-modified alkali rice straw, citric acid-modified kenaf core fibers, sulfuric acid-

treated parthenium and EDTA-modified rice husk given in Table 6.

Advantages over other biosorbents for SKOH may also include a less energy-intensive treatment process. For example, modification by the alkali treatment process is less energy intensive than biosorbent treatments that require high temperatures to produce activated carbon (e.g., activated carbon from apricot [40] and sago waste carbon [41]). Less intensive processing requirements for SKOH suggests potentially lower expenses in the use of this biosorbent.

## Adsorption kinetics

A kinetic study of adsorption is necessary as it provides the information about the adsorption mechanism, which is

**Table 5** Comparison of parameters for $R^2$ and error functions using different isotherm models

| Adsorbent | Models | $R^2$ | EABS | EERSQ/SSE | ARE | $\chi$ |
|---|---|---|---|---|---|---|
| SW | Langmuir | 0.9917 | 0.204 | 0.003 | 0.210 | 0.030 |
| | Freundlich | 0.9307 | 0.469 | 0.020 | 0.524 | 0.102 |
| | Temkin | 0.9724 | 1.471 | 0.130 | 2.267 | 1.814 |
| | Dubinin–Radushkevich (D–R) | 0.6407 | 0.850 | 0.059 | 0.985 | 0.650 |
| SKOH | Langmuir | 0.9979 | 0.152 | 0.013 | 0.204 | 0.060 |
| | Freundlich | 0.9025 | 0.614 | 0.040 | 0.637 | 0.134 |
| | Temkin | 0.9864 | 0.233 | 0.007 | 0.548 | 0.101 |
| | Dubinin–Radushkevich (D–R) | 0.6701 | 1.055 | 0.136 | 1.153 | 0.770 |
| SHP | Langmuir | 0.9488 | 0.101 | 0.002 | 0.200 | 0.030 |
| | Freundlich | 0.9217 | 0.121 | 0.001 | 0.384 | 0.021 |
| | Temkin | 0.9274 | 0.105 | 0.001 | 0.329 | 0.015 |
| | Dubinin–Radushkevich (D–R) | 0.7133 | 0.267 | 0.007 | 0.559 | 0.131 |

**Table 6** Comparison of the maximum monolayer adsorption capacities, $q_{max}$ (mg/g) of MB dye on various biomass-based adsorbents

| Adsorbent | Adsorption capacity (mg/g) | Reference |
|---|---|---|
| Untreated waste | | |
| Untreated (*Metroxylon sagu*) waste | 83.48 | This study |
| Pine cone biomass | 109.89 | [10] |
| Cucumber peels | 111.11 | [11] |
| Watermelon seed hulls | 57.14 | [11] |
| Grape pulp | 153.85 | [15] |
| Peanut husk | 72.13 | [46] |
| Spent rice biomass | 8.3 | [47] |
| Treated waste | | |
| Potassium hydroxide-treated (*Metroxylon sagu*) waste | 212.8 | This study |
| Phosphoric acid treated (*Metroxylon sagu*) waste | 36.82 | This study |
| Citric and/or EDTA-modified alkali rice straws | 62.90–135.10 | [22] |
| Activated carbon from apricot | 136.98 | [40] |
| Citric acid-modified rice straw | 270.3 | [48] |
| Citric acid-modified kenaf core fibers | 131.6 | [18] |
| Sulfuric acid-treated parthenium | 88.29 | [49] |
| EDTA-modified rice husk | 46.30 | [23] |
| Sago waste carbon | 4.51 | [41] |
| HCL-treated meranti sawdust | 120.48 | [12] |
| TA-modified bagasse | 69.93 | [13] |
| NaOH-modified durian leaf powder | 125 | [14] |

**Table 7** Kinetic model equations

| Kinetic models | Equation | Plot |
|---|---|---|
| Pseudo-first-order model | $\text{Log}(q_e - q_t) = \text{Log}\,q_e - \frac{K_1}{2.303}t$ | $\text{Log}(q_e - q_t)$ vs $t$ |
| Pseudo-second-order model | $\frac{t}{q_t} = \frac{1}{k_2 q_e^2} + \frac{1}{q_e}t$ | $\frac{t}{q_t}$ vs $t$ |
| Elovich model | $q_t = \left(\frac{1}{b}\right)\ln(ab) + \frac{1}{b}\ln t$ | $q_t$ vs $\ln t$ |
| Intra-particle diffusion model | $q_t = k_{id}t^{1/2} + C$ | $q_t$ vs $t^{1/2}$ |

$q_e$ and $q_t$ are the amounts of MB adsorbed (mg g$^{-1}$) at equilibrium and at time $t$ (min), respectively; $k_1$ (min$^{-1}$) is the adsorption rate constant; $k_2$ (g mg$^{-1}$ min$^{-1}$) is the rate constant of second-order adsorption; $k_1$ d (mg/g h) is the intraparticle diffusion rate constant and $C$ gives an idea about the thickness of the boundary layer; $a$ (mg/g h) is the initial sorption rate and $b$ (g/mg) is related to the extent of surface coverage and activation energy for chemisorption

crucial for the practicality of the process. The pseudo-first-order, pseudo-second-order, intraparticle diffusion and Elovich models were implemented to evaluate the rate constant of the adsorption process for SW, SHP and SKOH samples onto MB at various initial concentrations. The experimental data were fitted with the kinetic models linearized equations in Table 7 and the linear regression analyses and the constants were calculated by using Microsoft Excel, Version 2010. The experimental and calculated qe values from the related plots together with the model constants and correlation coefficient R$^2$ determined from the kinetic models for sorption of MB onto SW, SHP and SKOH samples at 30 °C are summarized in Table 8. The validity of the exploited models is verified by the experimental qe $_{exp}$ and correlation coefficient R$^2$.

In the case of the pseudo-first-order kinetic model values proposed by Lagergren [42] in Table 8, low linear regression correlation coefficients ($R^2 < 0.9$) indicated poor agreement between the model and experimental data. The experimental equilibrium uptakes, $q_e$,exp (mg/g), do not concur with the calculated $q_e$,cal (mg/g) values from the pseudo-first-order model which reflect that experimental data obtained for sorption of the MB under investigation fails to predict the sorption process for the entire region of contact time. The rate constant, $k_1$, obtained for the pseudo-first-order model do not show a consistent trend with increasing concentration range for MB studied as most $R^2$ values are relatively small.

However, the linear form of pseudo-second-order equation proposed by Ho and McKay [43] was found to be

**Table 8** Kinetic model constants and correlation coefficients for adsorption of MB dye using various adsorbents at different initial concentrations

| Concentration (ppm) | Adsorbent | Exp. uptake capacity $q_{e,Exp}$ (mg/g) | Pseudo-first-order model | | | Pseudo-second-order model | | | Intraparticle diffusion model | | | Elovich model | | |
|---|---|---|---|---|---|---|---|---|---|---|---|---|---|---|
| | | | $R^2$ | $q_{e,calc}$ (mg/g) | $K_1$ (min$^{-1}$) | $R^2$ | $q_{e,calc}$ (mg/g) | $K_2$ g/mg(min) | $R^2$ | C | $K_{id}$ | $R^2$ | $\alpha$ (mg/g min) | $\beta$ (g/mg) |
| 300 | SW | 85.90 | 0.9447 | 7.46 | 0.031 | 0.9993 | 96.15 | $8.78 \times 10^{-6}$ | 0.9373 | 21.11 | 9.155 | 0.9059 | 19.19 | 0.062 |
| | SKOH | 146.07 | 0.6876 | 5.90 | 0.014 | 0.9999 | 144.93 | $4.28 \times 10^{-8}$ | 0.7738 | 135.82 | 1.388 | 0.8919 | 135.37 | 0.390 |
| | SHP | 55.88 | 0.8619 | 14.07 | 0.033 | 0.9990 | 57.14 | $7.99 \times 10^{-6}$ | 0.7827 | 36.43 | 2.952 | 0.8819 | 34.56 | 0.178 |
| 150 | SW | 64.16 | 0.5879 | 16.73 | 0.008 | 0.9993 | 64.52 | $3.48 \times 10^{-6}$ | 0.6768 | 44.42 | 2.952 | 0.8553 | 41.55 | 0.168 |
| | SKOH | 75.10 | 0.8552 | 5.93 | 0.021 | 0.9999 | 74.63 | $6.64 \times 10^{-7}$ | 0.7938 | 65.65 | 1.368 | 0.9318 | 65.11 | 0.392 |
| | SHP | 26.61 | 0.6859 | 7.03 | 0.038 | 0.9994 | 27.70 | $9.30 \times 10^{-5}$ | 0.7067 | 13.21 | 2.101 | 0.9184 | 10.99 | 0.233 |
| 75 | SW | 30.89 | 0.7467 | 2.01 | 0.019 | 0.9999 | 30.77 | $9.51 \times 10^{-6}$ | 0.7039 | 28.77 | 0.304 | 0.9028 | 26.27 | 0.833 |
| | SKOH | 35.64 | 0.4407 | 0.92 | 0.015 | 0.9999 | 35.46 | $1.99 \times 10^{-6}$ | 0.5817 | 32.88 | 0.406 | 0.7657 | 32.61 | 1.246 |
| | SHP | 22.09 | 0.6250 | 2.83 | 0.027 | 0.9998 | 22.42 | $9.47 \times 10^{-5}$ | 0.8839 | 16.89 | 0.791 | 0.9449 | 16.67 | 0.720 |

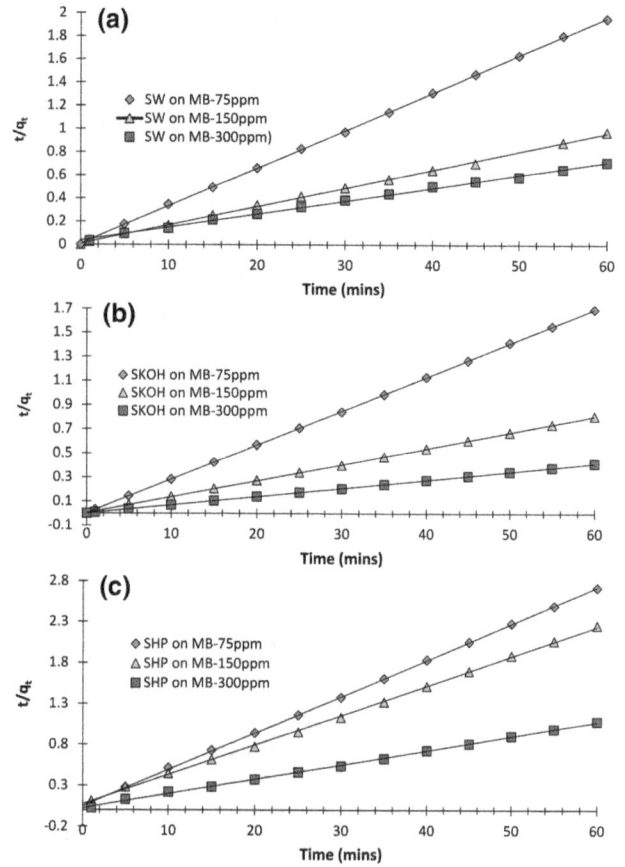

**Fig. 9** Pseudo-second-order kinetics model for adsorption of MB on **a** SW, **b** SKOH and **c** SHP

able to predict the behavior of the sorption process for all the range of concentrations studied here. The linear plots of $t/q_t$ against $t$ (min) gives $1/q_e$(cal) as the slope and $1/k_2 q_e^2$ as the intercept, where $k_2$ (g/mg-min) is the rate constant of the second-order adsorption as shown in Table 8 and Fig. 9. It is observed that the regression lines are almost superimposed by the experimental data. $R^2$ values between 0.9990 and 0.9999 and calculated $q_e$ values similar to those determined experimentally indicated that MB adsorption obeyed pseudo-second-order kinetics for all adsorbents, whereby an assumption of the model is that the rate of chemisorption is a significant factor in determining adsorption rates.

The values of the intraparticle diffusion model by Weber and Morris [43] constants ($k_{id}$ and $C$) obtained for all adsorbent system together with the $R^2$ values obtained are presented in Table 8. Positive $C$ values indicated that intraparticle diffusion was not solely the rate-determining step and that other factors contribute to the rate of adsorption. This is particularly the case at higher MB concentrations where larger $C$ values indicate a greater degree of boundary layer control and deviation from the intraparticle diffusion model. $R^2$ values also indicated a poor agreement of experimental data with the pseudo-second-order kinetic model. Contributors to adsorption rates that lead to positive $C$ values may include other diffusion mechanisms (e.g., film diffusion) or chemisorption, as found earlier by the agreement with pseudo-second-order kinetics. However, their contributions to the adsorption rate cannot be deduced from this model. Linear plots with reasonable $R^2$ values indicate agreement with chemisorption processes contributing significantly to adsorption rates.

Elovich equation, described by Chien and Clayton [45], is another most frequently used model for depicting chemisorption process and is expressed by Equation in Table 7. It is observed that the values of $\alpha$ (mg g$^{-1}$ min$^{-1}$) increase with the increase of initial concentration range studied. The values of the regression coefficient ($R^2 = 0.79$–$0.98$) of the Elovich kinetic model suggest that kinetic data did not follow the Elovich model. However, the higher values of the Elovich constants, $\alpha$ (mg g$^{-1}$ min$^{-1}$) and $\beta$ (g mg$^{-1}$), as shown in Table 8 are suggestive of an increased rate of chemisorption. However, experimental data again showed better agreement with the pseudo-second-order kinetic model where the $R^2$ values obtained for other three kinetics models are less than 0.9, and the majority of the data do not fall on a straight line, indicating that these models are inappropriate.

## Conclusions

The chemical treatment of adsorbents has been shown here to have significant effects on the adsorption capacity of sago waste powder on basic dyes such as methylene blue. Acid treatment resulted in a lower adsorbent surface area that subsequently reduced the adsorption capacity. However, alkaline treatment demonstrated the effectiveness of using chemical treatment to enhance the adsorption capacity of biosorbents. The treatment process increased the porosity and surface area of the sago waste powder, thereby increasing the number of adsorption sites. The adsorption capacity and kinetics for the adsorbents investigated were well described by the monolayer adsorption model of the Langmuir isotherm and the chemisorption model of pseudo-second-order kinetics. An improvement in the maximum adsorption capacity was observed for the alkali-treated sago waste with an adsorption capacity more than double that exhibited by the untreated sago waste. Maximum adsorption capacities of 83.5, 212.8 and 36.8 mg/g were found for untreated, alkali-treated and acid-treated sago wastes, respectively. The alkali-treated sago waste also showed significant improvement over a wide range of biosorbents previously reported. The sago waste used in this investigation is a low-cost agricultural waste found abundantly in the Asia–Pacific region. The alkali treatment was a simple process that requires few procedural steps, avoids energy-intensive heat treatment and shows potential for implementation on wastewater treatment. We propose that alkali-treated sago waste represents an effective and economically feasible material for the treatment of dye-containing effluents.

**Acknowledgments** The authors thank the Government of Brunei Darussalam and the Universiti Brunei Darussalam (UBD) for their financial support and the award of a PhD scholarship to JOA. The authors are grateful to the Bioenvironmental Engineering Research Unit (BERU), Faculty of Engineering, International Islamic University Malaysia for providing the laboratory support and facilities.

**Compliance with ethical standards**

**Conflict of interest** The authors have no conflicts of interest to declare.

## References

1. Chatterjee S, Lee DS, Lee MW, Woo SH (2009) Enhanced adsorption of congo red from aqueous solutions by chitosan hydrogel beads impregnated with cetyl trimethyl ammonium bromide. Bioresour Technol 100(11):2803–2809
2. Yao Z, Wang L, Qi J (2009) Biosorption of methylene blue from aqueous solution using a bioenergy forest waste: *Xanthoceras sorbifolia* seed coat. Clean Soil Air Water 37(8):642–648
3. Nigam P, Armour G, Banat IM, Singh D, Marchant R (2000) Physical removal of textile dyes from effluents and solid-state fermentation of dye-adsorbed agricultural residues. Bioresour Technol 72(3):219–226
4. Tan IAW, Hameed BH, Ahmad AL (2007) Equilibrium and kinetic studies on basic dye adsorption by oil palm fibre activated carbon. Chem Eng J 127:111–119
5. Ahmad A, Mohd-Setapar SH, Chuong CS, Khatoon A, Wani WA, Kumar R, Rafatullah M (2015) Recent advances in new generation dye removal technologies: novel search for approaches to reprocess wastewater. RSC Adv 5(39):30801–30818
6. El-Latif MA, Ibrahim AM, El-Kady M (2010) Adsorption equilibrium, kinetics and thermodynamics of methylene blue from aqueous solutions using biopolymer oak sawdust composite. J Am Sci 6(6):267–283
7. Ghaedi M, Hassanzadeh A, Kokhdan SN (2011) Multiwalled carbon nanotubes as adsorbents for the kinetic and equilibrium study of the removal of alizarin red S and morin. J Chem Eng Data 56(5):2511–2520
8. Vimonses V, Lei S, Jin B, Chow CWK, Saint C (2009) Kinetic study and equilibrium isotherm analysis of Congo Red adsorption by clay materials. Chem Eng J 148:354–364
9. Mohammad M, Maitra S, Ahmad N, Bustam A, Sen T, Dutta BK (2010) Metal ion removal from aqueous solution using physic seed hull. J Hazard Mater 179(1):363–372
10. Sen TK, Afroze S, Ang H (2011) Equilibrium, kinetics and mechanism of removal of methylene blue from aqueous solution by adsorption onto pine cone biomass of *Pinus radiata*. Water Air Soil Pollut 218(1–4):499–515
11. Akkaya G, Güzel F (2014) Application of some domestic wastes as new low-cost biosorbents for removal of methylene blue: kinetic and equilibrium studies. Chem Eng Commun 201(4):557–578

12. Ahmad A, Rafatullah M, Sulaiman O, Ibrahim M, Hashim R (2009) Scavenging behaviour of meranti sawdust in the removal of methylene blue from aqueous solution. J Hazard Mater 170(1):357–365

13. Low LW, Teng TT, Rafatullah M, Morad N, Azahari B (2013) Adsorption studies of methylene blue and malachite green from aqueous solutions by pretreated lignocellulosic materials. Sep Sci Technol 48(11):1688–1698

14. Hussin ZM, Talib N, Hussin NM, Hanafiah MA, Khalir WK (2015) Methylene blue adsorption onto NaOH modified durian leaf powder: isotherm and kinetic studies. Am J Environ Eng 5(3A):38–43

15. Saygili H, Akkaya Saygili G, Güzel F (2014) Using grape pulp as a new alternative biosorbent for removal of a model basic dye. Asia Pac J Chem Eng 9(2):214–225

16. Vakili M, Rafatullah M, Salamatinia B, Abdullah AZ, Ibrahim MH, Tan KB, Gholami Z, Amouzgar P (2014) Application of chitosan and its derivatives as adsorbents for dye removal from water and wastewater: a review. Carbohydr Polym 113:115–130

17. Thakur VK, Thakur MK (2014) Recent advances in graft copolymerization and applications of chitosan: a review. ACS Sustain Chem Eng 2(12):2637–2652

18. Sajab MS, Chia CH, Zakaria S, Jani SM, Ayob MK, Chee KL, Khiew PS, Chiu WS (2011) Citric acid modified kenaf core fibres for removal of methylene blue from aqueous solution. Bioresour Technol 102(15):7237–7243

19. Rafatullah M, Sulaiman O, Hashim R, Ahmad A (2010) Adsorption of methylene blue on low-cost adsorbents: a review. J Hazard Mater 177(1):70–80

20. Han R, Zhang L, Song C, Zhang M, Zhu H, Zhang L (2010) Characterization of modified wheat straw, kinetic and equilibrium study about copper ion and methylene blue adsorption in batch mode. Carbohydr Polym 79(4):1140–1149

21. Azlan K, Wan Saime WN, Lai Ken L (2009) Chitosan and chemically modified chitosan beads for acid dyes sorption. J Environ Sci 21(3):296–302

22. Fathy NA, El-Shafey OI, Khalil LB (2013) Effectiveness of alkali-acid treatment in enhancement the adsorption capacity for rice straw: the removal of methylene blue dye. ISRN Phys Chem 2013:15. doi:10.1155/2013/208087

23. Ong S-T, Keng P-S, Lee C-K (2010) Basic and reactive dyes sorption enhancement of rice hull through chemical modification. Am J Appl Sci 7(4):447–452

24. Prasad RN, Viswanathan S, Devi JR, Rajkumar J, Parthasarathy N (2008) Kinetics and equilibrium studies on biosorption of CBB by coir pith. Am Eurasian J Sci Res 3(2):123–127

25. Vijayaraghavan K, Yun Y-S (2008) Bacterial biosorbents and biosorption. Biotechnol Adv 26(3):266–291

26. Awg-Adeni D, Abd-Aziz S, Bujang K, Hassan MA (2010) Bioconversion of sago residue into value added products. Afr J Biotechnol 9(14):2016–2021

27. Kaewprasit C, Hequet E, Abidi N, Gourlot JP (1998) Quality measurements. J Cotton Sci 2:164–173

28. Linggang S, Phang L, Wasoh M, Abd-Aziz S (2012) Sago pith residue as an alternative cheap substrate for fermentable sugars production. Appl Biochem Biotechnol 167(1):122–131

29. Tan W-f, Lu S-j, Liu F, Feng X-h, J-z He, Koopal LK (2008) Determination of the point-of-zero charge of manganese oxides with different methods including an improved salt titration method. Soil Sci 173(4):277–286

30. Goertzen SL, Theriault KD, Oickle AM, Tarasuk AC, Andreas HA (2010) Standardization of the Boehm titration. Part I. $CO_2$ expulsion and endpoint determination. Carbon 48(4):1252–1261

31. Özsgn G (2011) Production and Characterization of Activated Carbon from Pistachio-Nut Shell. Middle East Technical University. Master Thesis

32. El-Sonbati AZ, El-Deen IM, El-Bindary MA (2016) Adsorption of Hazardous Azorhodanine Dye from an Aqueous Solution Using Rice Straw Fly Ash. J Disper Sci Technol 37(5):715–722

33. Chakraborty S, Chowdhury S, Das Saha P (2011) Adsorption of crystal violet from aqueous solution onto NaOH-modified rice husk. Carbohydr Polym 86(4):1533–1541

34. Shahryari Z, Goharrizi AS, Azadi M (2010) Experimental study of methylene blue adsorption from aqueous solutions onto carbon nano tubes. Int J Water Res Environ Eng 2(2):16–28

35. Mahmoud DK, Salleh MAM, Karim WAWA, Idris A, Abidin ZZ (2012) Batch adsorption of basic dye using acid treated kenaf fibre char: equilibrium, kinetic and thermodynamic studies. Chem Eng J 181–182:449–457

36. Garg V, Kumar R, Gupta R (2004) Removal of malachite green dye from aqueous solution by adsorption using agro-industry waste: a case study of Prosopis cineraria. Dyes Pigm 62(1):1–10

37. Gimbert F, Morin-Crini N, Renault F, Badot P-M, Crini G (2008) Adsorption isotherm models for dye removal by cationized starch-based material in a single component system: error analysis. J Hazard Mater 157(1):34–46

38. Chien SH, Clayton WR (1980) Application of Elovich equation to the kinetics of phosphate release and sorption in soils. Soil Sci Soc Am J 44:265

39. Zwain HM, Vakili M, Dahlan I (2014) Waste material adsorbents for zinc removal from wastewater: a comprehensive review. Int J Chem Eng 2014:13. doi:10.1155/2014/347912

40. Basar CA (2006) Applicability of the various adsorption models of three dyes adsorption onto activated carbon prepared waste apricot. J Hazard Mater 135:232–241

41. Kadirvelu K, Kavipriya M, Karthika C, Radhika M, Vennilamani N, Pattabhi S (2003) Utilization of various agricultural wastes for activated carbon preparation and application for the removal of dyes and metal ions from aqueous solutions. Bioresour Technol 87(1):129–132

42. Lagergren S (1898) About the theory of so-called adsorption of soluble substances. Kungliga Sven Vetensk Handl 24(4):1–39

43. Ho Y-S, McKay G (1998) Kinetic models for the sorption of dye from aqueous solution by wood. Process Saf Environ Prot 76(2):183–191

44. Weber WJ, Morriss JC (1963) Kinetics of adsorption on carbon from solution. J Sanit Eng Div Am Soc Civil Eng 89:31–60

45. Chien S, Clayton W (1980) Application of Elovich equation to the kinetics of phosphate release and sorption in soils. Soil Sci Soc Am J 44(2):265–268

46. Song J, Zou W, Bian Y, Su F, Han R (2011) Adsorption characteristics of methylene blue by peanut husk in batch and column modes. Desalination 265(1):119–125

47. Rehman MSU, Kim I, Han J-I (2012) Adsorption of methylene blue dye from aqueous solution by sugar extracted spent rice biomass. Carbohydr Polym 90(3):1314–1322

48. Gong R, Zhong K, Hu Y, Chen J, Zhu G (2008) Thermochemical esterifying citric acid onto lignocellulose for enhancing methylene blue sorption capacity of rice straw. J Environ Manage 88(4):875–880

49. Lata H, Garg V, Gupta R (2007) Removal of a basic dye from aqueous solution by adsorption using Parthenium hysterophorus: an agricultural waste. Dyes Pigm 74(3):653–658

# 17

# Response surface optimization of Rhodamine B dye removal using paper industry waste as adsorbent

Anita Thakur[1] · Harpreet Kaur[1]

**Abstract** The present investigation describes the conversion of waste product into effective adsorbent and its application for the treatment of wastewater, i.e., chemically modified solid waste from paper industry has been tested for its adsorption ability for the successful removal of Rhodamine B dye from its aqueous solution. The adsorption isotherm, kinetics and thermodynamic parameters of process have been determined by monitoring the different parameters, such as effect of pH, amount of adsorbent dose, concentration, contact time and temperature. The equilibrium data has been well described on the basis of various adsorption isotherms, namely Langmuir, Freundlich and Temkin adsorption isotherm. From Langmuir isotherm, the maximum monolayer adsorption capacity has been found to be 6.711 mg g$^{-1}$ at 308 K temperature. The kinetics of adsorption has been studied using pseudo-first order, pseudo-second order and intra-particle diffusion model and the results show that kinetics has been well described by pseudo-second order. Thermodynamic parameters, such as free energy change ($\Delta G$), enthalpy change ($\Delta H$) and entropy change ($\Delta S$), have been evaluated. The free energy has been obtained as $-11.9452$ kJ mol$^{-1}$ for 75 mg L$^{-1}$ concentration at 308 K temperature. Desorption and recycling efficiency of adsorbent has been studied and the adsorbent shows good recycling efficiency.

**Keywords** Paper industry waste · Rhodamine B · Adsorption · Kinetics · Isotherms

## Abbreviations

| | |
|---|---|
| CMSW | Chemically modified solid waste |
| BET | Brunauer–Emmett–Teller |
| SEM | Scanning electron microscope |
| FTIR | Fourier transformation infrared spectroscopy |
| EDAX | Energy dispersive X-ray spectroscopy |

## List of symbols

| | |
|---|---|
| $q_e$ | Adsorption capacity |
| $C_0$ | Initial equilibrium concentration |
| $C_e$ | Final equilibrium concentration |
| $V$ | Volume of the solution |
| $W$ | Weight of adsorbent |
| $q_m$ | Maximum adsorption capacity |
| $b_L$ | Energy of adsorption |
| $R_L$ | Dimensionless constant |
| $K_f$ | Freundlich constant |
| $1/n$ | Heterogeneity factor |
| $R^2$ | Regression coefficient |
| $B$ | Intensity of adsorption |
| $K_T$ | Constant related to adsorption capacity |
| $K_2$ | Pseudo-second order coefficient |
| $t$ | Time |
| $K_{ipd}$ | Intra-particle diffusion rate constant |
| $\Delta S$ | Entropy change |
| $\Delta H$ | Enthalpy change |
| $\Delta G$ | Free energy change |

## Introduction

India ranks third among the leading textile-producing countries in the world behind China and European nations, and more than 95 million peoples got engaged in textile and related sectors in India [1]. But despite of

✉ Harpreet Kaur
preetjudge@yahoo.co.in

[1] Department of Chemistry, Punjabi University, Patiala 147002, India

significance, the textile industries are the main source of pollution due to discharge of hazardous effluent containing colours and organic chemicals used for bleaching, dyeing, printing and other finishing processes [2]. Globally, about 10–15% of total dyestuff (equivalent to 280,000 tonnes) is released annually into the environment during the manufacturing of textile products, which leads to the contamination of water reservoirs, and thereby affects human and animal health [3, 4].

One of the most commonly used dyes in industries is Rhodamine B dye. Rhodamine B is synthetically prepared xanthene cationic dye and widely used for paper printing and as a colourant in textile and food stuff [5]. It is harmful to both human beings and animals, because if this dye is swallowed it can cause irritation to skin, eyes and respiratory track [6]. It has been medically proven that drinking water contaminated with Rhodamine B dye is highly carcinogenic, neurotoxin and chronic [7, 8]. Thus, the wastewater contaminated with Rhodamine B dye must be treated carefully before discharged into water streams [9].

A number of conventional physical, chemical and biological methods, such as ion-exchange [10], coagulation/flocculation [11], reverse osmosis [12], membrane filtration [13], electrochemical oxidation [14], electrochemical degradation [15], photodegradation [16], and heterocatalytic Fenton oxidation [17], have been used for the removal of dyes. The serious drawbacks of these methods are low efficiency, disposal of waste, low sensitivity, etc. [18, 19]. Among all these, adsorption has been found to be very simple and innovative method for treating dye wastewater even at very low concentration of dyes [20]. In adsorption process, adsorbate adhered on the surface of adsorbent by physical, chemical or electrostatic forces [21]. Activated carbon has been the most widely used adsorbent for the wastewater treatment due to its high surface area and high adsorption capacity [22]. Though the removal of dyes through activated carbon is very effective, but sometimes its use is restricted due to its high cost and difficulties associated with regeneration [23]. The removal of hazardous dyes through adsorption technique using industrial waste materials, such as blast furnace dust, sludge, slag from steel plant and carbon slurry from fertilizer plant [24], chitosan [25], bottom ash [26], and agriculture wastes, such as date palm [27], coconut tree flowers [28] have been already reported.

The paper industries produce a large amount of sludge every year, which can be used as an adsorbent for the removal of dyes. Thus, this study aimed to investigate the potential use of CMSW for the removal of hazardous dye Rhodamine B.

## Experimental

### Materials and methods

*Preparation of dye solution*

Rhodamine is a basic dye having IUPAC name [9-(2-carboxyphenyl)-6-diethylamino-3-xanthenylidene]-diethylammonium chloride has been purchased from S.D. Fine chemicals, Mumbai, India. Stock solution of dye (500 mg L$^{-1}$) has been prepared by dissolving 0.5 g of dye in 1000 mL of deionised water. Another solution of desired concentration has been prepared by successive dilutions of the stock solution. Concentration of the dye after adsorption has been determined using Shimadzu—1800 UV Visible Spectrophotometer at 553 nm wavelength.

*Preparation of adsorbent*

The waste material (sludge) from paper industry has been used as an adsorbent for the removal of dye. The sludge has been washed with deionised water and dried (under sunlight) and then kept in the oven at 100 °C for 3 days. The dried material has been grounded into fine powder. The finely powdered sludge has been mixed with sulphuric acid and kept overnight and then washed with deionised water to remove residue acid. The material has been dried at 100 °C for 24 h and then grounded, sieved and kept in air tight container for further uses.

*Adsorption studies*

Batch adsorption studies of removal of Rhodamine B dye onto CMSW has been carried out as a function of initial dye concentration, contact time, adsorbent dose and pH. All the adsorption experiments have been conducted by shaking 100 mL of solution of definite concentration of dye along with fixed amount of adsorbent at room temperature (308 K) and pH (4.40) at constant speed on mechanical shaker. 5 mL of solution has been withdrawn at pre-determined time intervals. The concentration of Rhodamine B dye in solution has been determined using UV–Visible spectrophotometer. During adsorption, equilibrium has been established between adsorbed dye on active sites of adsorbent and unadsorbed dye in the solution. The percentage of dye adsorbed and adsorption capacity at equilibrium has been calculated by the following formula:

$$\text{Percentage adsorption} (\%) = \frac{(C_0 - C_e)}{C_0} \times 100$$

$$q_e = \frac{(C_0 - C_e)}{W} V$$

where $C_0$ and $C_e$ represent the initial and final equilibrium concentrations (mg L$^{-1}$), $V$ is the volume of solution and $W$ is the weight of adsorbent.

## Effect of contact time

The influence of contact time on the adsorption process has been studied for different intervals of time, i.e., 10, 20, 30, 45, 60, 90 and 120 min. The initial dye concentration and adsorbent dose chosen for this study were 50 mg L$^{-1}$ and 2.0 g, respectively.

## Effect of initial dye concentration

The effect of initial dye concentration (25, 50, 75, 100 and 125 mg L$^{-1}$) on percentage adsorption has been analysed by agitating 100 mL of dye solution along with 2.0 g of adsorbent for equilibrium time, i.e., 60 min.

## Effect of adsorbent dose

Variable amount of CMSW dose (0.5, 1.0, 1.5, 2.0 and 2.5 g) has been agitated along with 100 mL of dye solution (50 mg L$^{-1}$) for different intervals of time as described above.

## Effect of pH and ionic strength

To investigate the effect of solution pH on the colorant adsorption, the pH values of solutions has been adjusted to pH 2.40, 4.40, 8.40 and 10.40 using 1 N sodium hydroxide and 1 N hydrochloric acid. The pH of solution has been monitored with the help of pH-meter. The effect has been studied by stirring dye solution of concentration 50 mg L$^{-1}$ along with adsorbent (2.0 g) for 60 min.

## Effect of temperature

To study the effect of temperature on the adsorption of Rhodamine B by CMSW, the experiments have been performed at three different temperatures, i.e., 308, 313 and 318 K. The concentration of dye taken is 50 mg L$^{-1}$ and CMSW dose is 2.0 g.

## Effect of surfactant

The effect of surfactant has been studied by agitating 100 mL of dye (50 mg/L) solution along with 10 mg of sodium dodecyl sulphate and 2.0 g of CMSW.

## Desorption studies

For the desorption studies, the adsorbent collected after adsorption has been dried and divided into three parts. One part is dissolved in water, other in 1 N acetic acid and remaining in 1 N hydrochloric acid for 24 h and then washed gently with water to remove any unadsorbed dye. To study the recycling efficiency, 2.0 g of adsorbent collected after desorption with water, acetic acid and hydrochloric acid has been agitated separately with 100 mL of dye solution of 50 mg L$^{-1}$ concentration for 60 min. The solutions after adsorption have been subjected to UV–Visible spectrophotometer to determine the amount of dye adsorbed.

## Characterization of adsorbent

### BET, SEM, FT-IR and EDAX studies

The physical parameters, such as surface area, total pore volume and mean pore diameter of CMSW has been determined using (Belstrop mini Japan) Brunauer, Emmett and Teller (BET) N$_2$ sorption procedure with liquid N$_2$ at $-195.72$ °C. For the BET analysis, the material has been degassed. The sample materials is placed in a vacuum chamber at a very low constant temperature ($-195.72$ °C) and it is operated at a wide range of pressure. The surface area, mean pore volume and mean pore diameter of CMSW has been found as 1600 cm$^2$ g$^{-1}$, 0.1083 cm$^3$ g$^{-1}$ and 27.058 nm, respectively. As compared with the surface area of other adsorbents, such as bottom ash (870.5 cm$^2$ g$^{-1}$) and deoiled soya (728.6 cm$^2$ g$^{-1}$), CSMW shows a very good surface area [29].

Scanning electron microscopy has been used as a primary source for characterizing the surface morphology and fundamental physical properties of the adsorbent. Figure 1a indicates that before adsorption the surface is rough and porous, so there is a good possibility for the dye to be adsorbed into these pores. It is clear from Fig. 1b that after adsorption the surface becomes smooth, which indicates that the surface of adsorbent is covered with dye molecules.

CMSW has been characterized using Fourier transformation infrared, i.e., FT-IR analysis. The FT-IR spectrum of CMSW before and after adsorption has been shown in Fig. 2, in which the lower one is unloaded CMSW and the upper one is loaded CMSW with Rhodamine B dye. The spectrum of unloaded CMSW shows weak absorption band at 3675 cm$^{-1}$ corresponds to hydroxyl group (–OH) stretching. An absorption band at 2915 cm$^{-1}$ corresponding to C–H stretching of the CH$_2$ groups, which indicates the presence of various amino groups. The spectrum shows weak absorption band at 1620 cm$^{-1}$, which may be due to –C=O stretching. The peak around 1260 cm$^{-1}$ may be due the presence of lignin [30].The weak absorption bands at

**Fig. 1 a** Scanning electron micrographs (SEM) of CMSW before adsorption. **b** Scanning electron micrographs (SEM) of CMSW after adsorption

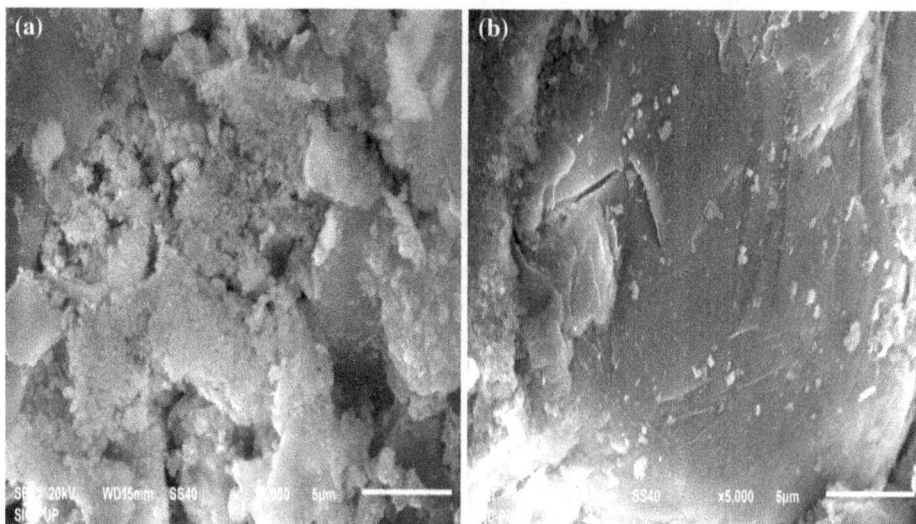

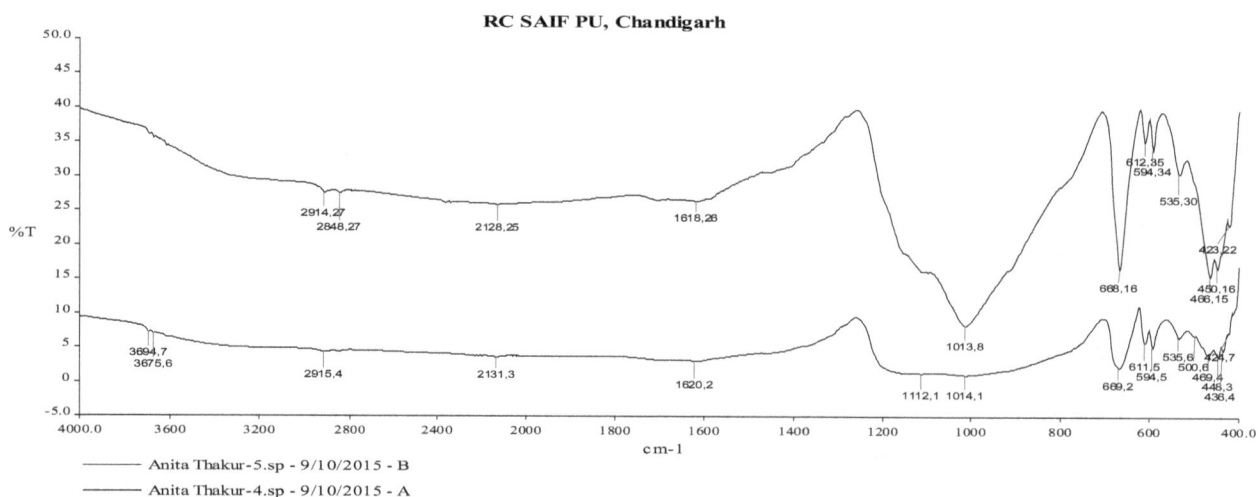
Fig. 2 FTIR spectra of loaded and unloaded CMSW

1121 and 1014 cm$^{-1}$ may be attributed to –C=N and C–O stretching of polysaccharide like substances. The stretching vibration in the region 700–600 cm$^{-1}$ may be assigned to C–S linkage and peak due to brominated compounds may be appeared in the region of 600–500 cm$^{-1}$. The absorption band due to –OH and –C=N stretching is missing after adsorption, which shows that these may be involved in the adsorption process. There is slight shifting of peaks of adsorbent after adsorption. No new peak has been observed, which indicates that no chemical bond is formed between adsorbate and adsorbent after adsorption, i.e., FT-IR data supports that adsorption of dye on adsorbent is due to physical forces.

The chemical composition of adsorbent has been determined using EDAX analysis. Figure 3 shows the elemental percentage composition of O, C, Si, S, Mg and Al in CMSW adsorbent. The oxygen content has been

found to be maximum in CMSW, i.e., 51.26%. The carbon content has been found to be 32.06%. The other contents, such as silicon, sulphur, magnesium and aluminium have been found to be 6.86, 4.61, 4.48 and 0.73%, respectively. Higher oxygen contents indicate that metal ions must be present in oxide form.

## Results and discussion

### A batch adsorption study

*Effect of contact time*

The adsorption potential of CMSW towards Rhodamine B dye as a function of contact time has been shown in

**Fig. 3** EDAX spectra of CMSW

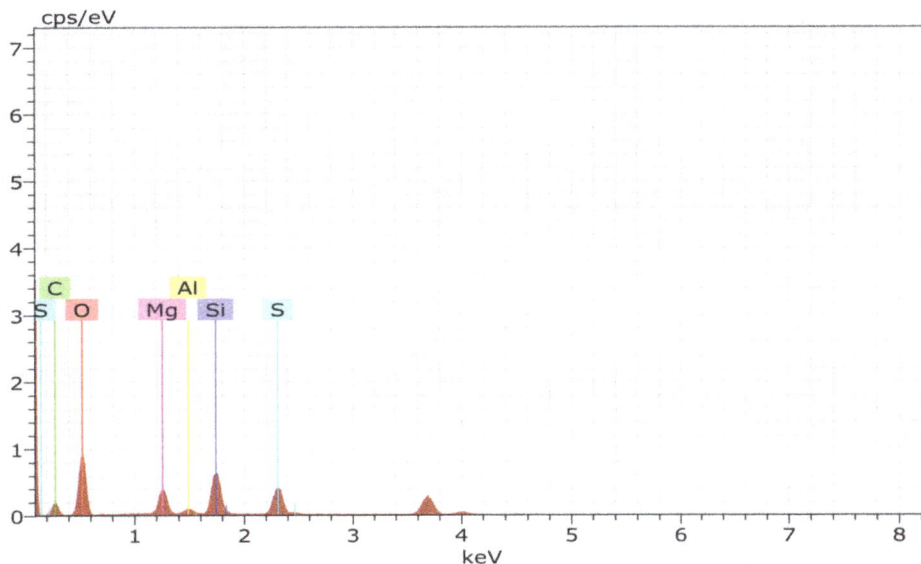

**Fig. 4 a** Effect of contact time and initial dye concentration on % removal of dye. Initial dye concentration = 50 mg L$^{-1}$, contact time = 60 min, adsorbent dose = 2.0 g. **b** Effect of contact time and initial dye concentration on adsorption capacity of dye. Initial dye concentration = 50 mg L$^{-1}$, contact time = 60 min, adsorbent dose = 2.0 g

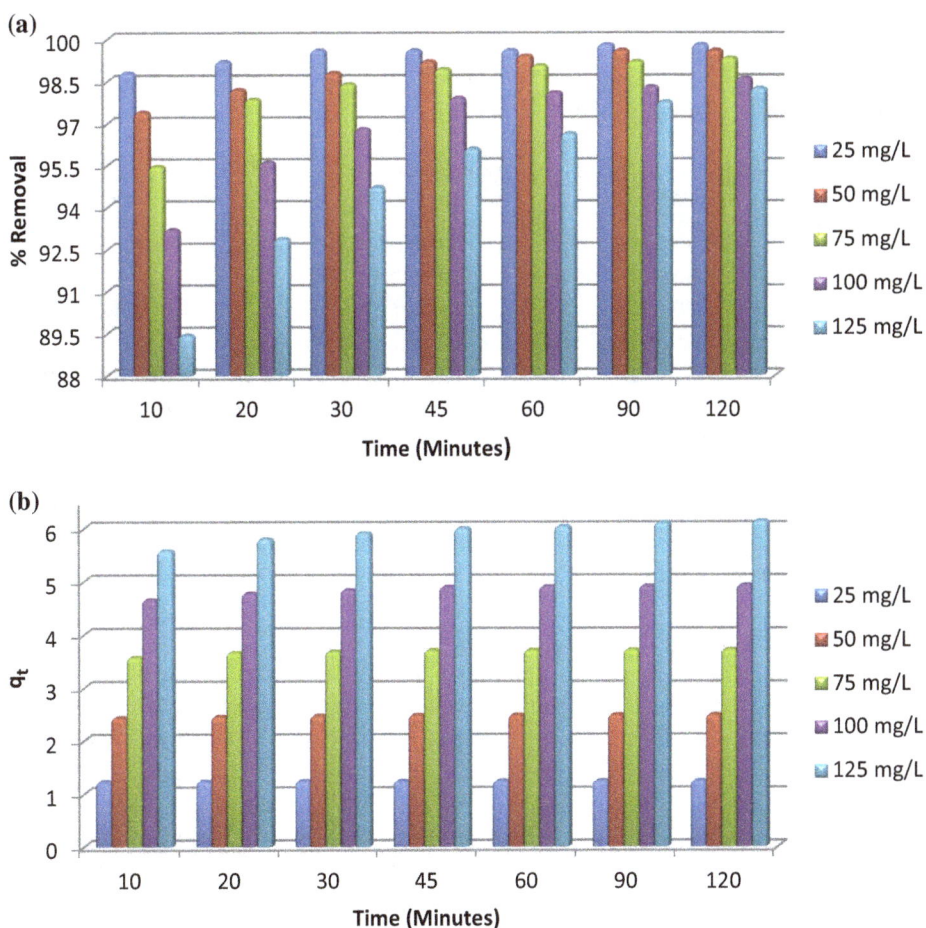

Fig. 4a, b and it is evident from figures that percentage removal of Rhodamine B dye has been increased with increase in contact time. The percentage removal has been found to be rapid in early stages of adsorption and remained almost constant after 60 min. This is due to the reason that at initial stages, all the active sites are free for adsorption, but after 60 min equilibrium is established between dye in solution and dye on adsorbent, i.e., there is electrostatic hindrance or repulsion between the adsorbed dye onto the adsorbent surface [6, 31]. Approximately

**Table 1** Contact time for Rhodamine B adsorption on various adsorbents

| Adsorbents | Equilibrium contact time (min) | References |
|---|---|---|
| Walnut shell | 80 | [32] |
| *Casuarina equisetifolia* needles | 180 | [33] |
| Rise husk | 180 | [34] |
| Coconut shell activated carbon | 180 | [35] |
| *TyphaAngustata L* plant materials | 210 | [36] |
| Walnut shell charcoal | 300 | [37] |
| CMSW | 60 | Present study |

99% of dye has been removed within 10 min at all initial concentrations, which shows that CMSW is a good adsorbent. A comparison of contact time for the adsorption of Rhodamine B dye onto CMSW with other adsorbents (Table 1) shows that CMSW takes lesser contact time for adsorption.

*Effect of initial dye concentration*

The data indicate that percentage of dye removed decreases with increase in the initial concentration of dye. As at lower concentration, maximum dye particles in solution occupy available binding sites on adsorbent, which results in better adsorption [38]. But at higher concentration, the available sites on the adsorbent become limited and there is no further adsorption. In case of adsorption capacity, the adsorption capacity increases with increase in initial dye concentration because the increase in initial dye concentration enhances the interaction between dye and adsorbent [35, 37, 39].

*Effect of adsorbent dose*

In adsorption process, the amount of adsorbent dose is an important parameter because it determines the potential of adsorbent to remove the dye at a particular given concentration. It has been observed that percentage of dye removed increases from 79.20 to 99.80% and adsorption capacity decreases from 7.92 to 1.996 mg g$^{-1}$ as amount of CMSW increased from 0.5 to 2.5 g. The increase in percentage removal at higher adsorbent dose is attributed to the fact that by increasing the amount of adsorbent dose, the adsorptive surface area increases, due to which the number of available sites increases and results in increase in percentage removal [29, 32, 33]. But the adsorption capacity decreases with increase in adsorbent dose because there is a split in concentration gradient between the concentration of dye in solution and that on the surface of adsorbent [40] (Fig. 5).

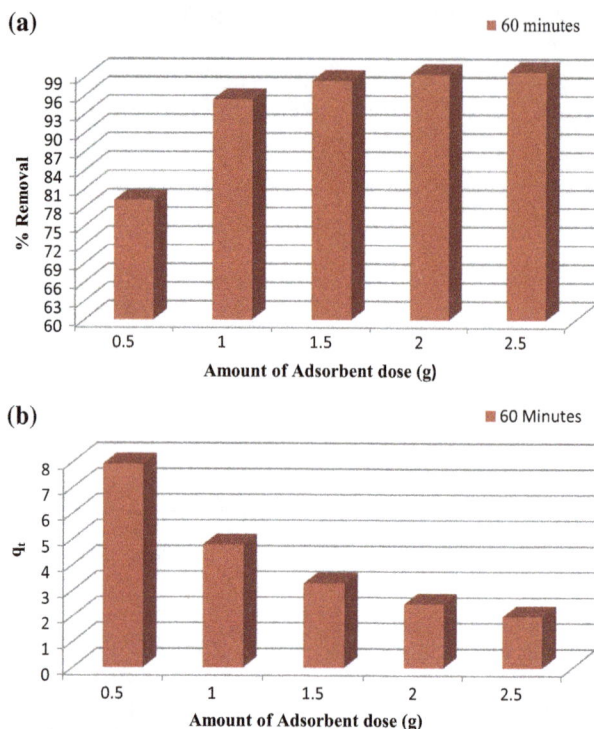

**Fig. 5 a** Effect of adsorbent dose on percentage removal of dye. Contact time = 60 min. Initial dye concentration = 50 mg L$^{-1}$. **b** Effect of adsorbent dose on adsorption capacity of dye. Initial dye concentration = 50 mg L$^{-1}$, contact time = 60 min

*Effect of pH and ionic strength*

The pH of solution plays an important role in adsorption process because it directly affects the dissociative and adsorptive ability of dye on the surface of adsorbent [41]. Figure 6 shows that removal of dye is higher in acidic medium than alkaline medium. It may be explained on the bases that change in pH of the solution results in the formation of different ionic species and different carbon surface charges. When the pH is lower, the Rhodamine B dye exists in cationic and monomeric form and is able to easily enter in the pores of adsorbent. But as the pH increases, the zwitterionic form of Rhodamine B in water may lead to the aggregation of dye molecules to dimmers [42]. Due to large size at high pH, dye molecules are enabling to fit, and this results in decrease in percentage removal at higher pH.

The effect of ionic strength is also important because it verify the attraction between the non-polar groups of dye and adsorbent, i.e., hydrophobic–hydrophobic interactions. It has been observed that adsorption has been increased with increase in ionic strength, i.e., with the addition of NaCl (0.1 mol L$^{-1}$ NaCl). This may be due to the fact that with increase in ionic strength, there is a partial neutralization of the positive charge on the adsorbent surface. The high ionic strength enhances the hydrophobic–hydrophobic interactions by compression of electric double layer that

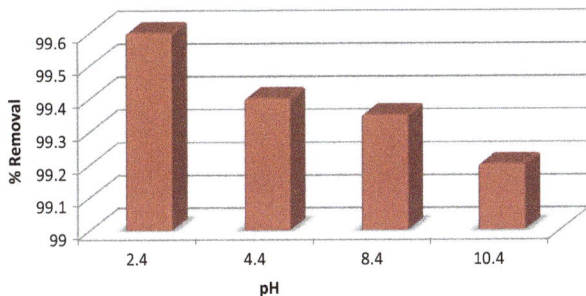

**Fig. 6** Effect of pH on percentage removal of dye. Initial dye concentration = 50 mg L$^{-1}$, contact time = 60 min, adsorbent dose = 2.0 g

moves particles much closer, which leads to increase in dye adsorption [43].

### Effect of temperature

Since adsorption is a temperature dependent process. Thus, the removal of dye has been studied at three different temperatures, i.e., 308, 313 and 318 K. The extent of adsorption of dye has been found to be slightly increased with increase in temperature (Fig. 7), indicating the endothermic nature of the process [36, 44].

### Effect of surfactant

The adsorption of cationic dye onto CMSW has been studied in the presence of anionic surfactant sodium dodecyl sulfate (SDS). The result indicates that 100% of dye has been removed using SDS along with the adsorbent. This can be explained on the fact that Rhodamine B is cationic dye and SDS is anionic surfactant, so there is more adsorption of ionic solute in the presence of oppositely charged surfactant, i.e., electrostatic attraction between adsorbate and adsorbent increases (Fig .8).

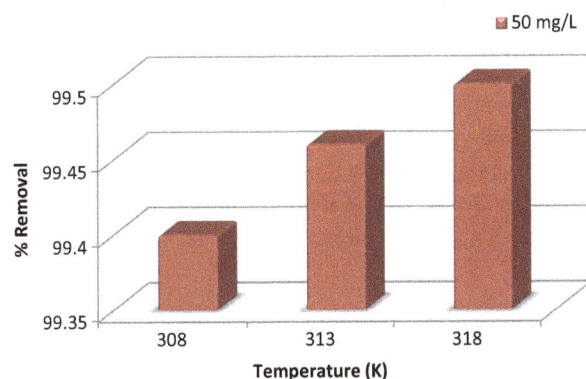

**Fig. 7** Effect of temperature percentage removal of dye. Initial dye concentration = 50 mg L$^{-1}$, contact time = 60 min, adsorbent dose = 2.0 g

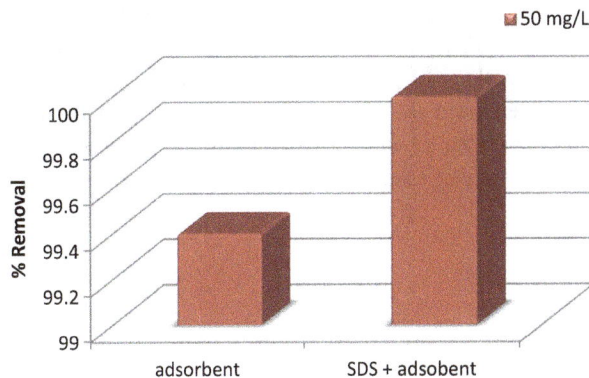

**Fig. 8** Effect of surfactant on percentage removal of dye. Initial dye concentration = 50 mg L$^{-1}$, contact time = 60 min, adsorbent dose = 2.0 g, SDS dose = 100 mg

### Desorption studies

Desorption studies help to elucidate the nature of interaction existing between adsorbate and adsorbent and the recycling of adsorbent. It is evident from Fig. 9 that the adsorbent which is treated with hydrochloric acid desorbed to maximum extent, i.e., why a large amount of dye has been removed using hydrochloric acid treated desorbed adsorbent. It indicates that hydrochloric acid has good regenerating power and CMSW shows good recycling efficiency.

### Adsorption isotherms

The data of adsorption studies has been tested with Langmuir, Freundlich and Temkin adsorption isotherms.

### Langmuir adsorption isotherm

The isotherm is based on the assumption that the adsorption takes place at specific homogeneous sites on the adsorbent surface and is monolayer in nature.

The linear equation for Langmuir isotherm model is given below [45]:

$$\frac{C_e}{q_e} = \frac{C_e}{q_m} + \frac{1}{q_m \cdot b_L}$$

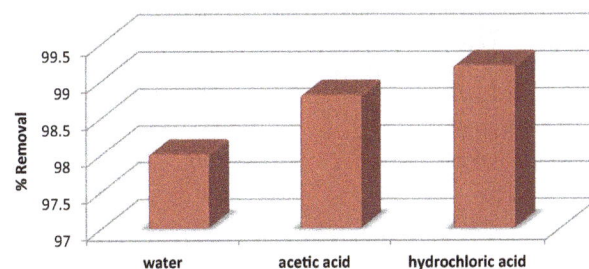

**Fig. 9** Desorption studies using various solvents

where, $q_m$ and $b_L$ are the Langmuir constants related to the maximum adsorption capacity (mg g$^{-1}$) and energy of adsorption (L mg$^{-1}$). The values of $q_m$ and $b_L$ have been determined from the slope and intercept of plot between $C_e/q_e$ versus the $C_e$ and are listed in Table 3. The essential characteristics of Langmuir isotherm can be expressed by a dimensionless constant called equilibrium parameter $R_L$, which is defined by equation:

$$R_L = \frac{1}{(1 + b_L \cdot C_0)}$$

The value of $R_L$ indicated the type of Langmuir isotherm to be irreversible ($R_L = 0$), favourable ($0 < R_L < 1$), linear ($R_L = 1$), or unfavourable ($R_L > 1$). The $R_L$ was found to be 0.010, 0.009 and 0.007 for 50 mg L$^{-1}$ concentration of Rhodamine B dye at 308, 313 and 318 K temperatures, respectively, which indicates the favourable adsorption.

A comparison of adsorbent capacity of CMSW with other adsorbents (Table 2) shows that CMSW has a better adsorption capacity than others (Fig. 10).

*Freundlich adsorption isotherm*

Freundlich adsorption isotherm is an empirical adsorption isotherm describing the adsorption on heterogeneous surface. This isotherm does not predict any saturation of the adsorbent by the adsorbate, indicating multilayer adsorption.

Freundlich isotherm can be described by the equation given below [56]:

**Table 2** Comparison of adsorption capacities of different waste adsorbents for Rhodamine B removal

| Waste materials | Adsorption capacity (mg g$^{-1}$) | References |
|---|---|---|
| Fly ash | 2.330 | [46] |
| Iron chromium oxide (ICO) | 2.980 | [47] |
| Tamarind fruit shell Activated carbon | 3.940 | [48] |
| Coir pith | 2.560 | [49] |
| Raw orange peel | 3.230 | [50] |
| Natural diatomite | 8.130 | [51] |
| Mimusops Elengi activated carbon | 1.700 | [52] |
| Mango leaf powder | 3.310 | [30] |
| Pigeon dropping | 8.550 | [6] |
| Walnut shell | 1.541 | [32] |
| Coconut shell carbon | 2.330 | [35] |
| Akash Kinari coal | 1.183 | [53] |
| Mango leaf powder | 3.310 | [54] |
| Raw Flint Clay | 1.488 | [55] |
| Exhausted coffee ground powder | 5.255 | [7] |
| Paper industry waste sludge | 6.711 | Present study |

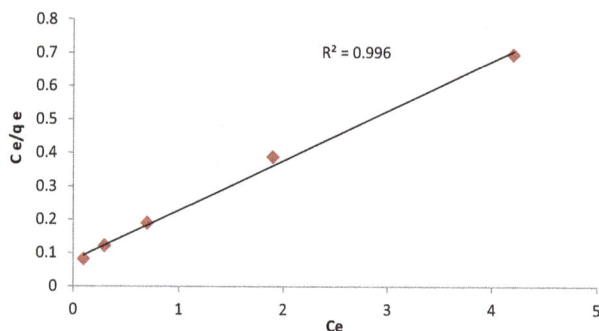

**Fig. 10** Langmuir adsorption isotherm for Rhodamine B adsorption at 308 K

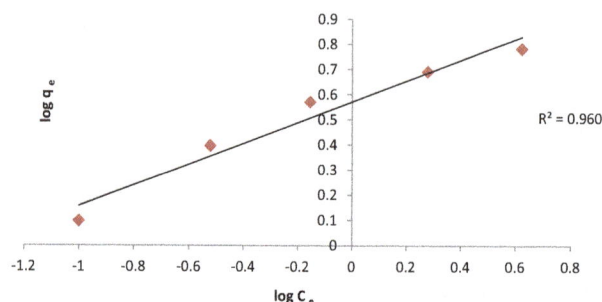

**Fig. 11** Freundlich adsorption isotherm for Rhodamine B adsorption at 308 K

$$\log q_e = \log K_f + \frac{1}{n}\log C_e$$

where $K_f$ is Freundlich constant and $1/n$ is the heterogeneity factor. It is evident from Fig. 11 that data fit well to Freundlich adsorption isotherm with regression coefficient $R^2 = 0.960$. The values of $K_f$ and $1/n$ have been calculated from intercept and slope of this straight line (listed in Table 3).

*Temkin adsorption isotherm*

The linear form of Temkin isotherm model is given by the following equation (by Temkin and Pyzhev)

$$q_e = B \ln K_T + B \ln C_e$$

where, $K_T$ and $B$ are the constants related to adsorption capacity and intensity of adsorption, respectively. A linear plot between $q_e$ verses $\ln C_e$ shows that adsorption follows Temkin isotherm. The values of $K_T$ and $B$ have been evaluated from slope and intercept of the plot (Fig. 12).

*Adsorption kinetics*

The data of adsorption of Rhodamine B dye has been applied to pseudo-first order, pseudo-second order and intra-particle diffusion models to determine the kinetics of adsorption process.

**Table 3** Langmuir, Freundlich and Temkin isotherms and their constants at different temperatures

| Temperature | Langmuir constants | | | | Freundlich constants | | | Temkin constants | | |
|---|---|---|---|---|---|---|---|---|---|---|
| Temp (K) | $q_m$ (mg/g) | $b_L$ (L/mg) | $R^2$ | $R_L$ | $n$ | $K_f$ | $R^2$ | $B$ | $K_T$ | $R^2$ |
| 308 | 6.711 | 1.886 | 0.996 | 0.01 | 2.40 | 3.715 | 0.960 | 2.962 | 24.871 | 0.998 |
| 313 | 6.757 | 2.145 | 0.992 | 0.009 | 2.43 | 3.971 | 0.980 | 2.916 | 30.683 | 0.990 |
| 318 | 6.757 | 2.552 | 0.985 | 0.007 | 2.53 | 4.236 | 0.994 | 2.763 | 43.531 | 0.972 |

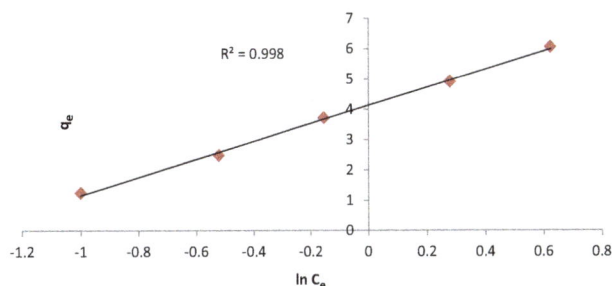

**Fig. 12** Temkin adsorption isotherm for Rhodamine B adsorption at 308 K

*Pseudo-first order kinetic model*

The data is subjected to Lagergren's first order equation. It has been found that it does not fit to straight line.

*Pseudo-second order kinetic model*

The integrated linear form of pseudo-second order kinetic model is given below [57]

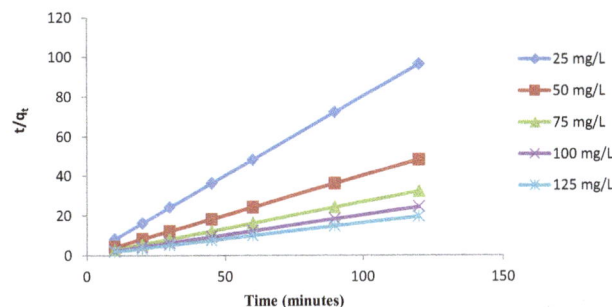

**Fig. 13** Pseudo-second order kinetics for Rhodamine B adsorption

$$\frac{t}{q_t} = \frac{1}{(K_2 q_e^2)} + \frac{1}{q_e}t$$

where $K_2$ is the pseudo-second order rate constant (g mg$^{-1}$ min$^{-1}$). For pseudo-second order kinetic model, the linear plot between $t/q_t$ verses $t$ shown in Fig. 13. The values of $K_2$ and $R_2$ have been calculated from the plot, which are represented in Table 4.

*Intra-particle diffusion model*

In adsorption process, the adsorbed species are most probably transported from the bulk of the solution into the solid phase through intra-particle diffusion, which is the rate limiting step. In addition, there is a possibility of the adsorbate to diffuse into the interior pores of the adsorbent. Weber and Morris proposed linear equation for intra-particle diffusion model, which is given in the following form [58]

$$q_t = K_{ipd}t^{1/2} + C$$

where $K_{ipd}$ is the intra-particle diffusion rate constant (mg g$^{-1}$ min$^{-1}$) and $C$ is the constant (mg g$^{-1}$). The intra-particle diffusion rate constant $K_{ipd}$ and $C$ have been calculated from the slope and intercept of the plot between $q_t$ verses $t^{1/2}$ which are listed in Table 4 (Fig. 14).

*Thermodynamic parameters*

Thermodynamic parameters, such as free energy change ($\Delta G$), enthalpy change ($\Delta H$) and entropy change ($\Delta S$) have important role for the determination of spontaneity and heat change of the adsorption process. The free energy

**Table 4** Pseudo-second order and intra-particle diffusion values for adsorption of Rhodamine B dye

| $C_0$ (mg L$^{-1}$) | Pseudo-second order calculated | | | Intra-particle diffusion parameters | | |
|---|---|---|---|---|---|---|
| | $K_2$ (g mg$^{-1}$ min$^{-1}$) | $q_e$ (mg/g) | $R^2$ | $K_{ipd}$ (mg g$^{-1}$ min$^{-1}$) | $C$ (mg g$^{-1}$) | $R^2$ |
| 25 | 7.1276 | 1.2484 | 1.000 | 0.001 | 1.234 | 0.733 |
| 50 | 1.344 | 2.5000 | 1.000 | 0.006 | 2.425 | 0.833 |
| 75 | 0.6989 | 3.7453 | 1.000 | 0.015 | 3.582 | 0.677 |
| 100 | 0.2992 | 4.9751 | 1.000 | 0.031 | 4.633 | 0.780 |
| 125 | 0.1184 | 6.2110 | 1.000 | 0.065 | 5.496 | 0.870 |

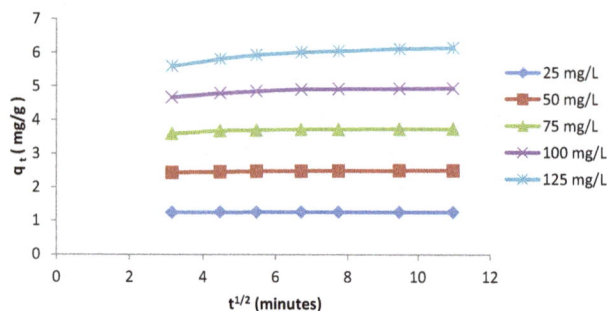

**Fig. 14** Intra-particle diffusion kinetics for Rhodamine B adsorption

**Table 5** Thermodynamic parameters for adsorption of Rhodamine B dye

| $C_o$ (mg L$^{-1}$) | $-\Delta G$ (KJ mol$^{-1}$) | | | $\Delta H$ (KJ mol$^{-1}$) | $\Delta S$ (KJ mol$^{-1}$K$^{-1}$) |
|---|---|---|---|---|---|
| | 308 K | 313 K | 318 K | | |
| 25 | 14.1287 | 15.1091 | 16.4252 | 52.12 | 0.2150 |
| 50 | 13.0849 | 13.5732 | 13.9947 | 13.88 | 0.0876 |
| 75 | 11.9452 | 12.3335 | 12.7442 | 11.70 | 0.0768 |
| 100 | 10.2198 | 10.2733 | 11.0634 | 17.83 | 0.0903 |
| 125 | 8.6012 | 9.3078 | 9.9829 | 31.46 | 0.1301 |

change ($\Delta G$) has been calculated from the thermodynamic equilibrium constant $K_0$ using the following equation:

$$\Delta G = -RT \ln K_o$$

where $R$ is the universal gas constant; $T$ is the absolute temperature in Kelvin.

Enthalpy change $\Delta H$ and entropy change $\Delta S$ have been evaluated from the Van't Hoff equation

$$\ln K_o = \Delta S/R - \Delta H/RT$$

$\Delta H$ and $\Delta S$ has been calculated from the slope and intercept of the plot between $\ln K_o$ verses $1/T$. Thermodynamic parameters obtained from the adsorption of Rhodamine B dye onto CMSW are given in Table 5. The negative value of $\Delta G$ confirms the feasibility of adsorption process. The increase of values of $\Delta G$ with temperature indicates that adsorption process is more favourable at higher temperatures, probably as a result of the increased mobility of dye species in solution. The positive values of $\Delta H$ confirm endothermic nature of adsorption process. The lower values of $\Delta S$ indicate that entropy decreases at solid–liquid interface.

## Conclusion

This study investigated the adsorption of a basic dye Rhodamine B onto CMSW as a function of adsorbent dose, initial dye concentration, pH and temperature. From the results it has been concluded that

1. CMSW is an efficient adsorbent for the removal of Rhodamine B dye in a smaller contact time, i.e., 60 min. Approximately 90% of dye has been removed within 10 min for each initial concentration.
2. The equilibrium adsorption data has been analysed by Langmuir, Freundlich and Temkin adsorption isotherms. The value of regression coefficient $R^2$ indicates that Langmuir, Freundlich and Temkin isotherms well describes the process. The monolayer adsorption capacity is 6.711, 6.757 and 6.757 mg g$^{-1}$ at 308, 313 and 318 K temperature.
3. Kinetic studies showed that data is best described by pseudo-second order kinetics with very good regression coefficient value equal to unity.
4. The negative value of thermodynamic parameters $\Delta G$ indicates the spontaneity, where as positive values of $\Delta H$ confirms the endothermic behaviour of the adsorption process. The low value of enthalpy change shows that it is a case of physio-sorption.
5. Desorption of the dye can be successfully carried out using different solvents. The adsorbent treated with hydrochloric acid show maximum recycling efficiency.

Taking into account the results of this study it has been concluded that CMSW can be considered as promising, eco-friendly adsorbent with low-cost production for the removal of dyes from wastewater in small time.

**Compliance with ethical standards**

**Conflict of interest** The authors declare that they have no competing interests.

## References

1. Kothari DD (2008) Rupee-value appreciation—calculating the crisis. Mod Text 3(1):26–29
2. Kant R (2012) Textile dyeing industry an environmental hazard. Nat Sci 4:22–26
3. Gonawala KH, Mehta MJ (2014) Removal of colour from different dye wastewater using ferric oxide as an adsorbent. Int J Eng Res Appl 4(5):102–109
4. Errais E, Duplay J, Darragi F (2010) Textile dye removal by natural clay—case study of Fouchana Tunisian clay. Environ Technol 31:373–380
5. Das SK, Ghosh P, Ghosh I, Guha AK (2008) Adsorption of Rhodamine B on *Rhizopus oryzae*: role of functional groups and cell wall components. J Colloids Surf B 65:30–34
6. Kaur H, Kaur R (2014) Removal of Rhodamine-B dye from aqueous solution onto Pigeon dropping: adsorption, kinetic,

equilibrium and thermodynamic studies. J Mater Environ Sci 5(6):1830–1838

7. Shen K, Gondal MA (2013) Removal of hazardous Rhodamine B dye from water by adsorption onto exhausted coffee ground. J Saudi Chem Soc. doi:10.1016/j.jscs.2013.11.005

8. Oliveira EGL Jr, Rodrigues JJ, de Olivrira HP (2011) Influence of surfactant on the fast photodegradation of Rhodamine B induced by $TiO_2$ dispersions in aqueous solution. Chem Eng J 172:96–101

9. Ahmad A, Setapar SHM, Chuonq CS, Khatoon A, Wani WA, Kumar A, Rafatullah M (2015) Recent advances in new generation dye removal technologies: novel search for approaches to reprocess wastewater. RSC Adv 5(39):30801–30818

10. Raghu S, Basha CA (2007) Chemical or electrochemical techniques followed by ion exchange, for recycle the textile dye wastewater. J Hazard Mater 149:324–330. doi:10.1016/j.hazmat.2007.03.087

11. Dragan ES, Dinu IA (2008) Removal of azo dyes from aqueous solution by coagulation/flocculation with strong polycations. Res J Chem Environ 12(3):5–11

12. Asl MK, Bahrami F (2014) Removal of vat dyes from coloured wastewater by reverse osmosis process. Bull Georg Natl Acad Sci 8(1):260–267

13. Wu J, Eiteman M, Law S (1998) Evaluation of membrane filtration and ozonation processes for treatment of reactive-dye wastewater. J Environ Eng 124(3):272–277

14. Babu BR, Parande AK, Kumar SA, Bhanu SU (2011) Treatment of dye effluent by electrochemical and biological processes. Open J Saf Sci Technol 1:12–18

15. Rajkumar K, Muthukumar M, Mangalaraja RV (2015) Electrochemical degradation of C.I. Reactive Orange 107 using Gadolinium ($Gd^{3+}$), Neodymium ($Nd^{3+}$) and Samarium ($Sm^{3+}$) doped cerium oxide nanoparticles. Int J Ind Chem 6:285–295

16. Gupta VK, Jain R, Nayak A, Agarwal S, Shrivastva M (2011) Removal of hazardous dye tartrazine by photodegradation on titanium dioxide surface. Mater Sci Eng C 31(5):1062–1067

17. Karthikeyan S, Gupta VK, Boopthy R, Titus A, Sekaran G (2012) A new approach for the degradation of high concentration of aromatic amine by heterocatalytic Fenton oxidation: kinetic and spectroscopic studies. J Mol Liquids 35:153–163

18. Kaur H, Thakur A (2014) Adsorption of Congo red dye from aqueous solution onto Ash of Cassia Fistula seeds: kinetic and thermodynamic studies. Chem Sci Rev Lett 3(11S):159–169

19. Saleh TA, Gupta VK (2014) Processing methods, characteristics and adsorption behaviour of tire derived carbons: a review. Adv Colloid Interface Sci 211:93–101

20. Hermawan AA, Bing T, Salamatinia B (2015) Application and optimization of using recycled pulp for Methylene Blue removal from wastewater: a response surface methodology approach. Int J Environ Sci Dev 6(4):267–274

21. Suteu D, Malutan T (2013) Industrial cellolignin waste as adsorbent for removal of Methylene blue dye from aqueous solution. Bioresources 8(1):427–446

22. Mittal A, Mittal J, Malviya A, Gupta VK (2010) Removal and recovery of Chrysoidine Y from aqueous solution by waste material. J Colloid Interface Sci 344:497–507

23. Mathivan M, Saranathan ES (2015) Sugarcane Bagasse—a low cost adsorbent for removal of Methylene Blue from aqueous solution. J Chem Pharm Res 7(1):817–822

24. Jain AK, Gupta VK, Bhatnagara A, Suhas S (2003) A comparative study of adsorbent prepared from industrial waste for removal of dyes. Sep Sci Technol 38(2):463–481

25. Vakili M, Rafatullah M, Salamatinia B, Abdullah AZ, Ibrahim MH, Tan KB, Gholami Z, Amouzgar P (2014) Application of chitosan and its derivatives as adsorbents for dye removal from water and wastewater: a review. Carbohydr Polym 113:115–130

26. Gupta VK, Mittal A, Jhare D, Mitaal J (2012) Batch and bulk removal of hazardous colouring agent Rose Bengal by adsorption technique using Bottom Ash as adsorbent. RCS Adv 2:8381–8389

27. Ahmad T, Danish M, Rafatullah M, Ghazali A, Sulaiman O, Hashim R, Nasir M, Ibrahim M (2012) The use of date palm as a potential adsorbent for wastewater treatment: a review. Environ Sci Pollut Res 19(5):1464–1484

28. Senthilkumar S, Kalaamani P, Subburaam CV (2006) Liquid phase adsorption of crystal violet onto activated carbon derived from male flowers of coconut tree. J Hazard Mater 136(3):800–808

29. Mittal A, Mittal J, Malviya A, Kaur D, Gupta VK (2010) Decoloration treatment of a hazardous triarylmethane dye, Light Green SF (Yellowish) by waste material adsorbents. J Colloid Interface Sci 342:518–527

30. Abdullah R, Ishak CF, Kadir WR, Bakar RA (2015) Characterization and feasibility assessment of recycled paper mill sludge for land application in relation to the environment. Int J Environ Res Public Health 12:9314–9329

31. Rangabhashiyam S, Selvaraju N (2015) Efficacy of unmodified and chemically modified *Swietenia Mahogani* shells for the removal of hexavalent chromium from simulated wastewater. J Mol Liquids 209:487–497

32. Shah J, Jan MR, Haq A, Khan Y (2013) Removal of Rhodamine B dye from aqueous solution and wastewater by Walnut shells: kinetic, equilibrium and thermodynamic studies. Front Chem Sci Eng 7(9):428–436

33. Kooh MRR, Dahri MK, Lim LBL (2016) The removal of Rhodamine B dye from aqueous solution using *Casuarina equisetifolia* needles as adsorbent. Cogent Environ Sci. doi:10.1080/23311843.2016.1140553

34. Jain R, Mathur M, Sikarwar S, Mittal A (2007) Removal of hazardous dye Rhodamine B through photocatalytic and adsorption treatments. J Environ Manag 85:956–964

35. Balasubramani K, Sivarajasekar N (2014) Adsorption studies of organic pollutants onto activated carbon. Int J Innov Res Sci Eng Technol 3(3):10575–10581

36. Santhi M, Kumar PE (2015) Adsorption of Rhodamine B from an aqueous solution: kinetic, equilibrium and thermodynamic studies. Int J Innov Res Sci Eng Technol 4:497–510

37. Sumanjit WT, Kansal I (2008) Removal of Rhodamine B by adsorption on walnut shell charcoal. J Surf Sci Technol 24(3–4):179–193

38. Rangabhashiyam S, Nakkeeran E, Anu N, Selvaraju N (2015) Biosorption potentials of a novel ficus auriculata leaves powder for the sequestration hexavalent chromium from aqueous solutions. Res Chem Intermed 41(11):8405–8424

39. Ponnusamy SK, Subramaniam R (2013) Process optimization studies of Congo red dye adsorption onto Cashew nut shell using Response surface methodology. Int J Ind Chem 4(17):2–10

40. Kavitha K, Sentamilselvi MM (2015) Removal of Malachite Green from aqueous solution using low cost adsorbent. Int J Curr Res Acad Rev 3(6):97–104

41. Mittal A, Kaur D, Malviya A, Mittal J, Gupta VK (2009) Adsorption studies on the removal of coloring agent phenol red from wastewater using waste materials as adsorbents. J Colloid Interface Sci 337:345–354

42. Venkatraman BR, Gayathri U, Elavarasi S, Arivoli S (2012) Removal of Rhodamine B dye from aqueous solution using the acid activated *Cynodondactylon* carbon. Der Chem Sin 3(1):99–113

43. Hema M, Arivoli S (2009) Rhodamine B adsorption by activated carbon: kinetic and equilibrium studies. Ind J Chem Technol 16:38–45

44. Mohan D, Singh KP, Singh G, Kumar K (2002) Removal of dyes from wastewater using Flyash, a low cost adsorbent. Ind Eng Chem Res 41:3688–3695

45. Langmuir I (1918) The adsorption of gases on plane surfaces of glass, mica and platinum. J Am Chem Soc 40:1361–1403

46. Khan TA, Imran A, Singh VV, Sharma S (2009) Utilization of flash as low-cost adsorbent for the removal of Methylene Blue, Malachite Green and Rhodamine B dyes from textile wastewater. J Environ Prot Sci 3:11–22

47. Kannan N, Murugavel S (2007) Column studies on the removal of dyes Rhodamine-B, Congo red and Acid violet by adsorption on various adsorbents. EJEAFChem 6:1860–1868

48. Vasu AE (2008) Studies on the removal of Rhodamine B and Malachite Green from aqueous solution by activated carbon. J Chem 5(4):844–852

49. Namasivayam C, Radhika R, Suba S (2001) Uptake of dyes by a promising locally available agriculture solid waste: coir pith. Waste Manag 21(4):381–387

50. Namasivayam C, Muniasamy N, Gayatri K, Rani M, Ranganathan K (1996) Removal of dyes from aqueous solution by cellulosic waste orange peel. Bioresour Technol 57:37–43

51. Koyuncu M, Kul AR (2014) Thermodynamics and adsorption studies of dye (Rhodamine B) onto natural diatomite. Physicochem Probl Miner Process 50(2):631–643

52. Gurunathan V, Gowthami P (2016) The effective removal of Rhodamine B dye by activated carbon (*Mimusops Elengi*) by adsorption studies. Int J Res Inst 3(21):575–581

53. Khan TA, Singh VV, Kumar D (2004) Removal of some basic dyes from artificial textile wastewater by adsorption onto Akash Kinari coal. J Sci Ind Res 863:355–364

54. Khan TA, Sharma S, Ali I (2011) Adsorption of Rhodamine B dye from aqueous solution onto acid activated Mango (*Magnifera indica*) leaf powder: equilibrium, kinetic and thermodynamic studies. J Tozicol Environ Health Sci 3(10):286–297

55. Kareem SH, Al-Hussien EABD (2012) Adsorption of Congo red, Rhodamine B and Disperse blue dyes from aqueous solution onto Raw Flint Clay. Baghdad Sci J 9(4):680–688

56. Freundlich H, Hellen W (1993) The adsorption of *cis*- and *trans*-azobenzene. J Am Chem Soc 61:2–28

57. Ho YS, McKay G (1999) Pseudo-second order model for sorption processes. Process Biochem 34:451–465

58. Weber WH, Morris JC (1963) Kinetics of adsorption on carbon from solution. J Sanit Eng Div Am Soc Civ Eng 89(2):31–60

# 18

# Adsorption of hydrogen sulphide over rhodium/silica and rhodium/alumina at 293 and 873 K, with co-adsorption of carbon monoxide and hydrogen

Claire Gillan[1] [iD] · Martin Fowles[2] · Sam French[2] · S. David Jackson[1]

**Abstract** In this study, we have examined the adsorption of hydrogen sulphide and carbon monoxide over rhodium/silica and rhodium/alumina catalysts. Adsorption of hydrogen sulphide was measured at 293 and 873 K and at 873 K in a 1:1 ratio with hydrogen. At 293 K, over Rh/silica, hydrogen sulphide adsorption capacity was similar to that of carbon monoxide; however, over Rh/alumina, the carbon monoxide adsorption capacity was higher, probably due to the formation of $Rh^1(CO)_2$. Over Rh/silica, the primary adsorbed state was HS(ads), in contrast to Rh/alumina, where the $H_2$:S ratio was 1:1 indicating that the adsorbed state was S(ads). Competitive adsorption between CO and $H_2S$ over Rh/silica and Rh/alumina revealed adsorption sites on the metal that only adsorbed carbon monoxide, only adsorbed hydrogen sulphide or could adsorb both species. At 873 K, hydrogen sulphide adsorption produced the bulk sulphide $Rh_2S_3$; however, when a 1:1 $H_2$:$H_2S$ mixture was used formation of the bulk sulphide was inhibited and a reduced amount of hydrogen sulphide was adsorbed.

**Keywords** Rhodium · Hydrogen sulphide · Carbon monoxide · Adsorption

✉ S. David Jackson
david.jackson@glasgow.ac.uk

[1] Centre for Catalysis Research, WestCHEM, School of Chemistry, University of Glasgow, Glasgow G12 8QQ, Scotland, UK

[2] Johnson Matthey Plc, Belasis Ave, Billingham TS23 1LB, UK

## Introduction

Catalyst poisoning is the strong chemisorption of a species on a site otherwise available for catalysis. Whether a species is a poison depends upon its adsorption strength relative to other species competing for active sites. For many reactions, such as methanation, methanol synthesis and Fischer–Tropsch synthesis, over group 8–11 metal catalysts, sulphur is a known poison [1]. To be able to interpret quantitatively the extent and nature of poisoning by sulphur, it is essential to know the structure and bonding of sulphur to metal atoms at the surface. There are two types of sulphides that form on a catalyst, 2-D surface sulphides and 3-D bulk sulphides, the formation of which requires the metal cation to diffuse through the adsorbed sulphide layer [2] forming a new metal sulphide layer on the outer surface. This phenomenon of segregation is strongly exothermic and is therefore favoured by a reduction in temperature. Surface sulphide formation is simply the adsorption of sulphur on the surface of the metal. Pt, Ni, Ru and Rh all have lower free energies of formation of their bulk sulphides than their surface sulphides, hence, large hydrogen sulphide concentrations are required for stable bulk sulphides to exist.

It has also been inferred that an SH surface species is present as an intermediate in the dissociation of hydrogen sulphide. For example, over $Pt/Al_2O_3$, it was observed that at increasing sulphur coverages, dissociated hydrogen is gradually desorbed but that a percentage spends a significant lifetime on the catalyst [3], and can participate in reactions. Also, on Pt/alumina two types of adsorbed hydrogen sulphides were detected, different due to strengths of adsorption and three different adsorption sites. These include: a site which bonds sulphur strongly and will not exchange, a site which bonds sulphur weakly and is removed under vacuum and a site which will allow exchange between gas and adsorbed phases. These were determined from radioactive labelling experiments [4], in which it was also found that the S:Pt$_{(surface)}$ ratio was 1:1 on Pt/SiO$_2$ but only 0.6:1 on Pt/Al$_2$O$_3$. There has been limited research on the adsorption of sulphur species on Rh catalysts; however, some work has been conducted on Rh single crystal faces. Hedge and White [5] studied the chemisorption and decomposition of H$_2$S on Rh(100) at 100 K using auger electron spectroscopy (AES) and obtained results suggesting a saturation coverage near 0.5 monolayer. On heating to 600 K, the sulphur coverage increased, which the authors inferred was due to physisorbed H$_2$S, which is consistent with results for H$_2$S adsorption on Pt and Ni. The thermal desorption spectra of molecular H$_2$S from Rh(100) exhibit low- and high-temperature peaks, hence, Hedge and White [5] assigned the low temperature peak as physisorbed H$_2$S. It was also found that a decreasing fraction of H$_2$S dissociated as the coverage of H$_2$S increased [5]. The similarities between H$_2$S adsorption on Rh(100) and Pt(111), Ru(110) and Ni(100) were noted. In all these cases, there is complete dissociative adsorption at low temperatures and low coverages with hydrogen remaining on the surface. At low temperatures and higher coverages on Pt(111), Ru(110) and Ni(100), first SH and then H$_2$S were observed.

In catalytic systems, the sulphur compound is not there in isolation, rather it is there coincidentally with the reactants. However, studies where competitive adsorption has been examined are rare and over rhodium rarer still. The interaction of CO and H$_2$S over supported Pt catalysts was studied in detail by Jackson et al. [6], who found that when H$_2$S was pre-adsorbed on Pt/silica no subsequent CO adsorption was detected due to the adsorption of H$_2$S being dissociative, so there was no mechanism by which sulphur could desorb, and hence no sites can be liberated for CO adsorption. When CO was pre-adsorbed on Pt/silica, the amount of H$_2$S adsorbed was reduced by 81% in comparison to a fresh surface, though it was suggested that 20% of the H$_2$S was able to adsorb onto the silica support, indicating that CO had completely suppressed H$_2$S adsorption on the Pt sites. However, when the same experiment was carried out over Pt/alumina [6], there was no reduction in adsorptive capacity for H$_2$S on a CO saturated surface indicating that CO did not block H$_2$S adsorption on the metal sites, and therefore must be related to the effect of the support. It has previously been reported that CO$_2$ is produced from the reaction of adsorbed CO with hydroxyl groups from the alumina support [6], therefore, it was proposed that CO may be able to desorb via this route liberating sites for H$_2$S adsorption. When CO and H$_2$S were co-fed over Pt/silica the amount of H$_2$S adsorbed decreased by 78%, whereas the amount of CO adsorbed increased by 67%. The enhancement in CO adsorption was explained by

the adsorption of H$_2$S and its displacement by CO. This caused desorption of residual hydrogen from the reduction procedure, possibly by surface reconstruction, which had been found to have a deleterious effect on CO adsorption [6].

A similar study examining the interaction of CO and H$_2$S over Rh/silica catalysts was carried out [7], but unlike Pt, it was found that CO could adsorb onto samples that had been saturated with sulphur. Displacement of H$_2$S was also evident, but this was dependent on the metal precursor used. It was only found to occur on an oxide-derived catalyst, and since the desorption of sulphur requires hydrogen, it was proposed that H$_2$S only partially dissociates on the oxide catalyst to produce an HS-* species, which would provide a source of hydrogen to allow for desorption [7]. The effect of passing H$_2$S over CO pre-covered surfaces was the displacement of CO and the adsorption of H$_2$S, i.e. similar to Pt/alumina. It was speculated that the CO displaced, reflected the different modes of adsorbed CO, and this was also found to be dependent on the metal precursor used. For example, a chloride-derived catalyst appeared to displace bridge-bonded Rh$_2$–CO [7].

This work follows on from a study over Pt/alumina and Pt/silica where the adsorption of hydrogen sulphide was examined in relation to its effect on steam reforming [8]. In this study, the adsorption of hydrogen sulphide and methanethiol over Rh/silica and Rh/alumina at 293 K will be examined and the amount compared with that found from carbon monoxide chemisorption. In this way, the total amount of hydrogen sulphide adsorption can be measured and related to the amount of surface rhodium. Competitive adsorption of carbon monoxide and hydrogen sulphide will be examined as it is rare that a poison is present in the absence of another species (reactant or product). By comparing the competitive adsorption, an assessment can be made as to the relative strength of adsorption between carbon monoxide and hydrogen sulphide. The extent hydrogen sulphide may adsorb in the presence of carbon monoxide can give an insight into sulphur poisoning of carbon monoxide hydrogenation over rhodium. The adsorption of hydrogen sulphide will be examined at high temperature (873 K) in the absence and presence of hydrogen, to understand whether hydrogen is effective at reducing the amount of sulphur adsorbed. Le Chatelier's principle indicates that the amount of sulphur adsorbed should be reduced; however this will be the first experimental verification. This will inform our understanding of sulphur poisoning in systems such as steam reforming.

## Experimental

Two catalysts were prepared on alumina and silica, 1.2% w/w Rh/alumina and 1% Rh/silica. Both catalysts were prepared by incipient wetness of the two supports (θalumina, surface area 101 m$^2$ g$^{-1}$; silica, surface area 220 m$^2$ g$^{-1}$) using Rh(NO$_3$)$_3$ hydrate (Aldrich) as the precursor salt. The precursor salt was dissolved in a volume of distilled water equal to the support pore volume (0.6 cm$^3$ g$^{-1}$ for alumina and 1 cm$^3$ g$^{-1}$ for silica) using 100 g of support. The catalysts were dried and calcined at 773 K for 4 h. The rhodium weight loading was confirmed by atomic adsorption.

Chemisorption studies were performed in a dynamic mode using a pulse-flow microreactor system in which the catalyst sample was placed on a sintered glass disc in a vertical tube (8 mm id, down flow) inside a furnace (Fig. 1).

The reactant pulses were introduced into the gas stream immediately before the catalyst bed using a fixed volume sample loop. Using this system, the catalysts (typically 0.50 g) were reduced in situ in a flow of hydrogen (40 cm$^3$ min$^{-1}$) by heating to 673 K at a rate of 10 K min$^{-1}$. The catalyst was held at this temperature for 2 h. The catalyst was then purged with argon (40 cm$^3$ min$^{-1}$) for 30 min and the catalyst was cooled in flowing argon to 293 K. The adsorbate gases were admitted by injecting pulses of known size (typically 24 μmol) into the argon carrier-gas stream, and hence onto the catalyst. The residence time of the pulse in the catalyst bed was ~1.5 s. In all cases, the whole pulse was analysed by on-line GC. For co-adsorption studies, the gases were mixed in the gas manifold prior to injection into the carrier gas. The amount of gas adsorbed, from any pulse, was determined from the difference between the peak area of a calibration pulse sent directly to the GC from the sample volume, and the peak area obtained following the injection of pulses of comparable size onto the catalyst. The detection limit for adsorption was 0.3 μmol g$^{-1}$. Adsorption measurements

were typically repeated three times and the values reported are the average. Standard deviation on the amount adsorbed was typically less than ±8%. Adsorptions were followed using a gas chromatograph fitted with a thermal conductivity detector and Molecular Sieve 5A and Porapak Q columns.

Both the helium (BOC, 99.997%) and the 5% hydrogen in dinitrogen (BOC) were further purified by passing through Chrompack Gas-Clean Oxygen filter to remove any oxygen impurity, and a bed of Chrompack Gas-Clean Moisture filter to remove any water impurity. Carbon monoxide (99.99% Research Grade) and hydrogen sulphide (>99%) were used as received.

## Results

Carbon monoxide and hydrogen sulphide adsorption was examined over the high weight-loading rhodium catalysts. As described in the "Experimental" section, multiple pulses of each gas were passed over the catalysts until no further adsorption was detected. Using this methodology, the pressure of the pulse is always 1 bar and only strongly bound species are detected. As expected, no adsorption of carbon monoxide was detected on the alumina or silica supports in the absence of the metal component. Carbon monoxide adsorption gave 143.8 μmol g$^{-1}$ for Rh/alumina and 65.1 μmol g$^{-1}$ for Rh/silica (this translates to metal dispersions of 123% for Rh/alumina and 67% for Rh/silica assuming a Rh:CO ratio of 1:1). The silica support did not adsorb hydrogen sulphide, but hydrogen sulphide did adsorb on the alumina, hence, the adsorption data for the Rh/alumina catalyst has had the support contribution subtracted from the total adsorption. The hydrogen sulphide adsorption data are reported in Table 1.

The adsorption of hydrogen sulphide and carbon monoxide was also studied by co-adsorption, carbon monoxide pre-adsorbed before hydrogen sulphide and vice versa. The results are shown in Table 2. For co-adsorption, pulses of the mixed gases at a 1:1 ratio were passed over the catalyst. For the sequential adsorptions, one gas was

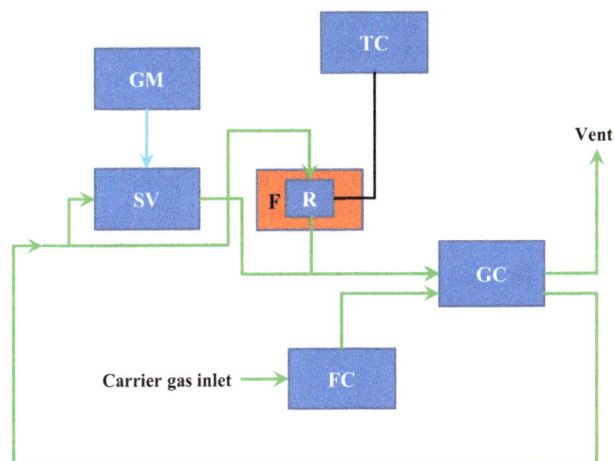

Fig. 1 Schematic of pulse-flow microreactor. *GM* gas manifold, *SV* standard volume, *F* furnace, *R* reactor, *TC* temperature controller, *FC* flow controller, *GC* gas chromatograph with a thermal conductivity detector

Table 1 Hydrogen sulphide adsorption at 293 K

| Catalyst | H$_2$S adsorbed[a] | H$_2$:S(ads)[b] | Dispersion[c] (%) |
|---|---|---|---|
| Rh/alumina | 112.7 | 1.0:1 | 97 |
| Rh/silica | 61.2 | 0.7:1 | 63 |

[a] μmol g$^{-1}$
[b] H$_2$ evolved during adsorption relative to H$_2$S adsorbed
[c] Calculated on the basis of the hydrogen sulphide adsorption, assuming a 1:1 S:Rh

**Table 2** Sequential and co-adsorption of carbon monoxide and hydrogen sulphide at 293 K ($\mu$mol g$^{-1}$)

**Rh/alumina**

| H$_2$S pre-adsorbed | CO adsorbed after H$_2$S adsorption | | |
|---|---|---|---|
| 112.7 | 14.7 | | |
| CO pre-adsorbed | H$_2$S adsorbed after CO adsorption | | H$_2$:S(ads)[a] |
| 143.8 | 25.0 | | 0.3:1 |
| Co-adsorption | CO adsorbed | H$_2$S adsorbed | H$_2$:S(ads) |
| | 64.9 | 30.1 | 0.3:1 |

**Rh/silica**

| H$_2$S pre-adsorbed | CO adsorbed after H$_2$S adsorption | | |
|---|---|---|---|
| 61.2 | 7.7 | | |
| CO pre-adsorbed | H$_2$S adsorbed after CO adsorption | | H$_2$:S(ads) |
| 65.1 | 9.0 | | 0.6:1 |
| Co-adsorption | CO adsorbed | H$_2$S adsorbed | H$_2$:S(ads) |
| | 21.3 | 17.4 | 0.7:1 |

[a] H$_2$ evolved during adsorption relative to H$_2$S adsorbed

**Table 3** Hydrogen sulphide adsorption at 873 K

| Catalyst | H$_2$S adsorbed[a] | H$_2$:S(ads)[b] | Rh:S(ads)[c] |
|---|---|---|---|
| Rh/alumina | 166.5 | 0.8:1 | 1:1.4 |
| Rh/silica | 152.5 | 1.0:1 | 1:1.5 |

[a] $\mu$mol g$^{-1}$

[b] Ratio of hydrogen evolved to sulphur adsorbed

[c] Ratio of Rh to sulphur adsorbed

**Table 4** Hydrogen sulphide adsorption from a 1:1 H$_2$S:H$_2$ mixture at 873 K

| Catalyst | H$_2$S adsorbed[a] | H$_2$:S(ads)[b] | Rh:S(ads)[c] |
|---|---|---|---|
| Rh/alumina | 90.4 | 0.7:1 | 1:0.8 |
| Rh/silica | 98.3 | 0.8:1 | 1:1.0 |

[a] $\mu$mol g$^{-1}$

[b] Ratio of hydrogen evolved to sulphur adsorbed

[c] Ratio of Rh to sulphur adsorbed

adsorbed to saturation (so if CO was pre-adsorbed the catalyst would be saturated with CO) before the second gas was passed over the catalyst.

The adsorption of hydrogen sulphide was also studied at 873 K in the absence and presence of hydrogen and the results are shown in Tables 3 and 4, respectively. When the hydrogen sulphide was adsorbed in the presence of hydrogen, the ratio of the mixture was 1:1. The amount of hydrogen sulphide adsorbed on the alumina at 873 K has been subtracted from the Rh/alumina adsorption in Tables 3 and 4. No adsorption took place on the silica.

## Discussion

The adsorption of hydrogen sulphide on rhodium has only been studied sparingly [5, 7, 9, 10]. Nevertheless, there is good agreement about what is expected from hydrogen sulphide adsorption at room temperature over highly dispersed Rh/silica catalysts. Our value of 0.6:1 S:Rh is typical for sulphur adsorption on Rh/silica [7, 9], as is the $\sim$1:1 correspondence between hydrogen sulphide adsorption and carbon monoxide adsorption. Carbon monoxide adsorption over rhodium can be described by three adsorbed states, linear Rh–CO, bridge-bonded Rh$_2$–CO, and gem-dicarbonyl Rh$^I$(CO)$_2$. The ratio of CO:Rh obtained with the silica catalyst suggests principally bridged and linear sites, which implies a similar bonding model for sulphur. This is in accordance with the work of Sachtler et al. [10]. Over Rh/alumina, a S:Rh ratio of $\sim$1:1 is obtained, which is higher than that observed with the silica catalyst. This is the opposite of what is found with carbon monoxide adsorption and hydrogen sulphide adsorption over Pt/alumina, where the alumina-supported catalyst gave a lower ratio [8]. The greater than 1:1 CO:Rh ratio was expected over the alumina supported catalyst, as it is known that the gem-dicarbonyl Rh$^I$(CO)$_2$ species can be formed leading to a greater than 1:1 ratio. Interestingly, the hydrogen sulphide adsorption follows the same trend as the carbon monoxide adsorption. It is normally assumed that sulphur occupies a multiply bonded site such as a threefold hollow [10] on a Rh(111) face, but the commonality in adsorption ratios between carbon monoxide and hydrogen sulphide suggests that on highly dispersed supported metals there may be a number of adsorption modes.

Further information on the mode of hydrogen sulphide adsorption can be obtained from the H$_2$:S ratio. The H$_2$:S ratio over the silica catalyst was 0.7:1, this implies that two adsorbed states are present, S(ads) and HS(ads), and we can obtain a ratio of these adsorbed states from a stoichiometric equation:

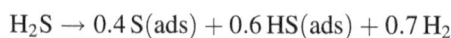

$$H_2S \rightarrow 0.4\,S(ads) + 0.6\,HS(ads) + 0.7\,H_2$$

Therefore, over Rh/silica, the primary adsorbed state is HS(ads), which is not the case for Rh/alumina, where the H$_2$:S ratio is 1:1 indicating that the only adsorbed state is S(ads) (Tables 1, 2). This is different from the Pt/alumina and Pt/silica systems where both catalysts had a H$_2$:S ratio of $\sim$0.7:1 [8], indicating that dissociation is enhanced over Rh/alumina.

When carbon monoxide was pre-adsorbed over the Rh/silica at 293 K and then hydrogen sulphide adsorbed over the same catalyst, the amount of hydrogen sulphide adsorbed was reduced by 85%. There was no evidence of carbon monoxide being displaced suggesting that the hydrogen sulphide was accessing sites that were unavailable to carbon monoxide. The extent of hydrogen evolution indicates that the mode of hydrogen sulphide adsorption had changed, such that 80% of the adsorbed sulphur was in the form HS(ads). When hydrogen sulphide was pre-adsorbed and then carbon monoxide adsorbed over the same catalyst, the behaviour observed was similar to the CO/$H_2$S couple, but with the carbon monoxide adsorption reduced by 88%. The two sequential adsorption experiments suggest that there are three sites on the surface, one that adsorbs both carbon monoxide and hydrogen sulphide, one that adsorbs only hydrogen sulphide and one which only adsorbs carbon monoxide. These results are in accordance with the literature [7], where multiple sites were also proposed. However, when we consider the co-adsorption experiment, we find a much lower total adsorption (Table 2). The nature of the hydrogen sulphide adsorption appears similar in that the amount of hydrogen liberated during the adsorption is the same as for the single-gas adsorption. Nevertheless, the adsorption capacity has reduced. It is not clear why such a reduction takes place although both adsorbates may cause restructuring of the metal crystallites.

The sequential and competitive adsorptions over Rh/alumina are similar to those found with Rh/silica confirming the absence of a support effect and reinforcing the validity of the results. Once again, under co-adsorption the total amount of gas adsorbed is reduced (Table 2). As has been noted for the Rh/silica catalyst, both adsorbates can cause restructuring; however, when hydrogen and carbon monoxide were co-adsorbed [12] above 373 K, the hydrogen inhibited the formation of $Rh^I(CO)_2$ but enhanced sintering. This behaviour at least suggests a mechanism, whereby the total amount adsorbed could be reduced. Also, in the co-adsorption, there is a significant change in the amount of hydrogen retained by the adsorbed hydrogen sulphide. When hydrogen sulphide is adsorbed first, the amount of hydrogen desorbed is such that the adsorbed species is S(ads). However, when hydrogen sulphide is co-adsorbed with carbon monoxide, or when it is adsorbed after carbon monoxide pre-adsorption, the adsorbed sulphur retains $\sim$2/3rd of the hydrogen, altering the adsorbed state to principally HS(ads) with the remainder adsorbed as non-dissociated hydrogen sulphide:

$$H_2S \rightarrow 0.4\,H_2S(ads) + 0.6\,HS(ads) + 0.3\,H_2$$

Therefore, the carbon monoxide appears to inhibit hydrogen sulphide dissociation.

It is noticeable that the amount of hydrogen sulphide adsorbed at 873 K is greater than that adsorbed at 293 K, which is the reverse of what would be expected. However, the adsorption of hydrogen sulphide at 873 K (Table 3) reveals a higher degree of dissociation approximating to the loss of all hydrogen and a S:Rh ratio of 1.4:1 for Rh/

alumina and 1.5:1 for Rh/silica. This suggests the formation of the bulk sulphide, $Rh_2S_3$. When hydrogen was co-fed with hydrogen sulphide at 873 K (Table 4), the amount of sulphur adsorbed decreased significantly (S:Rh $\sim$ 1:1) and the degree of dissociation also decreased, but not by as much as found with platinum catalysts, where the $H_2$:S(ads) ratio decreased from 1:1 to 0.44 for Pt/alumina and 0.25 for Pt/silica [8]. This is to be expected, as we are now displacing the following equilibria to the left-hand side:

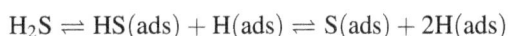

$$H_2S \rightleftharpoons HS(ads) + H(ads) \rightleftharpoons S(ads) + 2H(ads)$$

However, the extent of the displacement will be dependent upon thermodynamic factors, and under these conditions $PtS_2$ is unstable, so we may expect a larger move to the left. Nevertheless, it is clear that the addition of hydrogen was sufficient to inhibit the formation of the bulk sulphide and brings the system back closer to an adsorbed state. Indeed for Rh/silica, the adsorption gives $H_2S \rightarrow 0.6\,S(ads) + 0.4\,HS(ads) + 0.8\,H_2$, which is similar to that found for room temperature adsorption.

## Conclusions

In this study, we have examined the adsorption of hydrogen sulphide over rhodium/silica and rhodium/alumina catalysts at 293 and 873 K. At 293 K, over Rh/silica, hydrogen sulphide adsorption capacity was similar to that of carbon monoxide; however, over Rh/alumina, the carbon monoxide adsorption capacity was higher, probably due to the formation of $Rh^I(CO)_2$. Over Rh/silica, the primary adsorbed state was HS(ads), which was not the case for Rh/alumina, where the $H_2$:S ratio was 1:1 indicating that the adsorbed state was S(ads). Sequential adsorption between CO and $H_2$S over Rh/silica and Rh/alumina at 293 K revealed adsorption sites on the metal that only adsorbed carbon monoxide, only adsorbed hydrogen sulphide, or could adsorb both species. Co-adsorption of carbon monoxide and hydrogen sulphide resulted in a much reduced total adsorption (40% for Rh/silica, 33% for Rh/alumina). The reason for this is not clear, but may relate to restructuring/sintering of the systems. At 873 K, hydrogen sulphide adsorption produced the bulk sulphide $Rh_2S_3$; however, when a 1:1 $H_2$:$H_2$S mixture was used, formation of the bulk sulphide was inhibited. The Rh:S ratio was reduced to $\sim$1:1, and for Rh/silica a significant amount of HS(ads) was identified.

## References

1. Oudar J (1980) Sulfur adsorption and poisoning of metallic catalysts. Catal Rev 22:171–195
2. Bartholomew CH, Agrawal PK, Katzer JR (1982) Sulfur poisoning of metals. Adv Catal 31:135–242

3. Jackson SD, Leeming P, Grenfell J (1994) The effect of sulphur on the non-steady state reaction of propane over a platinum/alumina catalyst at 873 K. J Catal 150:170–176

4. Jackson SD, Leeming P, Webb G (1996) Supported metal catalysts; preparation, characterisation and function. Part IV. Study of hydrogen sulphide and carbonyl sulphide adsorption on platinum catalysts. J Catal 160:235–243

5. Hegde RI, White JM (1986) Chemisorption and decomposition of $H_2S$ on Rh(100). J Chem Phys 90:296–300

6. Jackson SD, Willis J, McLellan GD, Webb G, Keegan MBT, Moyes RB, Simpson S, Wells PB, Whyman R (1993) Supported metal catalysts; preparation, characterisation and function. Part II. Carbon monoxide and dioxygen adsorption on platinum catalysts. J Catal 139:207–220

7. Jackson SD, Brandreth BJ, Winstanley D (1987) The effect of hydrogen sulphide on the adsorption and thermal desorption of carbon monoxide over rhodium catalysts. J Chem Soc Faraday Trans 1 83:1835–1842

8. Gillan C, Fowles M, French S, Jackson SD (2013) Ethane steam reforming over a Pt/alumina catalyst: effect of sulphur poisoning. Ind Eng Chem Res 52:13350–13356

9. Laosiripojana N, Assabumrungrat S (2011) Conversion of poisonous methanethiol to hydrogen-rich gas by chemisorption/reforming over nano-scale $CeO_2$: the use of $CeO_2$ as catalyst coating material. Appl Catal B 102:267–275

10. Konishi Y, Ichikawa M, Sachtler WMH (1987) Hydrogenation and hydroformylation with supported rhodium catalysts. Effect of adsorbed sulfur. J Phys Chem 91:6286–6291

11. Rufael TS, Koestner RJ, Kollin EB, Salmeron M, Gland JL (1993) Adsorption and thermal decomposition of CH, SH on the Pt(111) surface. Surf Sci 297:272–285

12. Solymosi F, Pasztor M (1986) Infrared study of the effect of $H_2$ on CO-induced structural changes in supported Rh. J Phys Chem 90:5312–5317

# 19

# Rejection of far infrared radiation from the human body using Cu–Ni–P–Ni nanocomposite electroless plated PET fabric

Amir Khalili[1] · Amirreza Mottaghitalab[2] · Mahdi Hasanzadeh[1] · Vahid Mottaghitalab[1]

**Abstract** An experimental investigation was utilized to present the IR rejection performance of Cu–Ni–P–Ni composite double-layer electroless plated PET fabric compared to fabric samples composed of Cu–Ni–P metallic monolayer. Accordingly, the effect of a range of operational parameters was explored on the conductivity of electroless plated PET fabric. Results indicate higher conductivity and lower durability for Cu–Ni–P-coated samples compared to its counterpart with same sub-layer included with nickel on top layer. The SEM image of Cu–Ni–P particle on PET fabric shows a hexagonal non-homogenous morphology with nanoscale crack on its surface. However, the micrograph of the Cu–Ni–P–Ni electroless plated fabric shows an extremely compact and continuous phase with clear boundaries containing semispherical particles. The thermopile radiated sensing system was used as a sophisticated device to show the thermal energy absorption level. The acquired data indicate a 2.3 and 2.7 unit reduction in transmitted radiated power, respectively, for Cu–Ni–P and Cu–Ni–P–Ni conducting fabric. The captured thermal camera images of human body while keeping in front of a Cu–Ni–P conducting fabric revealed a nearly black and blue feature which proves the significant decrease in body radiated thermal energy. However, the thermal image of Cu–Ni–P–Ni conductive fabric shows almost black appearance in all areas. It can be presumably due to improving of the IR rejection performance and also formation of a massive barrier against body thermal radiation for promising camouflage applications.

**Keywords** Electroless plating · Nickel · Copper · Fabric · IR rejection

## Introduction

Deposition of metallic layers on fabric substrates in recent years has been area of interest for diverse applications such as electronic textiles and electromagnetic interference (EMI) shielding [1, 2]. The metalized textiles also are a preferred choice for light and radiant heat protection according to high surface reflection [3, 4]. The protection against intense radiant heat in short duration was reported for technical clothing [3]. This category of protective clothing is generally applicable for firefighter and others who exposed to high temperatures including workers in steel mills [5, 6]. Also, it can be mentioned that the metalized surface reflects solar heat during the summer and radiant heat back into the room during the winter. The textile metallization can also significantly improve the electrical conductivity and antistatic properties through conducting electricity and reducing static shock on insulated textiles [7, 8]. Moreover, the electromagnetic interference (EMI) is becoming the fourth kind of public space pollution in addition to noise, water and air pollution [1, 9]. The EMI is a well-known problem for commercial and scientific electronic instruments, antenna systems and military electronic devices. The high EMI shielding effectiveness (SE) of a typical metal is due to its high reflectance [10–12]. The previously published study indicated the high level of EMI SE of 80–90 dB for copper-plated PET fabric implying that these metallized fabrics

✉ Vahid Mottaghitalab
motaghitalab@guilan.ac.ir

[1] Textile Engineering Department, Faculty of Engineering, The University of Guilan, P. O. BOX 3756, Rasht, Guilan, Iran

[2] Chemical Engineering Faculty, Sahand University of Technology, Tabriz, Iran

can be used in the advanced electronic products and defense applications [13]. In the meantime, based on the particular specifications of the metalized textiles, promising practical potential as thermal camouflage has been reported in the literature [14, 15]. The infrared reflective materials composed of metal elements categorize in periodic table groups such as IB(Cu, Ag, Au), IIB(Zn, Cd), IIA(Be, Mg), IVA(C, Si, Ge), VIIB(Mn) and VIIIB(Fe, Ni, Co). These materials act as IR reflective layer on most substrates including textile, polymeric films and some other flexible sub-layers. In most reflection applications, it is preferred to adopt a range of total thickness between 10–50 μm and a conductive layer of 100–300 nm [16]. The clotting coated with metallic material might be a powerful reflective surface against human thermal emission detectable by thermal cameras [17, 18]. The thermal camera is a passive system working based on body reflection in middle and infrared range. Such a system works by contrasting between target IR signature and its surrounding in a wavelength range between 0.8 and 14 μm with temperature sensitivity of a few centigrade [19]. The embedded IR sensor also detects the contrast between target thermal radiation and its own background, which extremely depends on fabric optical properties. Therefore, the control of surface optical properties through morphology manipulation and conductivity adjusting plays a key role for IR rejection performance [18, 20]. In general, the contrast temperature difference is limited to 4–5 °C that needs to preserve for any atmosphere. The human skin radiates the infrared spectra in two wavelength including 32% in spectral band of 8–12 μm and 1% in spectral band of 3.3–4.8 μm. The thermal camouflage prohibits the thermal leakage but possibly creates thermal shock for wearer and reduces the overall performance. One of the main routes to remove deficiency is using the low emissivity and high reflectivity covering for prohibition of thermal radiation. The fabric finishing through lamination of a conductive layer or pattern can be a promising potential candidate in commercial bulk production. Accordingly, the variety of metalizing techniques on textile substrate was utilized to promote their properties for technical applications. The metal coating on fabric substrate is possible to be carried out using range of techniques including spraying of conductive paint [21], sputter coating [22] and electroless plating [13, 22–25]. Among the cited coating techniques, electroless plating is of special interest due to its advantages such as uniform and coherent metal deposition, good electrical conductivity and efficient heat transport. It can also be applied to almost all fiber made substances at any form of textile products such as yarn, stock, fabric or clothing. The electroless deposition method uses a catalytic redox reaction between metal ions and dissolved reduction agent [24]. For instance, the electroless

copper plating is obtained through the usage of conventional sensitization and activation steps involving expensive activators such as palladium chloride (PdCl2). The great performance of Cu–Ni–P alloy electroless plated fabric on cotton fabrics makes it as an essential research area in textile finishing process. However, the oxidation process plays a disadvantageous role on conductivity and reflectivity versus time. The empirical observation reveals a significant decrease in conductivity upon exposure to atmosphere [13]. Therefore, an antioxidation layer with negligible impact on conductivity and other physical properties is extremely necessary to empower the environmental resilience of copper-coated textiles. The nickel substance is a most abundant metal potentially applicable as coating material because of its excellent corrosion and wear resistance, hardness, reflectivity and brightness [28]. Moreover, its good resistance needs to be considered against corrosion in the normal atmosphere. These properties change depending on the nickel and phosphorus levels in the deposit which in turns attributed to the composition and pH of the plating bath. The brightness and reflectivity of electroless nickel-plated surface can also significantly vary, depending on the specific formulation. The reflectivity of the deposit is also affected by the surface finishing of the substrate.

Current study focuses on the technical procedure for development of a highly durable composite conductive Cu–Ni–P–Ni layer on PET fabric by electroless plating method. The novel part of the work points out two main ideas. Firstly, the resilience promotion of previously developed material using Cu–Ni–P-coated fabric by inclusion of nickel as top layer. Then, the enhancement of thermal camouflage was highlighted using promoted technique. The fabric product after conductive finishing was characterized using scanning electron microscopy (SEM), energy dispersive X-ray (EDX) analysis, Raman spectrometer and wide-angle X-ray diffraction (WAXD). A detailed investigation was also conducted on developed samples using thermal energy detector and thermal camera.

# Materials and methods

## Chemical reagents

The polyethylene terephthalate (PET) (47 × 23 Counts/cm$^2$, 133.7 gr/m$^2$, taffeta woven) fabrics with 50 cm$^2$ (5 × 10) surface area was used in electroless plating process. The nickel (II) sulfate hexahydrate, sodium hypophosphite monohydrate, copper (II) sulfate pentahydrate, trisodium citrate and boric acid supplied by Sigma-Aldrich and used without further purification. The dioinized water with conductivity of less than 2 μs/cm was used in all preparation steps.

## The electroless process

The electroless plating for either copper or nickel plating involves five steps including etching, sensitization, activation, electroless plating and post-treatment. Table 1 represents the composition of nickel and copper electroless bath. The whole plating procedure is illustrated schematically in Fig. 1.

The untreated fabric firstly scours and etches in 15 gr/L NaOH solution (1L) at 70 °C for 5 min prior to use and cleans with deionized water thereafter. Subsequently, the surface sensitization carries out by immersing of treated fabric samples into an aqueous solution (1 L) containing SnCl₂ and HCL at fixed temperature. Next, the substrate rinsed in deionized water and afterward activated by soaking in an activator solution containing PdCl₂ and HCL at room temperature.

The samples were then rinsed in a large volume of deionized water for more than 5 min to prevent contamination of the plating bath. Table 1 lists the composition of copper electroless bath. The pH value of the baths is adjusted by adding sufficient quantity of 10% sodium hydroxide solution. In the post-treatment stage, the plated fabrics are rinsed with deionized water at 40 °C for 20 min right after metalizing reaction of electroless plating. The prepared copper-plated fabric is immersed in the electroless nickel bath immediately to form nickel layer on

**Table 1** Bath Composition and operation conditions of electroless plating

| Chemical | Role | Copper electroless | Nickel electroless |
|---|---|---|---|
| Nickel(II) sulfate hexahydrate (gL⁻¹) | Catalyst for copper electroless Nickel source for nickel electroless | 0.25 | 0.75 |
| Sodium hypophosphite monohydrate (gL⁻¹) | Reducing agent | 5.5 | 1.2 |
| Copper(II) sulfate pentahydrate (gL⁻¹) | Copper source | 5.5 | – |
| Trisodium citrate (gL⁻¹) | Accelerant ligand | 4 | 1.5 |
| Boric acid (gL⁻¹) | pH adjustor | 5.5 | 1.2 |

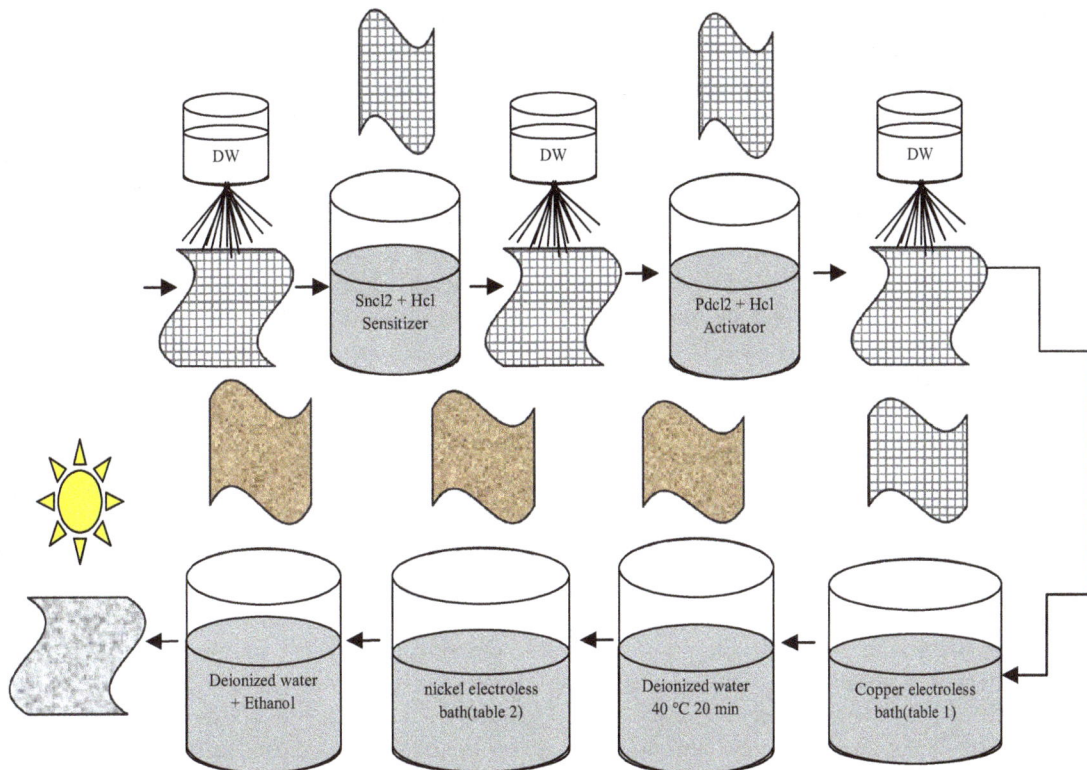

**Fig. 1** Graphical representation of Cu–Ni–P–Ni plating procedure

previously copper-plated fabric. After plating, the samples are carefully rinsed with deionized water and ultimately dried in an oven at 70 °C.

**The characterization techniques**

The surface and cross-sectional morphologies of coated fabric were observed using Philips XL 30 scanning electron microscope (SEM). The chemical composition of plated films was determined using an elemental analysis device incorporated with SEM instrument. The crystal structure of metalized fabric before and after nickel electroless plating was also investigated using X-ray diffraction (PW1840 diffractometer, Philips) with Cu K$\alpha$ radiation operating at 40 kV and 35 mA, and a scan rate of 0.1°/s. Raman spectra were characterized by Raman spectrometer (LabRam-1B, France, JY Co. Ltd). The spectra collected with a spectral resolution of 1.5 cm$^{-1}$ in the backscattering mode, using the 632.8 nm excitation line of a Helium/Neon laser. The nominal power of the laser polarized 500:1 was 20 mW. A Gaussian/Lorentzian-fitting function was used to obtain band position and intensity. The incident laser beam was focused onto the specimen surface through a 100× microscopic objective lens, forming a laser spot of approximately 5 $\mu$m in diameter, using a capture time of 50 s. The Raman signals obtained with the half wave plate were rotated at 170° with a confocal hole set at 1100 $\mu$m and the slit set at 300 $\mu$m. The electronic balance (HR200, Japan) was used to measure the percentage of the weight change for the coated fabric compared to neat fabric. The thickness of deposits qualitatively measures by weight gain method and further confirms by metallographic cross sections of the deposits by scanning or optical microscope. The color fastness of treated fabrics was evaluated separately based on standard testing method of the ISO 105-A02:1993 (color fastness to crocking under dry and wet conditions) and ISO 105-C06:1994 (color fastness to laundering). The surface resistances of each sample were measured five times for each sample with AATCC 76-1995 standard test method. The two 2 × 3 cm$^2$ copper electrodes separated by 1.5 cm were placed on the each specimen and a constant 10 N force applies onto the plated fabrics. The resistance was recorded with multimeter (ADM 552R, Arma Ltd., Iran). Surface resistivity ($R_s$) expressed in ohms/square is calculated by equation of $R = R_s (1 \times A/w)$ where $R$ is the resistance in ohms, $L$ (cm) is the distance between electrodes, $w$ (cm) is the width of the each electrode, and $A$ is the fabric area in cm$^2$.

The transmitted thermal radiated power detects and records by utilizing a home made system using SMTIR9901 silicone infrared thermopile sensor. The SMTIR is able to detect the radiated temperature and acquired thermal power without any contact in different range of temperature. The principal specification of SMTIR9901 is a high precision and sensitivity, proper signal-to-noise ratio and response time and cutoff wavelength of 5.5 $\mu$m. The detected variation in thermal power and temperature amplifies and translates to digital value for data storage and displaying.

Microcontroller using visual basic programming employed for displaying of gathered data. Data serially transfer to main storage system to configure and record in the specific database. A general data displaying scheme employs to show the temperature, power and time data. The acquired data were categorized in a Cartesian data sheet and graph. The experiment was carried out in dark room at constant temperature of 25 °C using a standard hot surface as thermal source. Every sample fixes on hot surface for 2 min before data acquisition. The distance between thermopile sensing system and 20 × 20 cm$^2$ fabric sample adjusts on 10 cm. Data acquisition time and the power and temperature data can be exported in variety of spread sheet for further analysis. Every sample was experimentally tested five times to prohibit error propagation. The thermal images were captured by hot finder SAT-IR thermal camera to validate the radiated thermal power recorded by thermopile system.

**Results and discussion**

**The mechanism of plating process**

Next, following sentences focus on the mechanism of electroless copper plating of polyester fabrics that utilize the PdCl$_2$ as activator. The neat fabric substrate passes a multistep preparation procedure before soaking smoothly in copper plating bath. The concentration and temperature of activator (Sncl$_2$) and sensitizer (PdCl$_2$) baths are selected as fixed parameters based on the previously published paper regarding to copper-plated fabric [13]. Similarly, the concentration of copper electroless bath needs to be adjusted based on latest finding [13]. However, an experimental investigation conducted to find the impact of pH and temperature on average conductivity of copper-plated fabric [23, 24]. After soaking the polyester fabrics into an acidic SnCl$_2$ bath, Sn$_2^+$ ions adsorb onto the particle surface, forming a uniform layer. The sensitized polyester fabric then placed in the activation bath, which is prepared by dissolving PdCl$_2$ in aqueous acidic solution. This leads to the formation of Pd$^{2+}$ ions in the activation bath, which adsorbs on the fabric surface following the addition of sensitized fabrics to the activation bath. However, Sn$^{2+}$ ions that are present on the sensitized fabric surface immediately reduce Cu$^{2+}$ to Cu$^0$. The deposited Ni$^{2+}$ then acts as a catalyst for the subsequent Cu deposition in the electroless plating bath. Under the catalytic action, The $Ni^{2+}$ metallic clusters adsorb on the fabric surface. The copper ions deposit onto the catalytic nickel surface by capturing electrons that are furnished by the reducing agent [24]. In the initiation reaction, the metallic nickel clusters act as nucleation sites for copper deposition. In aqueous alkaline solutions (pH 9–10), hypophosphite as reducing agent adsorb on catalytic nickel surfaces. It is easily oxidized to yield $HCOO^-$, the activated hydrogen atom (•H) and release electrons (anodic reaction). In the meantime, the copper ions in the plating bath reduce to metallic copper by the electrons generated through the oxidation of nickel (cathodic reaction) [26]. The combination of two activated hydrogen atoms will be responsible for part of the observed gas evolution. Once the copper deposition initiates, the deposited copper atoms themselves act as self-catalysts for further copper deposition [27]. As well known, the electroless copper plating reaction represents by following chemical mechanism (Scheme 1).

## Physical characteristics of Cu–Ni–P-coated fabric

The physical specifications of copper-coated fabric prepared in different pH and temperature were measured based on designated standard methods before and after color fastness tests (Table 2). According to data given in

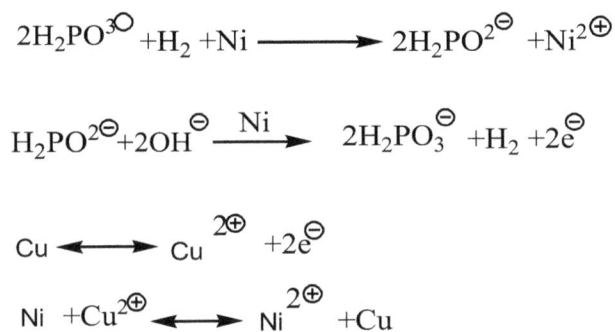

$$2H_2PO^{3\ominus} + H_2 + Ni \longrightarrow 2H_2PO^{2\ominus} + Ni^{2\oplus}$$

$$H_2PO^{2\ominus} + 2OH^{\ominus} \xrightarrow{Ni} 2H_2PO_3^{\ominus} + H_2 + 2e^{\ominus}$$

$$Cu \longleftrightarrow Cu^{2\oplus} + 2e^{\ominus}$$

$$Ni + Cu^{2\oplus} \longleftrightarrow Ni^{2\oplus} + Cu$$

**Scheme 1** Mechanism of copper electroless plating

Table 2, the weight of all fabric samples increases after copper electroless plating regarding to copper formation on fabric substrate. It is evident that the higher the pH value is, the heavier the sample produces. The monotonic increase of weight continues until the pH reach to 9.5 at a fixed temperature of 70 °C (Sample 5). Similarly, the influence of temperature was also investigated at fixed pH value. The low dependence was observed for fabric weight respect to temperature change from 70 to 85 °C. In another attempt, the color fastness and corresponding electrical resistance were evaluated for fabric samples. The comparison of sample 6 and 7 shows a small difference in fabric weight before fastness test. Meanwhile, the sample 7 shows slightly smaller resistance compared to sample 6, before and after color fastness tests.

## SEM and EDAX analysis of Cu–Ni–P-coated fabric

Figure 2 shows the SEM images of PET fabric before and after electroless plating from surface of the sample 5. The acquired image in low magnification clearly reveals the presence of small particles on fabric surface. The high-resolution scanning electron microscope and EDAX analysis utilized to focus on to the coating pattern and formation process. Figure 3 clearly demonstrates the smooth and homogenous plating of copper on PET fabric. The SEM image of Cu–Ni–P particle on PET fabric shows a semi-hexagonal non-homogenous morphology with nanoscale crack on its surface. EDAX analysis and the acquired image show a compacted bulk of particles on fabric surface leads to very high conductivity for sample 5. The EDAX analysis shows copper particle and other auxiliary material. Table 3 lists the weight percent of all chemical species which were formed on fabric surfaces. Results confirm the existence of nickel particle as second top most abundant element compared to other materials. The copper reduction using sodium hypophosphate was severally investigated in previous studies [13, 23, 24, 29]. The reduction process through sodium hypophosphate is more complicated

**Table 2** Effect of temperature and pH on coating weight and surface resistance before and after color fastness test

| Sample | pH | $T$ (°C) | Fabric weight before coating (gr) | Fabric weight after coating (gr) | Color fastness | Surface resistance ($\Omega$/sq) | Surface resistance after crocking ($\Omega$/sq) |
|---|---|---|---|---|---|---|---|
| 1 | 7.5 | 70 | $0.73 \pm 0.15$ | $0.76 \pm 0.13$ (4.0%↑) | 3 | $4.56 \pm 0.34$ | $12.36 \pm 0.87$ (171.05%↑) |
| 2 | 8 | 70 | $0.7 \pm 0.17$ | $0.76 \pm 0.11$ (8.7%↑) | 3–4 | $2.32 \pm 0.44$ | $4.27 \pm 1.21$ (84.05%↑) |
| 3 | 8.5 | 70 | $0.69 \pm 0.19$ | $0.78 \pm 0.15$ (13.0%↑) | 4 | $1.44 \pm 0.31$ | $2.22 \pm 0.87$ (54.16%↑) |
| 4 | 9 | 70 | $0.72 \pm 0.14$ | $0.84 \pm 0.17$ (16.6%↑) | 4 | $1.22 \pm 0.54$ | $1.87 \pm 0.54$ (53.27%↑) |
| 5 | 9.5 | 70 | $0.72 \pm 0.18$ | $0.85 \pm 0.18$ (8.05%↑) | 4 | $0.78 \pm 0.15$ | $1.28 \pm 0.65$ (64.10%↑) |
| 6 | 9.5 | 75 | $0.74 \pm 0.15$ | $0.87 \pm 0.16$ (17.0%↑) | 4–5 | $0.99 \pm 0.18$ | $1.43 \pm 0.32$ (44.44%↑) |
| 7 | 9.5 | 80 | $0.71 \pm 0.17$ | $0.84 \pm 0.14$ (18.0%↑) | 4–5 | $0.64 \pm 0.14$ | $0.92 \pm 0.22$ (43.75%↑) |
| 8 | 9.5 | 85 | $0.72 \pm 0.19$ | $0.83 \pm 0.19$ (15.0%↑) | 4 | $0.61 \pm 0.18$ | $0.97 \pm 0.25$ (59.01%↑) |

**Fig. 2** SEM image of **a** the neat PET fabric, **b** Cu–Ni–P electroless plated

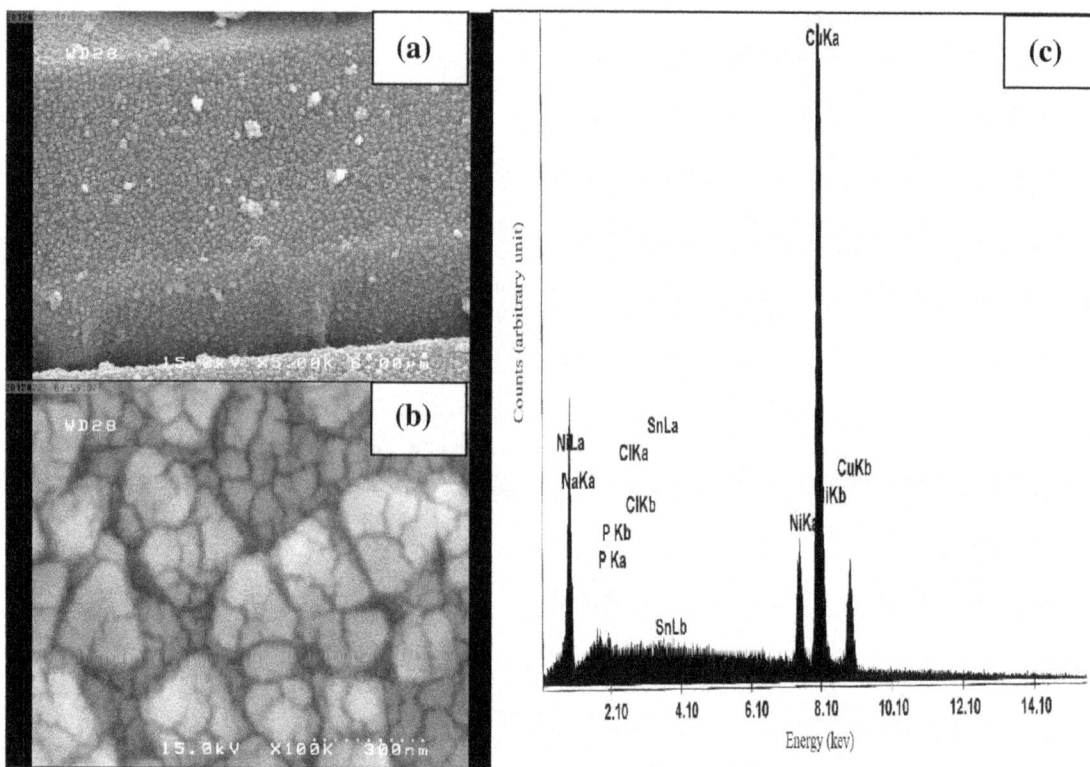

**Fig. 3** SEM image of copper electroless plated **a** 5000×, **b** 10,000×, **c** the EDAX analysis of Cu–Ni–P-coated layer

compared to other process using formaldehyde as reducing agent. Since copper is not strong enough for fast oxidation of sodium hypophosphate, therefore the nickel or other metallic ion assists the oxidation process as a catalyst. Consequently, existence of nickel particle in elemental analysis of copper-plated surface is inevitable. Moreover, according to previous studies, the nickel particle assists the copper-plated fabric for being a polished surface. The reduction of copper sulfate leads to oxidizing of sodium hypophosphite using nickel sulfate, which in turn sediment nickel, and phosphorus.

## The IR rejection of Cu–Ni–P-coated fabric

The IR rejection of polyester fabric coated by copper nanoparticles was characterized by thermopile sensor described in experimental section. The experiments were conducted based on the sample with high conductivity and durability, as they are most effective parameter in thermal camouflages. The detection probability of a target with a thermal imaging system is determined by the surface emissivity and temperature of the target, which is mainly controlled by the target's temperature, the radiant

**Table 3** Elemental analysis of deposited particle for copper electroless plating

| Element | Cu–Ni–P electroless plated | |
|---------|----------|----------|
| | Atomic % | Weight % |
| P | 1.72 | 0.72 |
| Na | 3.32 | 1.24 |
| Cl | 0.07 | 0.04 |
| Sn | 0.19 | 0.36 |
| Ni | 10.04 | 9.62 |
| Cu | 88 | 84.86 |

waveband, direction, the target's shape and surface area. The coating process not only influences on the microstructure of fabric surface, but also changes the surface emissivity or reflectance. Since textile materials are very susceptive to IR radiation, fabric's temperature rises quickly once it is exposed to thermal sources. It is well known that the metallic-coated fabrics belong to the lower emissivity materials.

Figure 4 depicts the recorded radiated thermal power from room condition, hot source with neat fabric and hot source with copper-coated polyester fabric. The recorded data reveal a meaningful reduction in radiated thermal power detected by thermopile sensor. The experiment shows nearly constant response with little fluctuation for different sample. It can be translated as proper and reliable recording time for data acquisition. The influence of texture density on radiated power was examined by testing of neat fabric and copper-coated fabric for dense and sparsely texture fabric (Fig. 5). The sparsely texture fabric coated by copper nanoparticle shows a range of resistivity around 0.71 $\Omega$/sq which is quite close to previous high-density sample. This leads to a 2.3 unit reduction in radiated thermal power compared to neat sparsely texture polyester fabric. However, the copper-coated dense fabric shows the lowest radiated power as it was logically expected. Figure 6 tries to explore the form and the thickness of copper

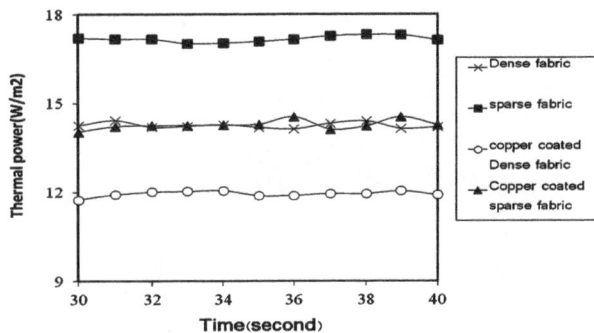

**Fig. 5** Detected radiated thermal power for conductive and neat fabric with different texture

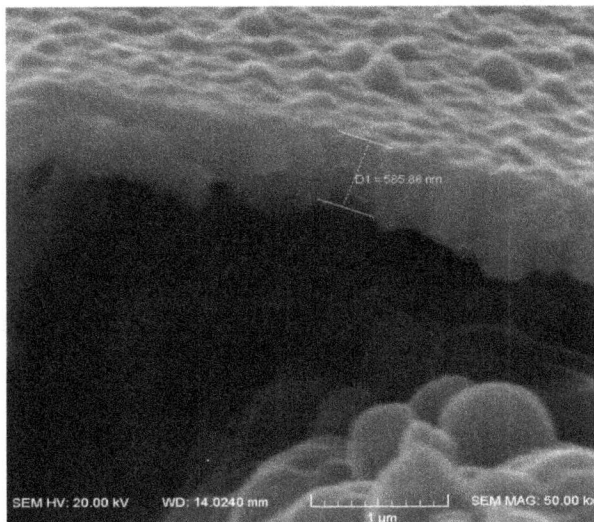

**Fig. 6** SEM picture from thickness of copper deposited layer on PET fabric

layer on fabric surface. The SEM image evidences the formation of a quite smooth layer with a thickness of 585 nm. In fact, the controlling of thickness is possible based on plating duration and proposed application. The captured picture by thermal camera from a human body shows blue and black regions while keeping a piece of copper-coated dense polyester fabric (Fig. 7). The temperature indicator in right panel shows a remarkable decrease in detected temperature from 36 to 28.8 °C. This is valuable decrease in body radiated thermal energy, which locates in infrared wavelength range. There is very close coherence between conductivity data, the radiated thermal power and the thermal images. The more conductivity, the less radiation and the lower detected temperature occur. Despite very clear impact of conductive copper-coated fabric on thermal camouflage, still there is some energy leakage, which is quite evident by a narrow gap between black and blue indicating about 0.5 °C temperature difference. The first part of experiments tried to show the conductivity and its stability and homogeneity of

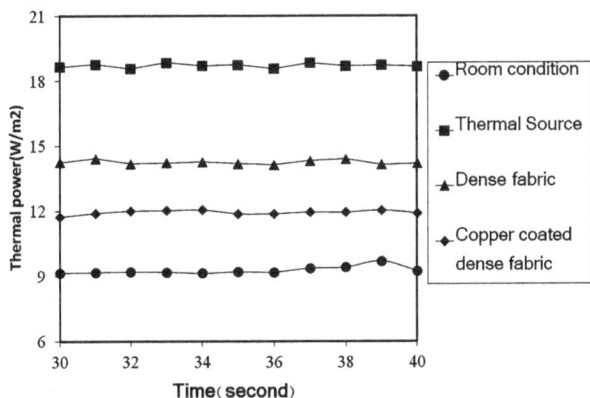

**Fig. 4** Recorded thermal radiation by thermopile sensor for different condition

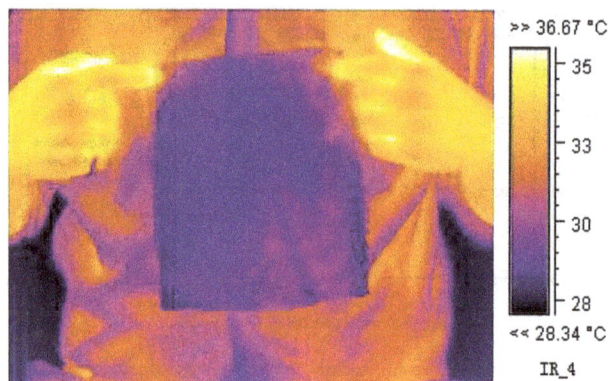

**Fig. 7** Captured thermal image from human body while keeping a piece of dense copper-coated fabric

the copper-plated fabric. Results indicate the less durability due to copper oxidation for copper-plated fabric and low and non-homogenous conductivity for nickel-plated fabric because of weak coherence between particles and fabric surface. These observations are in good agreement with previously published works [13, 23, 24, 29]. Moreover, the IR- rejection of body thermal energy using copper-coated fabric shows radiation leakage which is quite enough for thermal camera detection. This problem highlights more specifically for low detection limit thermal cameras.

## Physical characteristics of Cu–Ni–P–Ni-coated fabric

The rest of paper focuses on nickel coating as a top durable conductive layer on previously Cu–Ni–P composite

conductive layer. Table 4 lists the electrical properties and its durability after fastness test for Cu–Ni–P–Ni nanocomposite electroless plated PET fabric as a function of temperature and pH. It can be declared that the weight of all samples significantly increases after nickel plating compared to bare copper-plated fabric. The weight measurement of other samples indicates the nickel formation on copper electroless plated fabric. Choosing suitable pH and temperature is extremely critical. In such a manner, the fabric durability after fastness test needs to be evaluated. Results exhibits most acceptable change in conductivity and its absolute value after crocking for the samples 1 and 5. However, the detailed investigation demonstrates a lower conductivity for Cu–Ni–P–Ni nanocomposite electroless plated PET fabric compared to bare Cu–Ni–P-coated fabric. It was quite predictable since the Ni presence on top surface reduces surface conductivity compared to copper particle. According to given data, the pH and temperature, respectively, adjusted on 9 and 70 °C for a nickel concentration of 1.48 gr/l.

The other strategy explains another technical procedure in order to remove deficiencies of first protocol. It can be hypothesized that the nickel lamination on top of copper-plated fabric induces both durability and homogenous high level of conductivity to fabric.

## SEM and EDAX analysis of Cu–Ni–P-coated fabric

The SEM image of the nickel-plated surfaces of metallic fabric is shown in Fig. 8. It can be observed

**Table 4** Effect of temperature and pH on coating weight and surface resistance before and after color fastness test for nickel-coated copper electroless plated

| Sample | $T$ (°C) | pH | Fabric weight after coating (gr) | Ni coated on copper plated | Color fastness | Surface resistance ($\Omega$/sq) | Surface resistance after crocking ($\Omega$/sq) |
|---|---|---|---|---|---|---|---|
| 1 | 70 | 9.5 | 0.84 ± 0.12 | 0.95 ± 0.14 (13.09%↑) | 4–3 | 1.2 ± 0.3 | 2.3 ± 0.4 (88.43%↑) |
| 2 | 70 | 9 | 0.83 ± 0.15 | 0.9 ± 0.12 (8.43%↑) | 3 | 6.3 ± 0.5 | 13.2 ± 2.1 (109.49%↑) |
| 3 | 70 | 10 | 0.85 ± 0.17 | 0.96 ± 0.17 (12.94%↑) | 4–3 | 4.2 ± 0.8 | 7.6 ± 1.4 (78.92%↑) |
| 4 | 75 | 9.5 | 0.81 ± 0.18 | 0.89 ± 0.11 (9.87%↑) | 4 | 3.5 ± 0.9 | 5.8 ± 2.1 (66.19%↑) |
| 5 | 80 | 9.5 | 0.84 ± 0.11 | 0.93 ± 0.13 (10.71%↑) | 4–3 | 2.4 ± 0.2 | 4.6 ± 1.1 (89.3%↑) |

**Fig. 8** SEM image of **a** Nickel plating on copper electroless plated fabric at low magnification (×500), **b** the quasi spherical nickel particle morphology at higher magnification (×2500)

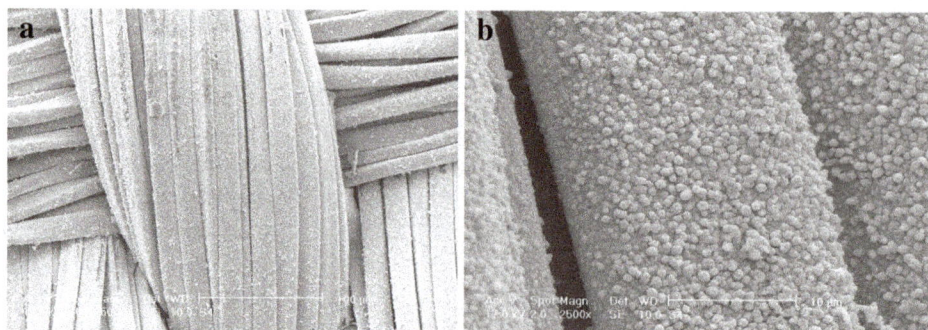

that the surfaces were covered with the dense particles and evenness layers. The compactness and exceptional coherence of particles in nanoscale lead to high electrical conductivity for Ni–Cu electroless plated fabric. Figure 9 shows the particle formation and its deposited area on fabric surface in high magnification. Figure 9a apparently shows an area covered by huge numbers of nucleated particle in nanoscale range between 100 and 500 nm. The metallic fabrics were subjected to elemental analysis by EDAX analysis, and the results are shown in Fig. 9c. EDAX studies revealed that the nickel-plated metallic fabric contained mainly of copper and nickel with small amounts of tin, chlorine, sodium and phosphorus. Table 5 shows the chemical composition of metallic coatings before and after nickel electroless plating. The results show that the percentage of copper and nickel in the deposited layer, respectively, decreases and increases after nickel coating on the copper-plated fabric. On the other hand, the nickel content increases up to a certain level of the coating. The presence of nickel substance in electroless copper plating plays a key role in formation of copper layers on the fabric surface [26]. After nickel plating on the copper-plated fabric, the nickel percentage increases from 9.62 to 26.73% (% w/w). On the other hand, the copper content decreases down to a certain percent, but it has highest value within other elements.

**Table 5** Elemental analysis of Cu–Ni–P–Ni plated fabric

| Element | Atomic percent (%) | Weight percent (%) |
| --- | --- | --- |
| P | 2.63 | 0.81 |
| Na | 2.17 | 0.82 |
| Cl | 0.21 | 0.12 |
| Sn | 0.11 | 0.21 |
| Ni | 27.67 | 26.73 |
| Cu | 68.21 | 71.31 |

## The IR rejection of Cu–Ni–P–Ni-coated fabric

Similar to first part of paper, the performance of conductive fabric is characterized by the level of deduction in transmitted power detects by thermopile sensor (Fig. 10). The Cu–Ni–P–Ni-coated fabric shows more homogenous surface compared to Cu–Ni–P-coated fabric. In addition, a thin layer around 300 nm containing nickel particle forms on previously deposited copper layer (Fig. 11). So, the Ni content is not the only reason for the IR rejection as it was tabulated in Tables 3 and 5.

The higher capability for IR ejection in Cu–Ni–P–Ni-coated fabric compared to Cu–Ni–P counterpart is a direct outcome of surface morphology. To illuminate the role of conductivity, a fabric sample with quite same range of

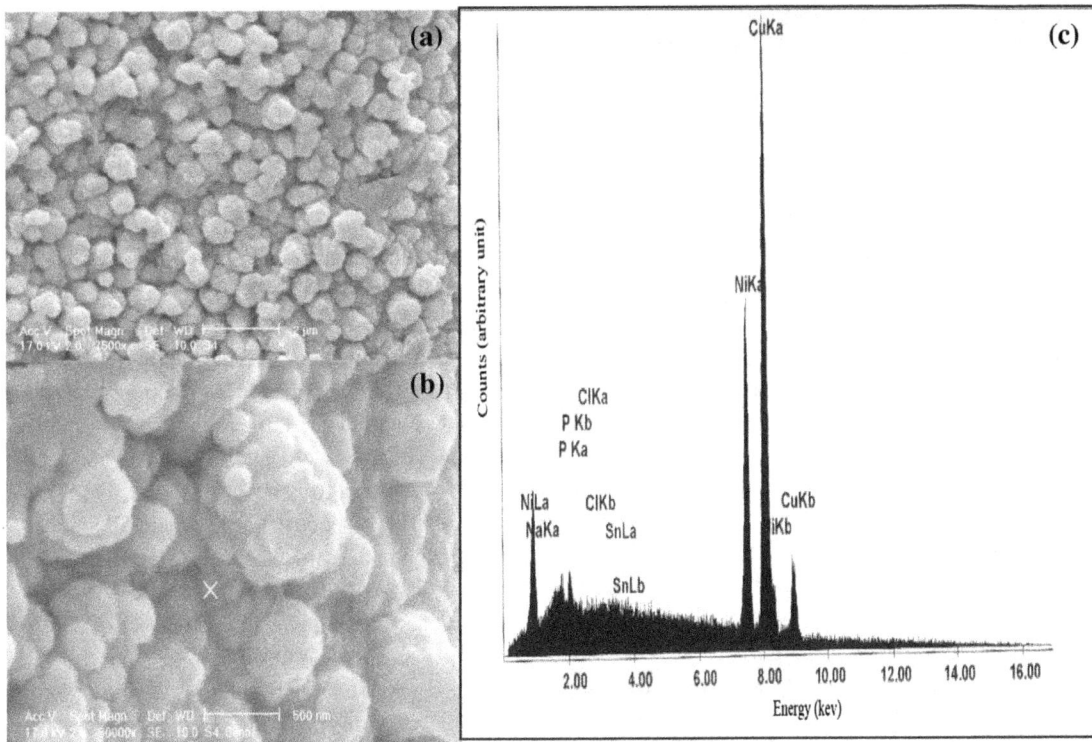

**Fig. 9** SEM image of Cu–Ni–P–Ni electroless plated PET fabric **a** ×7500, **b** ×80,000, **c** the EDAX analysis of Cu–Ni–P–Ni-coated layer

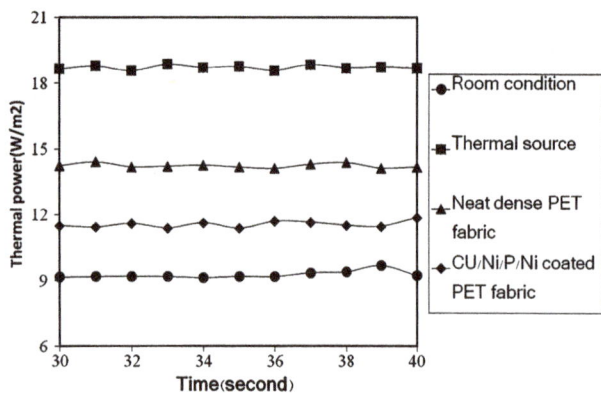

Fig. 10 Recorded thermal radiation by thermopile sensor for different condition

Fig. 12 Captured thermal image from human body while keeping a piece of dense Cu–Ni–P–Ni composite nanoparticle-coated fabric

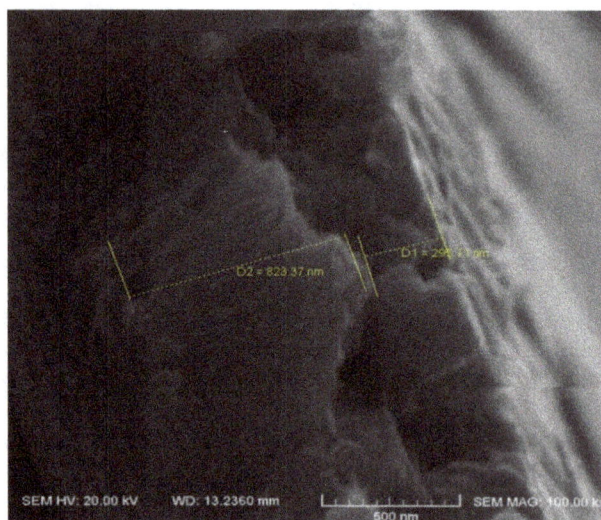

Fig. 11 SEM picture from thickness of nickel deposited layer on copper deposited PET fabric

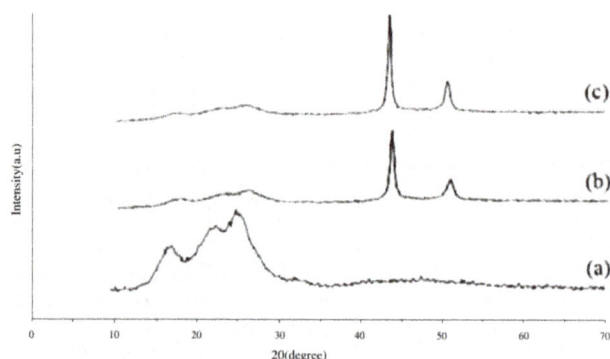

Fig. 13 XRD spectra of metallic fabrics: a untreated PET fabric, b Cu–Ni–P fabric I (1 1 1) = 13.13, I (2 0 0) = 4.75, c Cu–Ni–P–Ni fabric I (1 1 1) = 17.16, I (2 0 0) = 6.33

## XRD and Raman analysis

conductivity was characterized for detection of thermal power radiation. The thermopile sensor for Cu–Ni–P–Ni-coated fabric shows a 2.7 unit deduction compared to neat fabric. The Cu–Ni–P–Ni-coated fabric shows a higher amount of rejection about 0.4 units compared to the Cu–Ni–P. The radiation leakage in first protocol can be attributed to surface inconsistency, which is probably removes by inclusion of nickel crystals. In fact, the formation of spherical particles may improves the IR rejection performance and block most of body thermal radiation. The captured thermal image reveals nearly black appearance in most areas compared to Fig. 6, which shows a black and blue color (Fig. 12). The results indicate a significant decrease in detected temperature from 36 to 28 °C. It is pronounced as efficient blockage of leaking nodes through the fabric texture after conductive finishing.

Figure 13 shows the XRD patterns for metallic-coated fabrics before and after nickel plating. The strong diffraction peaks at $2\theta = 17.5°$, $22.8°$, $25.3°$ are characteristic peaks of untreated polyester fabrics. Comparing XRD patterns shown in Fig. 13a–c, the characteristic peak intensity of the plated PET substrate is decreased obviously. As a result, the covering degree of metallic coating is compact. The XRD patterns of copper-plated PET fabric and nickel–plated metallic fabric in Fig. 13b, c were superimposed as well as the peaks that appeared at $2\theta = 43.4°$ and $50.6°$ and correspond to (1 1 1), (2 0 0) crystal planes, respectively. The relative intensities of the diffraction peaks from the (1 1 1) and (2 0 0) planes are listed in Fig. 13. It can be seen that the crystalline structure of the copper coating was intensified with nickel plating on the copper surface of fabric.

Raman spectra of untreated PET fabric are shown in Fig. 14a. A strong peak at 347 cm$^{-1}$ is attributed to aliphatic chains (C–C) stretching vibration, and C–O–C

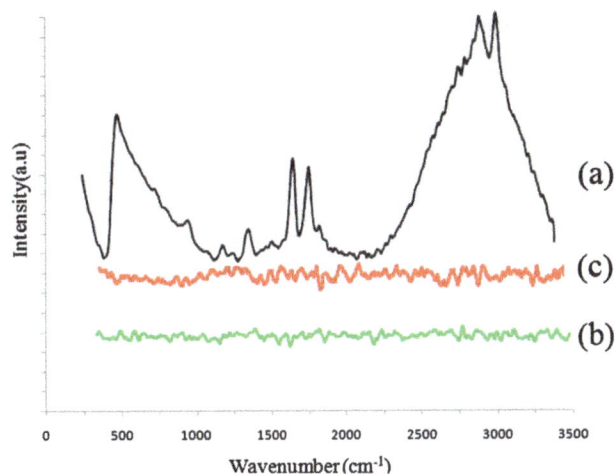

Fig. 14 Raman spectra of a untreated PET fabric, b Cu–Ni–P-coated fabric, c Cu–Ni–P–Ni composite nanoparticle-coated fabric

asymmetric stretching vibration is revealed at 1107 cm$^{-1}$. A peak at 1179 cm$^{-1}$ belongs to ester (C(O)–O) stretching vibration. The aromatic >C=C< stretching vibration is observed at 1294 cm$^{-1}$. A weak peak in 1466 cm$^{-1}$ is ascribed to the C–H stretching vibration of methylene group. Two strong peaks in 1620 and 1734 cm$^{-1}$ are characters of phenyl ring stretching vibration and carbonyl (C=O) stretching vibration, respectively. The successive peaks in region between 2800 and 3000 cm$^{-1}$ contribute to alkanes (C–H) stretching vibration. The alkenes (=C–H) stretching vibration developed at 3080 cm$^{-1}$.

Raman spectra of copper plating PET fabric and nickel-plating metallic fabrics are shown in Fig. 14b, c. Also, it can be seen that the intensity of Raman for copper-plated fabric increased after nickel plating. It can be seen that all characteristic peaks of PET fabrics are covered after metal plating. This indicates that the existence of metallic layer on the fabric surface causes a decrease in the scattering intensity by the reflection or blocking of Raman [24].

## Conclusions

The high-performance conductive durable fabric was developed for rejection of emitted radiated power from thermal source in the range of far infrared similar to body thermal energy. The electroless plating of PET fabric through Cu–Ni–P formulation leads to acceptable level of conductivity and moderate durability. The specific morphology provides a medium level of blocking layer against transmitted thermal wave. It was confirmed by both thermopile sensor and thermal camera. The fur electroless nickel plating on previously surface of Cu–Ni–P PET fabric was investigated. The properties of Cu–Ni–P–Ni fabric were enhanced with respect to SEM analysis due to

semispherical compact nanoparticle morphology, but there was a moderate decrease in the electrical conductivity. The nickel-coated copper-plated samples showed better properties and stable structure with uniformly distributed nickel particles. EDX shows that the chemical composition of coating still consisted mainly of copper even after nickel-plating copper-plated fabric. XRD patterns of copper-plated fabric shows that the peaks appear in the same place after nickel plating but the peaks intensity increase. Raman spectra indicate that peaks for untreated PET fabric are disappeared after plating with copper. Compared to copper-plated fabric, the Cu–Ni–P–Ni conductive fabric shows enhanced performance in terms of rejection of thermal radiation and also significant decrease in detected temperature by thermal camera.

Acknowledgements The Authors would like to express sincere appreciation from University of Guilan, Guilan Science and Technology Park (GSTP) and Iran nanotechnology initiative for their financial and technical supports.

## References

1. Chen H-C, Lee K-C, Lin J-H (2004) Electromagnetic and electrostatic shielding properties of co-weaving-knitting fabrics reinforced composites. Compos A Appl Sci Manuf 35:1249–1256
2. Narasimman P, Pushpavanam M, Periasamy VM (2011) Synthesis, characterization and comparison of sediment electro-codeposited nickel–micro and nano SiC composites. Appl Surf Sci 258:590–598
3. Granqvist CG (2007) Transparent conductors as solar energy materials: a panoramic review. Sol Energy Mater Sol Cells 91:1529–1598
4. Han EG, Kim EA, Oh KW (2001) Electromagnetic interference shielding effectiveness of electroless Cu-plated PET fabrics. Synth Met 123:469–476
5. Chung G-S, Lee DH (2005) A study on comfort of protective clothing. In: Yutaka T, Tadakatsu O (eds) Elsevier Ergonomics Book Series, Elsevier
6. Raimundo AM, Figueiredo AR (2009) Personal protective clothing and safety of firefighters near a high intensity fire front. Fire Saf J 44:514–521
7. Knittel D, Schollmeyer E (2009) Electrically high-conductive textiles. Synth Met 159:1433–1437
8. Textor T, Mahltig B (2010) A sol–gel based surface treatment for preparation of water repellent antistatic textiles. Appl Surf Sci 256:1668–1674
9. Cheng KB, Ramakrishna S, Lee KC (2000) Electromagnetic shielding effectiveness of copper/glass fiber knitted fabric reinforced polypropylene composites. Compos A Appl Sci Manuf 31:1039–1045
10. Huang C-Y, Wu C-C (2000) The EMI shielding effectiveness of PC/ABS/nickel-coated-carbon-fibre composites. Eur Polym J 36:2729–2737
11. Kim MS, Kim HK, Byun SW, Jeong SH, Hong YK, Joo JS, Song KT, Kim JK, Lee CJ, Lee JY (2002) PET fabric/polypyrrole composite with high electrical conductivity for EMI shielding. Synth Met 126:233–239
12. Lee H-C, Kim J-Y, Noh C-H, Song KY, Cho S-H (2006) Selective metal pattern formation and its EMI shielding efficiency. Appl Surf Sci 252:2665–2672

13. Afzali A, Mottaghitalab V, Motlagh MS, Haghi AK (2010) The electroless plating of Cu–Ni–P alloy onto cotton fabrics. Korean J Chem Eng 27:1145–1149

14. Singh S, Suhag N (2015) Camouflage textiles. Int J Enhanc Res Sci Technol Eng 4:351–359

15. Marin N, Buszka J (2013) In: Alternate Light Source Imaging: Forensic Photography Techniques (Forensic Studies for Criminal Justice).1st edn, Anderson Publishing Ltd., Boston

16. Li Y, Wu D-X, Hu J-Y, Wang S-X (2007) Novel infrared radiation properties of cotton fabric coated with nano Zn/ZnO particles. Colloids Surf A 300:140–144

17. Wang T, He J, Zhou J, Ding X, Zhao J, Wu S, Guo Y (2010) Electromagnetic wave absorption and infrared camouflage of ordered mesoporous carbon–alumina nanocomposites. Microporous Mesoporous Mater 134:58–64

18. Yin X, Chen Q, Pan N (2013) A study and a design criterion for multilayer-structure in perspiration based infrared camouflage. Exp Therm Fluid Sci 46:211–220

19. Boukhanouf R, Haddad A, North MT, Buffone C (2006) Experimental investigation of a flat plate heat pipe performance using IR thermal imaging camera. Appl Therm Eng 26:2148–2156

20. Yin X, Chen Q, Pan N (2012) Feasibility of perspiration based infrared Camouflage. Appl Therm Eng 36:32–38

21. Azim SS, Satheesh A, Ramu KK, Ramu S, Venkatachari G (2006) Studies on graphite based conductive paint coatings. Prog Org Coat 55:1–4

22. Yip J, Jiang S, Wong C (2009) Characterization of metallic textiles deposited by magnetron sputtering and traditional metallic treatments. Surf Coat Technol 204:380–385

23. Baskaran I, Narayanan TSNS, Stephen A (2006) Effect of accelerators and stabilizers on the formation and characteristics of electroless Ni–P deposits. Mater Chem Phys 99:117–126

24. Gan X, Wu Y, Liu L, Shen B, Hu W (2008) Electroless plating of Cu–Ni–P alloy on PET fabrics and effect of plating parameters on the properties of conductive fabrics. J Alloy Compd 455:308–313

25. Lien W-F, Huang P-C, Tseng S-C, Cheng C-H, Lai S-M, Liaw W-C (2012) Electroless silver plating on tetraethoxy silane-bridged fiber glass. Appl Surf Sci 258:2246–2254

26. Gan X, Wu Y, Liu L, Shen B, Hu W (2007) Electroless copper plating on PET fabrics using hypophosphite as reducing agent. Surf Coat Technol 201:7018–7023

27. Lu Y (2009) Electroless copper plating on 3-mercaptopropyltriethoxysilane modified PET fabric challenged by ultrasonic washing. Appl Surf Sci 255:8430–8434

28. DiBari GA (1995) Nickel plating. Met Finish 93:259–279

29. Choi Y-S, Yoo Y-H, Kim J-G, Kim S-H (2006) A comparison of the corrosion resistance of Cu–Ni–stainless steel multilayers used for EMI shielding. Surf Coat Technol 201:3775–3782

# Permissions

The contributors of this book come from diverse backgrounds, making this book a truly international effort. This book will bring forth new frontiers with its revolutionizing research information and detailed analysis of the nascent developments around the world.

We would like to thank all the contributing authors for lending their expertise to make the book truly unique. They have played a crucial role in the development of this book. Without their invaluable contributions this book wouldn't have been possible. They have made vital efforts to compile up to date information on the varied aspects of this subject to make this book a valuable addition to the collection of many professionals and students.

This book was conceptualized with the vision of imparting up-to-date information and advanced data in this field. To ensure the same, a matchless editorial board was set up. Every individual on the board went through rigorous rounds of assessment to prove their worth. After which they invested a large part of their time researching and compiling the most relevant data for our readers.

The editorial board has been involved in producing this book since its inception. They have spent rigorous hours researching and exploring the diverse topics which have resulted in the successful publishing of this book. They have passed on their knowledge of decades through this book. To expedite this challenging task, the publisher supported the team at every step. A small team of assistant editors was also appointed to further simplify the editing procedure and attain best results for the readers.

Apart from the editorial board, the designing team has also invested a significant amount of their time in understanding the subject and creating the most relevant covers. They scrutinized every image to scout for the most suitable representation of the subject and create an appropriate cover for the book.

The publishing team has been an ardent support to the editorial, designing and production team. Their endless efforts to recruit the best for this project, has resulted in the accomplishment of this book. They are a veteran in the field of academics and their pool of knowledge is as vast as their experience in printing. Their expertise and guidance has proved useful at every step. Their uncompromising quality standards have made this book an exceptional effort. Their encouragement from time to time has been an inspiration for everyone.

The publisher and the editorial board hope that this book will prove to be a valuable piece of knowledge for researchers, students, practitioners and scholars across the globe.

# List of Contributors

**R. Jothi Ramalingam**
Surfactants research chair, Chemistry department, College of Science, King Saud University, Riyadh, Kingdom of Saudi Arabia

**T. Radhika and Farook Adam**
School of Chemical Sciences, Universiti Sains Malaysia, Penang 11800, Malaysia

**Tarekegn Heliso Dolla**
College of Natural and Computational Sciences, Wolaita Sodo University, Wolaita Sodo, Ethiopia

**Juan Du, Ye Ying, Xiao-yu Guo, Chuan-chuan Li, Yiping Wu, Ying We and Hai-Feng Yang**
The Education Ministry Key Lab of Resource Chemistry, Department of Chemistry, Shanghai Normal University, Shanghai 200234, People's Republic of China

**E. A. Abdel-Galil, A. B. Ibrahim and M. M. Abou-Mesalam**
Atomic Energy Authority, Hot Labs. Center, P.O. 13759, Cairo, Egypt

**Ahmet Ozan Gezerman**
Department of Chemical Engineering, Faculty of Chemical and Metallurgical Engineering, Yildiz Technical University, Istanbul, Turkey

**Aaron Akah and Musaed Al-Ghrami**
Research and Development Center, Saudi Aramco, Dhahran 31311, Saudi Arabia

**Mian Saeed and M. Abdul Bari Siddiqui**
Center for Refining and Petrochemicals, Research Institute, KFUPM, P.O. Box 807, Dhahran 31261, Saudi Arabia

**Manoranjan Behera and Gitisudha Giri**
Silicon Institute of Technology, Bhubaneswar, Odisha, India

**Jonathan O. Babalola, Funmilayo T. Olayiwola, Joshua O. Olowoyo and Alimoh H. Alabi**
Department of Chemistry, University of Ibadan, 200284 Ibadan, Nigeria

**Emmanuel I. Unuabonah**
Environmental and Chemical Processes Research Laboratory, Department of Chemical Sciences, Redeemer's University, P.M.B. 230, Ede, Osun State, Nigeria

**Augustine E. Ofomaja**
Adsorption and Catalysis Research Laboratory, Department of Chemistry, Vaal University of Technology, Private Bag X021, Andries Potgieter Boulevard, Vanderbijlpark 1900, South Africa

**Martins O. Omorogie**
Adsorption and Catalysis Research Laboratory, Department of Chemistry, Vaal University of Technology, Private Bag X021, Andries Potgieter Boulevard, Vanderbijlpark 1900, South Africa
Environmental and Chemical Processes Research Laboratory, Department of Chemical Sciences, Redeemer's University, P.M.B. 230, Ede, Osun State, Nigeria

**Abir Tabaï, Ouahiba Bechiri and Mostefa Abbessi**
Laboratory of Environmental Engineering, Department of Process Engineering, Faculty of Engineering, University of Annaba, P.O. Box 12, 23000 Annaba, Algeria

**Atiya Banerjee, Devyani Varshney and Surendra Kumar**
Department of Chemical Engineering, Indian Institute of Technology Roorkee, Roorkee, Uttarakhand 247667, India

**Payal Chaudhary**
Centre for Transportation Systems, Indian Institute of Technology Roorkee, Roorkee, Uttarakhand 247667, India

**V. K. Gupta**
Department of Chemistry, Indian Institute of Technology Roorkee, Roorkee, Uttarakhand 247667, India

**Radhakrishnan Kannan, Sethuraman Lakshmi, Natarajan Aparna and Wilson Richard Thilagaraj**
Department of Biotechnology, School of Bioengineering, SRM University, Kattankulathur 603203, Tamil Nadu, India

**Sivaraman Prabhakar**
Department of Chemical Engineering, School of Bioengineering, SRM University, Kattankulathur 603203, Tamil Nadu, India

**Inkollu Sreedhar and Yandapalli Kirti Kishan**
Department of Chemical Engineering, BITS Pilani Hyderabad Campus, Hyderabad, India

**İsmet Kaya and Sebra Çöpür**
Polymer Synthesis and Analysis Laboratory, Department of Chemistry, C̣anakkale Onsekiz Mart University, 17020 C̣anakkale, Turkey

**Hatice Karaer**
Polymer Synthesis and Analysis Laboratory, Department of Chemistry, C̣anakkale Onsekiz Mart University, 17020 C̣anakkale, Turkey
Department of Chemistry, Faculty of Sciences, Dicle University, 21280 Diyarbakır, Turkey

**Meysam Sadeghi**
Young Researchers and Elite Club, Ahvaz Branch, Islamic Azad University, Ahvaz, Iran

**Sina Yekta and Esmaeil Babanezhad**
Department of Chemistry, Faculty of Basic Sciences, Qaemshahr Branch, Islamic Azad University, Qaemshahr, Iran

**Hamed Ghaedi**
Faculty of Engineering, Bushehr Branch, Islamic Azad University, Bushehr, Iran

**S. A. El-Molla and M. M. Ebrahim**
Department of Chemistry, Faculty of Education, Ain Shams University, Roxy, Cairo 11757, Egypt

**Sh. M. Ibrahim**
Department of Chemistry, Faculty of Education, Ain Shams University, Roxy, Cairo 11757, Egypt

Present Address: Department of Chemistry, Faculty of Science, Al-Qassim University, Al-Qassim, Buraidah 51452, Saudi Arabia

**Madhuchhanda Maiti, Ganesh C. Basak, Vivek K. Srivastava and Raksh Vir Jasra**
Reliance Technology Group, Vadodara Manufacturing Division, Reliance Industries Ltd., Vadodara, Gujarat 391346, India

**Jeminat O. Amode, Jose H. Santos, Aminul H. Mirza and Chan C. Mei**
Faculty of Science, Universiti Brunei Darussalam, Jalan Tungku Link, Gadong 1410, Negara Brunei Darussalam

**Zahangir Md. Alam**
Bioenvironmental Engineering Research Unit (BERU), Faculty of Engineering, International Islamic University Malaysia (IIUM), Kuala Lumpur, Malaysia

**Anita Thakur and Harpreet Kaur**
Department of Chemistry, Punjabi University, Patiala 147002, India

**Claire Gillan and S. David Jackson**
Centre for Catalysis Research, WestCHEM, School of Chemistry, University of Glasgow, Glasgow G12 8QQ, Scotland, UK

**Martin Fowles and Sam French**
Johnson Matthey Plc, Belasis Ave, Billingham TS23 1LB, UK

**Amir Khalili, Mahdi Hasanzadeh and Vahid Mottaghitalab**
Textile Engineering Department, Faculty of Engineering, The University of Guilan, P. O. BOX 3756, Rasht, Guilan, Iran

**Amirreza Mottaghitalab**
Chemical Engineering Faculty, Sahand University of Technology, Tabriz, Iran

# Index